Photoreactors in Advanced Oxidation Processes

Scrivener Publishing
100 Cummings Center, Suite 541J
Beverly, MA 01915-6106

Publishers at Scrivener
Martin Scrivener (martin@scrivenerpublishing.com)
Phillip Carmical (pcarmical@scrivenerpublishing.com)

Photoreactors in Advanced Oxidation Processes

The Future of Wastewater Treatment

Edited by

Elvis Fosso-Kankeu

Department of Electrical and Mining Engineering, the University of South Africa, Pretoria, South Africa

Sadanand Pandey

Particulate Matter Research Center, Research Institute of Industrial Science & Technology (RIST), South Korea

and

Suprakas Sinha Ray

DST-CSIR National Centre for Nanostructured Materials, Pretoria, South Africa

Scrivener Publishing

WILEY

This edition first published 2023 by John Wiley & Sons, Inc., 111 River Street, Hoboken, NJ 07030, USA and Scrivener Publishing LLC, 100 Cummings Center, Suite 541J, Beverly, MA 01915, USA

For more information about Scrivener publications please visit www.scrivenerpublishing.com.

Wiley Global Headquarters
111 River Street, Hoboken, NJ 07030, USA

For details of our global editorial offices, customer services, and more information about Wiley products visit us at www.wiley.com.

Library of Congress Cataloging-in-Publication Data

ISBN 978-1-394-16629-9

Cover image: Pixabay.Com
Cover design by Russell Richardson

Set in size of 11pt and Minion Pro by Manila Typesetting Company, Makati, Philippines

Printed in the USA

10 9 8 7 6 5 4 3 2 1

Contents

Preface

Global population growth along with ever-increasing industrialization are significantly compromising access to safe drinking water. Advanced oxidation processes (AOPs) for wastewater treatment are emerging as one of the most efficient, renewable green chemical technologies among those being discussed to address this problem. Recently, photochemical oxidation of organic and inorganic pollutants has become an attractive technique for water purification and wastewater treatment. In order to increase the water treatment efficiency, the selection of suitable biogenic and cheaper photo catalyst, as well as a light source and an oxidation system, are some of the key parameters required. It is well known that sufficient UV penetration into the radiated liquid (promotion of efficient conversion of incident photons to charge carriers) is of paramount importance, especially for an opaque environment, and UV radiation is effective when very close to the UV lamp surface. High mass transfer rates for efficient interaction between the pollutant and the photocatalyst and high oxygen uptake at the gas-liquid interface are vital requirements for practical applications. In this regard, designing a photoreactor for efficient wastewater treatment has been challenging. Many types of photoreactors have already been studied, reported on and patented in the literature.

In this book, we present the most up-to-date research on AOPs in order to make the argument that AOPs offer an eco-friendly method of wastewater treatment. In addition to an overview of the fundamentals and applications, it provides ample details of the reactive species involved in AOPs as well as reactor design concepts, thus providing readers with the necessary tools to better understand and implement these methods. Moreover, this book presents some conventional and novel photoreactors equipped with UV/vis lamps for working under solar radiation for wastewater treatment in a laboratory and on an industrial scale, which is an important focus of our book.

There have been numerous studies covering the modeling of light distribution in different photocatalytic reactors using simulation methods such as Six-Flux analysis and Monte Carlo analysis. These studies have reported novel methods of establishing the catalyst's optical properties and validating the models. Therefore, this book also reviews these recent developments with respect to the modeling of light distribution and reaction kinetics in photocatalysis reactors.

Since the main objective of this book is to critically discuss and evaluate the chemical oxidation applications for industrial wastewater presented in this up-to-date review, it will be of interest to scientists and engineers in academia or industry working on projects related to the removal of organic pollutants from wastewater.

The editors are grateful to the reviewers who have contributed to improving the quality of the book through their constructive comments. The editors also thank the publisher for including this book in their series.

Elvis Fosso-Kankeu, Sadanand Pandey and Suprakas Sinha Ray
December 2022

Part 1

ADVANCES IN PHOTOCATALYSTS SYNTHESIS

1

Advancement and New Challenges in Heterogeneous Photocatalysts for Industrial Wastewater Treatment in the 21st Century

Sadanand Pandey[1*], Tanushri Chatterji[2], Edwin Makhado[3], Abbas Rahdar[4], Elvis Fosso-Kankeu[5] and Misook Kang[1†]

[1]*Department of Chemistry, College of Natural Science, Yeungnam University, Daehak-Ro, Gyeongsan, Gyeongbuk, Republic of Korea*
[2]*School of Bioscience, IMS Ghaziabad (University Courses Campus), Uttar Pradesh, India*
[3]*Department of Chemistry, School of Physical and Mineral Sciences, University of Limpopo, Polokwane, Sovenga, South Africa*
[4]*Department of Physics, University of Zabol, Zabol, Iran*
[5]*Department of Mining Engineering, College of Science Engineering and Technology, University of South Africa, Florida Science Campus, South Africa*

Abstract
From the last few decades, heterogeneous photocatalysts have flourished significant consideration especially concerning energy and the environment. Heterogeneous photocatalysts play a vital role in the cleavage of solar water and in the removal of environmental pollutants, including organic and inorganic species from aqueous or gas phase systems in environmental remediation, drinking water treatment, industrial, and health care setups. The current chapter starts with a brief introduction on the background of industrial wastewater and the advancement of wastewater treatment processes through advanced oxidation processes (AOPs), comparing the importance of AOPs technology for water treatment. The recent development of heterogeneous photocatalysts for the treatment of minor pollutant concentrations in water/air is also reviewed. The chapter also focuses on

**Corresponding author*: sadanand.au@gmail.com
†*Corresponding author*: mskang@ynu.ac.kr

Elvis Fosso-Kankeu, Sadanand Pandey, and Suprakas Sinha Ray (eds.) Photoreactors in Advanced Oxidation Processes: The Future of Wastewater Treatment, (3–36) © 2023 Scrivener Publishing LLC

the mechanisms of heterogeneous photocatalysis, the impact of various designs of photoreactors with the review of the published literature, which includes various types and designs of photocatalytic reactors. It is our hope that readers will get an overview of the requirements guiding the usage of suitable photoreactors. Finally, the chapter ends with a discussion of the personal perspectives that can provide new insights into the future development and prospects of heterogeneous photocatalysts for industrial wastewater.

***Keywords*:** Photoreactors, heterogeneous photocatalysts, advanced oxidation processes, water treatment

1.1 Introduction

The existence of all living beings on this planet depends on water. It covered about 71% of the earth's surface but almost 2.5% is specified as freshwater. The limited amount of fresh water is used and then recycled to support the growing population. A rapid population growth, the increase in industrialization and material production inflicts the influx of anthropogenic pollutants into the water environment, engendering a potential threat to human health and the ecological environment. The increased usage of water by various industrial sectors has inescapably led to a rise in the generation of wastewater. Numerous modern industries, such as textile, paper printing, leather, food, mining, electroplating, cosmetics, and other chemical industries, discharge highly noxious chemicals into water sources. The main causes of water pollution lies in improper disposal and extensive usage of organic products that majorly include pharmaceuticals and personal care products, detergents, plasticizers, and dyes. Furthermore, hazardous substances are toxic, carcinogenic, and nonbiodegradable, making them a major threat to society. These classes of pollutants are becoming more complex and challenging to treat. Traditional methods for remediation of water gradually can no longer meet the requirement to treat complex contaminated water. For these reasons, researchers have focused on finding some emergent strategies to assist in removing these species of contaminants from wastewater.

Enormous attempts have been made to remediate organic products from wastewater, which include electrocoagulation/degradation process, membrane filtration [1], electrocoagulation [2], chemical coagulation [3], chemical precipitation [4], adsorption system [5], and advanced oxidation processes (AOPs) [6] have shown off a good performance in wastewater treatment and purification. The latter is a highly efficient treatment method owing to its fast reaction speed, simple technology and relatively no secondary pollution. Many factors like low efficiency, side product formation, and high-energy consumption are encouraging us to search for innovations in AOP.

Heterogeneous photocatalysis (HPC), the Fenton process, sonolysis, the ozonation process, and radiation-induced degradation are the AOPs, which exhibited great potential as a solution for decontaminating the aquatic environment. These techniques have shown enhanced efficacy for decaying, nonselective performance, and mineralizing organic toxins at relatively reduced concentrations without producing secondary pollution. The AOPs accomplish mineralization of organic compounds and sometimes inorganic compounds also to carbon dioxide and mineral acids [7]. For several years now, HPC has been one of the most promising approaches for the breakdown of organic compounds and metal ions in industrial wastewater. This process is based on aqueous phase hydroxyl radical chemistry and pair of lower-energy radiation or light source with semiconductors as photocatalysts. The technique has proven to be a viable alternative to solving environmental problems, overcoming many of the limitations of traditional industrial wastewater treatment methods. This emerging trend treatment promotes water purification, which includes decontamination, detoxification, discoloration, deodorization, and simultaneous degeneration of the pollutants.

The HPC is defined as the alteration in the rate of a chemical reaction or its onset, which is regulated by the action of ultraviolet, visible or infrared radiation in the presence of a substance called a photocatalyst. This photocatalyst consumes light and undergoes chemical conversion. The factors which accelerate the rate of photocatalysis are light intensity, pH, and modified photocatalyst [8]. The efficacy of HPC could be enhanced by the use of different semiconductors due to its advanced oxidation process. Preparation of HPCs by semiconductor oxides is one of the promising methods and acts for remediation of many organic and inorganic pollutants from water and air [9]. On the grounds of their unique combination of physical and chemical properties, and their low cost and photostability under irradiation [10], Titanium oxide (TiO_2) nanomaterial, provide a wide variety of possible applications. Few studies revealed its effectiveness studies related to air cleaning and water purification. For environmental applications, visible light-harvesting nanomaterials will be increasingly applied in combination with different advanced oxidative processes (AOPs) technologies [11].

The effectiveness of HPCs in the removal of organic compounds from polluted soil is quite remarkable. The stringent method is the action of TiO_2 under UV irradiation and solar light is noted. On the contrary, the difficulty of removing simple deposition of the photocatalyst on the soil is also observed. The reason behind this is that light cannot penetrate deeper to induce the process of photocatalysis. Hence, the degradation of pollutant is restricted to a maximum of 4cm in contaminated soil. To overcome, the polluted soil is missed with the photocatalyst followed by the exposure to irradiation light. In the previous studies, it was reported that heterogeneous photocatalytic degrade the pesticide Diuron (3-(3,4-dichlorophenyl)-1,1-dimethylurea), atrazine, 2-chlorophenol, and 2,7-dichlorodibenzodioxin present in the soil samples [12].

Nowadays, the process of HPCs is implemented for remediation of environmental problems including air, water, wastewater treatment [13], disinfection processes [14, 15]. Additionally, it is being also used for energy production by degrading biomass, hydrogen generation by water splitting, treating oil spills and chemical synthesis [16]. Recently, Shukla *et al.* (2021) reviewed the current advances in heterogeneous micro-photocatalytic reactors for wastewater treatment [17]. Contrary to previous chapters, this one discusses the advancements in wastewater treatment using HPCs. The photocatalytic degradation of organic pollutants by employing recently developed HPCs has been comprehensively discussed. Mechanism of HPCs, effects of different designs of photoreactors and the parameters required to conduct the photocatalytic process are discussed. Lastly, the future challenges for wastewater treatment via photocatalytic degradation are also considered.

1.2 Development of Heterogeneous Photocatalysts

Various advanced oxidation processes (AOPs) have been continually explored for the decomposition of organic pollutants, such as dyes, surfactants, phenolic substances, personal care products, pharmaceuticals, hydrocarbons, endocrine disruptors, fertilizers, and pesticides. Most of these organic contaminants are usually active at low concentrations. Several studies have been focused on the photocatalytic remediation method to eliminate the abovementioned organic contaminates because this approach is clean, sustainable, and it completely decomposes/degrade pollutants or converts them into nontoxic forms. In this direction, the degradation of these pollutants has spurred great interest in the remediation of wastewater, and this has led to a rise in the development of different HPCs. The benefit of HPCs lies in the fact that they can regulate optical band gaps; increase the sorption threshold via combining semiconductors having different bang gaps. Moreover, they can improve the

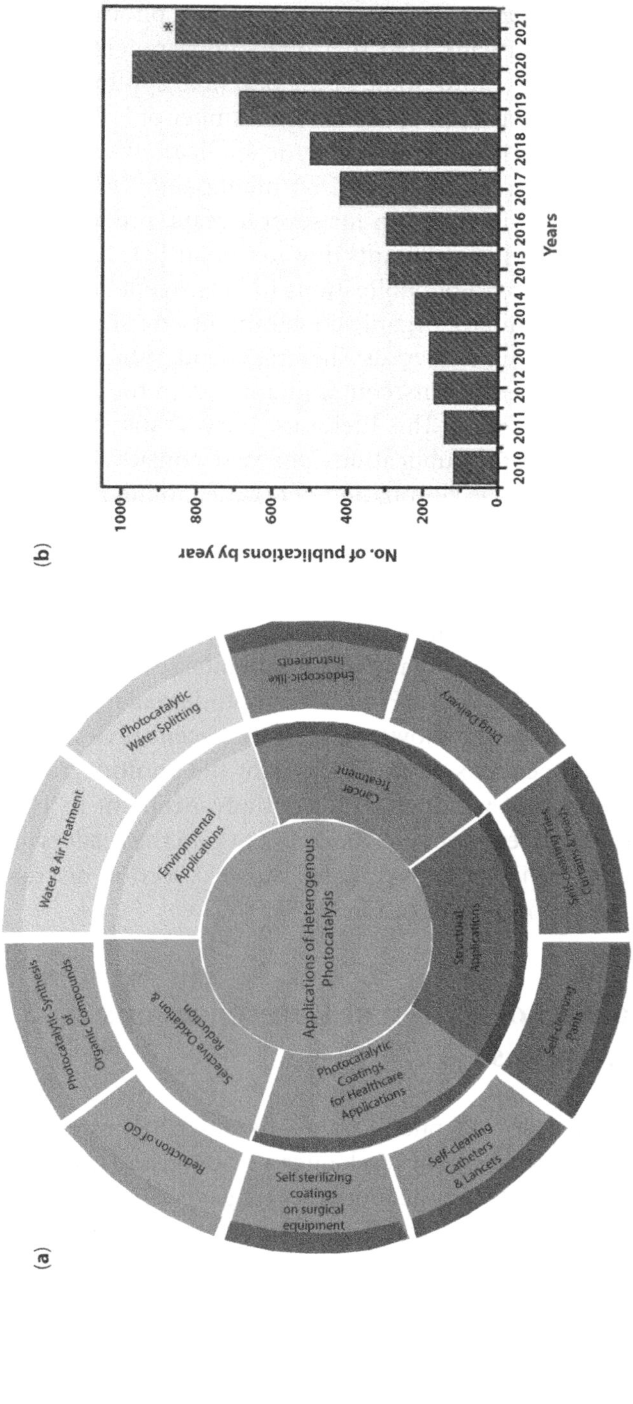

Figure 1.1 (a) Applications of heterogeneous photocatalysts (adopted from Ref [20]) and (b) The total number of publications per year on heterogeneous photocatalysts for degradation of organic pollutants/contaminants during the period 2010 to 2021 (using ScienceDirect database). *Data collected in June 2021.

charge separation under light irradiation and lower the recombination rate of electrons and holes, which may be appropriate for the oxidation-reduction process and leads to enhance the efficacy of photocatalysis process [18, 19]. HPCs is a wide field that offers great potential for many applications. Figure 1.1a shows some of the potential applications of heterogeneous photocatalysts [20]. A significant number of HPCs have been developed for a broad array of photocatalytic applications, among which water and wastewater treatment could be mentioned. HPCs have witnessed extensive scientific attention for several years, predominately for the treatment of low load of pollutants in water or air [21, 22]. In the past 20 years, a developing trend of publications in heterogeneous photocatalysts for the decomposition of organic contaminants are shown in Figure 1.1b. Search results using keywords "heterogeneous photocatalysts for degradation of organic pollutants/contaminants" from the ScienceDirect database on June 30, 2021. The literature survey statistics show an increase in the number of publications per year considering heterogeneous photocatalysts for the elimination of organic pollutants in aqueous environment. Significant growth is noticed in last few years (2017–2021). The usage of HPCs in treating wastewater has gained enormous interest from academics; researchers have been encouraged to write a review in the field of photocatalytic technology. Among many publications on the HPCs for degradation of organic contaminants, several reviews have published concerning the use of HPCs for photodegradation of various organic pollutants. Table 1.1 shows some of the recent reviews concerning the use of heterogeneous photocatalysts for the photodegradation of organic contaminants of wastewater. Recent advancement in HPCs for water and wastewater treatment are cited here to give the readers an overview background on the progress to date. The next section discusses the mechanism of action of photocatalysis.

1.3 Mechanism of Action of Heterogeneous Photocatalysis

The mechanism of action of photocatalysis is described as a combined action of light and catalyst, which accelerates a reaction and leads to chemical transformation [29, 30]. Semiconductors are generally used as catalysts for these purposes. The action of catalysis is modulated due to their specific electronic structure, which is characterized by a filled valence band (VB)

Table 1.1 Recent reviews on heterogeneous photocatalytic for degradation of organic contaminants in the aquatic environment.

Descriptions/titles	References
Recent advances on modelling of solar heterogeneous photocatalytic reactors applied for degradation of pharmaceuticals and emerging organic contaminants in water	[23]
An overview on nonspherical semiconductors for heterogeneous photocatalytic degradation of organic water contaminants	[24]
Recent developments in the use of metal oxides for photocatalytic degradation of pharmaceutical pollutants in water—a review	[25]
An overview of the recent advances of carbon quantum dots/metal oxides in the application of heterogeneous photocatalysis in photodegradation of pollutants towards visible-light and solar energy exploitation	[26]
Application of heterogeneous nano-semiconductors for photocatalytic advanced oxidation of organic compounds: a review	[27]
The application of heterogeneous visible light photocatalyst in organic synthesis	[28]
Heterogeneous photocatalysis and its potential applications in water and wastewater treatment: a review	[20]

and an empty conduction band [31]. The variation in the energy level between the conduction band (CB) and the valence band (VB) is called band gap energy. The band gap energy is normally relatively low approximately a few electron volts (eV). To activate the catalyst, a photon with sufficient energy could be used by transporting an electron from the filled valence band into the conduction band (Figure 1.2) [16]. Apart from the photocatalyst, the process of photocatalysis is a preferable as it does not require any additional reagents, and the catalytic reaction is initiated by light absorption. In addition, the semiconductor photocatalyst is activated by the consumption of a photon with ultra-band gap energy, ensuing in the advancement of an electron (e^-) from the valence band (VB) into the CB and the instantaneous formation of a positive hole (h^+) in the VB. This leads to the formation of e^- and h^+ pairs, known as charge separation. The combination of these products induces reduction and oxidation reactions with species adsorbed on the surface of photocatalyst. Elimination of pollutants from water bodies, e^- in the CB can interact with adsorbed O_2, creating a superoxide radical anion ($O_2\cdot$), whereas h^+ in the can react with water adsorbed on the photocatalyst surface and lead to the formation of hydroxyl radicals (OH·) [49].

The following are the four varied reactions, which are predicted for while the removal of pollutants photocatalytic degradation [32].

a. The reaction occurs while both species are adsorbed [33].

$$OH^* + R_1 \longrightarrow R_2^*;$$

b. A nonbound radical reacts with an adsorbed organic species [34].

$$OH^* + R_1 \longrightarrow R_2^*;$$

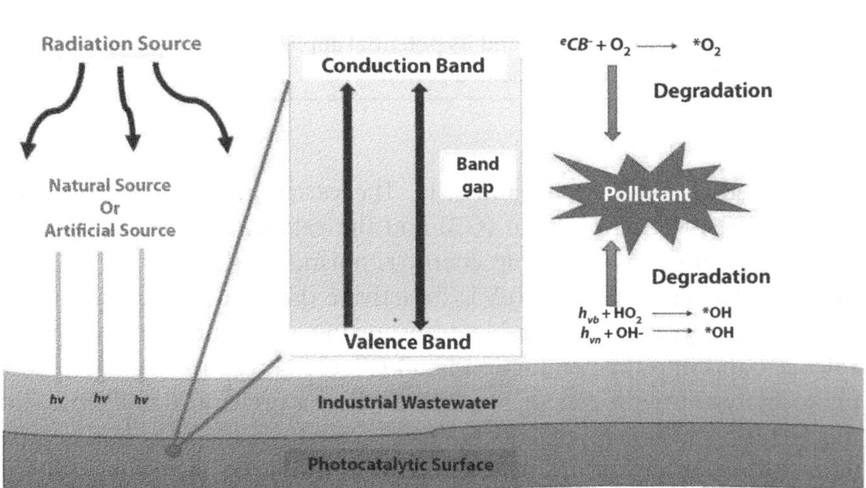

Figure 1.2 Mechanism of action of a photocatalyst.

c. An adsorbed radical reacts with a free organic species arriving at the catalyst surface [35].

$$OH^* + R_1 \longrightarrow R_2^*;$$

d. A reaction occurs between two free species in the bulk solution [36].

$$OH^* + R_1 \longrightarrow R_2^*;$$

Heterogeneous photocatalysis is a promising approach for treating polluted industrial wastewater possessing organic compounds. These new technologies are composed of advanced oxidation processes. A few of the major AOPs processes are H_2O_2/UV, O_3/UV, H_2O_2/O_3/UV, TiO_2/UV and VUV. Under the UV illumination of 254nm, H_2O_2/UV photocatalysis takes place and H_2O_2 is cleaved into hydroxyl radicals (OH^*). The hydroxyl radicals that are produced further degrade the organic compounds of wastewater by the mechanism of hydrogen abstraction, electrophilic addition and electron transfer. The rate of oxidation during photocatalysis is regulated by the hydroxyl radicals formed [37]. In the case of O_3/UV, oxidation of organic compounds present in industrial wastewater takes place. In this system, UV radiation (254 nm) is irradiated from aqueous solution saturated with ozone. It is observed that the use of ozone separately increase accelerates the oxidative degradation rates but the limitation is low solubility in water and consequent mass transfer [38]. The mechanism of action of O_3/H_2O_2/UV is similar to O_3/UV process. The rate of degradation increases by the photochemical generation of hydroxyl radicals by adding H_2O_2. The vacuum UV of 190nm is emitted by an excimer lamp as a light source. In VUV, the generation of hydroxyl radicals is mainly generated from H_2O. VUV photocatalysis is quite simple and advantageous as it requires no chemical usage.

1.4 Recent Advances in Heterogeneous Photocatalyst

Industrial wastewater contains a complex composition and dominant pollutants, such as ammonia, cyanide benzene, naphthalene, phenols, and cresols. Nowadays, the treatment of industrial wastewater has become a demanding topic. Therefore, various new technologies, such as membrane technology, electrochemical processes, and membrane bioreactor (MBR), for the treatment of industrial wastewater have been proposed and adopted [39]. In this context, the use of metal oxide-based photocatalysis is the method of choice. Since the discovery of HPCs by Fujishima and Honda in 1972, the use of semiconductors has received great attention [35]. The metal oxides used for photocatalysis are TiO_2, WO_3, ZnO, Fe_3O_4, V_2O_5, SnO_2, ZrO_2, Cu_2O, and Ta_2O_5. Sulfides (e.g. SnS, CnS, CDs) and ferrites are successfully used to break down organic pollutants. Principles of HPCs have been well documented in the literature, with the emphasis on the electronic structure of semiconductors [18, 40–42]. TiO_2, ZnO, and CdS are the main semiconductors that are used for heterogeneous photocatalysis.

The most common and frequently used semiconductor used in this process is TiO_2 [43]. The treatment has been chosen due to biocompatibility, abundance, low cost, and well suited for the design of efficient photocatalysis. During the photodegradation of varied organic compounds like aromatic and aliphatic chlorinated hydrocarbons using TiO_2 AOPs the following reactions occurs [43]:

$$CH_3COOH + 2O_2 \longrightarrow 2CO_2 + 2H_2O$$

$$CCl_4 + 2H_2O \longrightarrow CO_2 + 4H^+ + 4Cl$$

$$2CHCl_3 + 2H_2O + O_2 \longrightarrow 2CO_2 + 6H^+ + 6Cl$$

$$2HOC_6Cl_5 + 7O_2 \longrightarrow 4HCO_2H + 8CO_2 + 10HCl$$

Although semiconductors are ideal candidates for photocatalysis, they are associated with some inherent disadvantages that limit their photocatalytic performance, as well as their application prospects. Anatase TiO_2, for example, can only be activated under UV light, which is less than 5% of the solar spectrum. To take advantage of the abundant visible light in the solar spectrum, it is of paramount importance to produce photocatalysts that operate with high efficiency under visible light. In the case of conventional TiO_2, defects, such as quantum efficiency, low specific surface area, low use of visible light, recovery from the reaction medium, weak photoreductivity, limit its performance as photocatalysts. The HPCs have shown several advantages over the conventional homogeneous catalysts (Table 1.2). The ideal HPCs must have immediately recognizable properties, such as high activity, efficient recovery, low cost, photo stability, nontoxicity, chemical inertness, and high efficiency [44–48]. The applications of HPCs on a pilot scale are hampered by their poor consumption of visible light along with the high recombination rate of electron-hole pairs. Given the above challenges, the continuing exploration in the field of photocatalytic technology has led researchers to expand both the photoreaction and photoactivity of the visible light range in the solar spectrum.

The development of new semiconductors has received considerable attention for their application to eliminate organic pollutants from aqueous environments. In this context, efforts have been made to use semiconductors for various applications, including photocatalytic technology. For example, experiments, such as doping can be considered, since TiO_2 is highlighted, since it has a band gap in the range of 3.0 to 3.2 eV. The photocatalytic performance of TiO_2 could be improved by incorporating additional components into the semiconductor structure, which in turn promote the sensitivity of TiO_2 to visible light. This could be possible by changing its electronic as well as optical properties. The technique increases the VB edge of TiO_2 without changing the position of the CB edge, thereby lowering the band gap. The trending approach minimizes the electron-hole recombination method in some way and improves photocatalysis [49]. Other approaches include the formation of composites, precious metal deposition, nonprecious metal deposition, surface modification, and dye sensitization [50–53]. Several approaches have been tried to improve the inherited properties of conventional TiO_2. Among

Table 1.2 Advantages and disadvantages of HPCs.

Advantages	Disadvantages
Low-cost stability	Harvesting of visible light
Chemical inertness	Photocatalysis recovery from the mixture is not easy
High activity nontoxic	Difficult isolation
Stability in aqueous environment	High recombination rate of electron-hole pairs
Efficient recovery and reasonable recyclability	Poor electric adsorption and treatment of high concentration of organic pollutants

Table 1.3 Photodegradation efficiencies of some of HPCs on organic contaminants.

HPCs	Target pollutants	Reaction conditions	Time (min)	Degradation efficiency (%)	References
Nb_2O_5/ZnAL-LDH	Congo red	Visible light	390	85	[56]
Graphene-based TiO_2	Bisphenol A	Visible light	180	67.6	[57]
TiO_2/WO_3	Methylene blue Rhodamine B	Visible light	60 20	100 50	[58]
Sn/N-TiO_2	Zopiclone	UV-Vis light	120	91	[59]
ZnO-TiO_2	Azo dye	UV-Vis light	180	99	[60]
Sepiolite/BiOCl/ TiO_2	Tetracycline	Visible light	180	90	[61]
Sm(III), N, P-doped TiO_2	4-Chlorophenols	Visible light	120	100	[62]
C/N-doped TiO_2	Phenols	Visible light	150	87	[63]
Fe(III)-doped TiO_2	Nitrobenzene	Visible light	240	88	[64]
$ZnSnO_3$	Ciprofloxacin	Visible light	100	85.9	[65]

(Continued)

Table 1.3 Photodegradation efficiencies of some of HPCs on organic contaminants. (*Continued*)

HPCs	Target pollutants	Reaction conditions	Time (min)	Degradation efficiency (%)	References
MOF-@rGO	Methylene blue Rhodamine B Methyl Orange	Sunlight Sunlight Sunlight	20 20 20	93 97 92	[66]
TiO_2@LDH	Methylene blue Phenol	Visible light UV light	60 60	95 90	[67]
SiO_2-TiO_2	Phenol	Visible light	240	90	[68]
Cu-TiO_2	Chlorophenols	Visible light	144	98.9	[69]
Zr^{+4}-TiO_2	4-Chlorophenols	UV-Vis light	480	37.4	[70]
TiO_2@MIL-101	Methyl orange	UV light	30	99	[71]
Fe_2O_4@MIL-100(Fe)	Methylene blue Methylene blue	UV-Vis Visible light	40 200	100 100	[72]
ZnO/CdS@ZIF-8	Rhodamine B Methylene blue	Visible light	120	62 99.9	[73]

(*Continued*)

Table 1.3 Photodegradation efficiencies of some of HPCs on organic contaminants. (*Continued*)

HPCs	Target pollutants	Reaction conditions	Time (min)	Degradation efficiency (%)	References
Fe_3O_4@rGO@ZnO	Metformin	Visible light	60	100	[74]
Ag-$ZnFe_2O_4$	Oxytetracycline	Visible light	150	90.5	[75]
Ca/Zn-Al_2O_3	Caffeine	UV light	70	98.5	[76]
GO-based TiO_2	Bromothymol blue Rose bengal	UV-Vis light	80	86 90	[77]

the photocatalytic materials developed, ZnO is considered an alternative photocatalyst to TiO_2, as it has almost the similar band gap energy, but has a high absorption efficiency over a wide range of the solar spectrum compared to TiO_2 [54, 55]. In recent years, ZnO has attracted great interest as a potential photocatalyst for the decomposition of pollutants, including organic contents, due to its unique properties, such as long-term photo stability, excellent chemical stability, nontoxicity, and excellent charge transport binding energy at 60 meV. The modification of these semiconductors via the above approaches improves the surface structure to allow the absorption of light in the visible region of the solar spectrum. Table 1.3 shows some of the reported photocatalytic degradation efficiencies of the HPCs, with respect to organic pollutants. Overall, recently developed HPCs show a higher photocatalytic degradation performance for the elimination of organic pollutants. Coupling or modification of semiconductors has been shown to improve charge separation, reduce band gap energy, and reduce recombination rate.

1.5 Heterostructure Photocatalysts for the Degradation of Organic Pollutants

The HPCs can regulate the optical band gap, increase the absorption threshold, improve charge separation and lowers the recombination rate of electrons and holes, which can be preferable for oxidation and reduction reaction, and proved to be highly effective in photocatalytic degradation [18, 78, 79]. The heterostructure photocatalysts can be prepared by a combination of two semiconductors with varied energy levels, these photocatalysts show improved activity in photocatalysis as compared to pure photocatalysts. Semiconductors that are having low bandgap energy and negative CB are coupled with semiconductors that are having large bandgap, and then the transfer of electrons occurs between semiconductors. Based on several components, heterostructure photocatalysts are termed binary, ternary, etc. Heterostructure can be categorized by alignments of the semiconductors like Type-I, Type-II, Z-scheme, and S-scheme (Figure 1.3).

The positions of VB and CB of semiconductor A must be more positive and negative than those of semiconductor B. Under a suitable source of light, photogenerated electrons and holes of semiconductor A are simultaneously moved to semiconductor B. The recombination of electron-hole

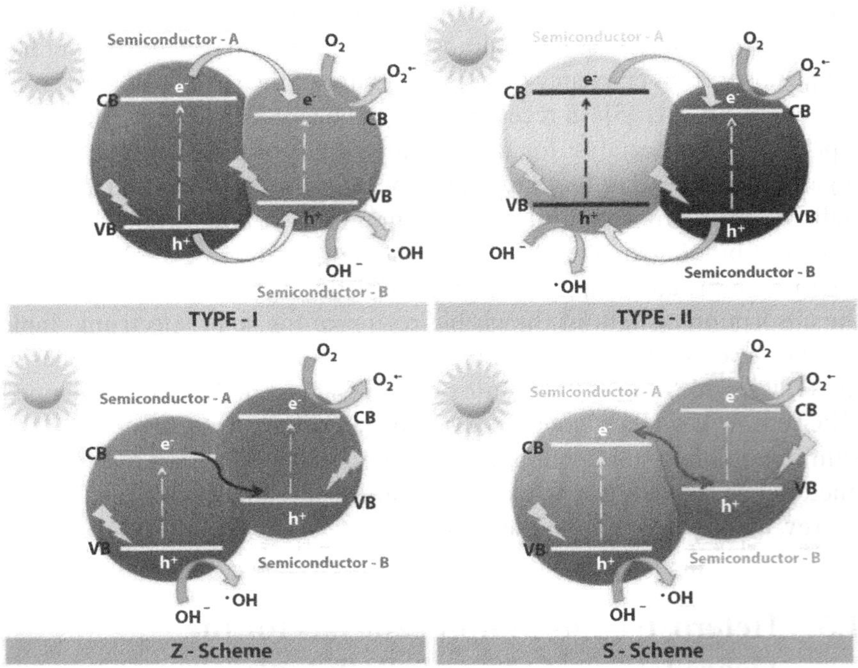

Figure 1.3 Types of heterostructure semiconductor photocatalysts (adopted from Ref [80]).

pairs within the similar semiconductor surface is high, this form of heterostructure is called type I. On the other hand, in the type II heterostructure, photoinduced electron-hole pairs are prepared within semiconductor A and semiconductor B in the presence of light. The photogenerated electrons are moved from semiconductor A to semiconductor B, whereas photogenerated holes are transferred in the opposite direction. This led to the aggregation of electrons on the semiconductor B, whereas semiconductor A receives holes. It promotes oxidation and reduction sites on semiconductor A or semiconductor A. For this reason, the photogenerated charges are spatially separated. In the Z-scheme heterostructure, holes in semiconductor A react with electron donors to form electron acceptors. Then the higher energy level electron in the CB of semiconductor A and holes in the VB of semiconductor B contribute in the photoreduction or oxidation reactions. The S-scheme heterostructure include combination of reduction and oxidation photocatalysts. Here, the strong electrons and holes in the

CB of the reduction photocatalyst or VB of the oxidation photocatalyst are reserved [80].

1.6 Photoreactors

The equipment used to carry out the photocatalytic process is known as a photoreactor. The photoreactors are designed accordingly to the reaction kinetics of photoreaction resulting in the production of intermediate products in a short period. The equipment is focused on the artificial or natural source of radiation for photocatalysis. The process of photocatalysis involves an advanced oxidation process that mineralizes the carbon containing compounds to water and carbon dioxide. Therefore, to carry out the advanced oxidation process, a semiconductor material is required. This semiconductor has a distinctive energy band gap level variating among its valence band and its conduction band, which is adequate to be overcome by the electrons excited by solar radiation [81].

Industrial wastewater is purified and degraded through the interaction of three phases. These three phases are the solid phase; the photocatalyst, the liquid phase; the contaminant and gas phase; oxygen. The intermediates and the radicals are short-tensioned. The photoreactors are designed to process the complex phenomena, a suitable interaction of the three phases under high turbulence, so that a proper reaction for industrial wastewater treatment takes place. The activation of the photocatalyst is controlled by the illumination of reactants and phases. Therefore, the management of the efficient operation of photoreactors is responsible for several reactions in several phases and the use of several phenomena occurring simultaneously [83].

The two most important parameters that influence the photocatalytic reactor are the kinetic reaction and the mass transport. Efficiency of the process enhanced by few of the factors, which include light source, intensity, pollutant concentration, humidity, temperature, surface area and activity of the photocatalyst [83]. In addition, approaches for photoreactor modeling and coupling of PC with other advanced oxidation processes (AOPs) using H_2O_2, O_3, and peroxydisulfate are preferred. Photocatalysis takes into account only a few targets, as mentioned below [82]:

a) To change the photoactivity of catalysts in the visible light range or to accelerate the degradation rate;
b) Use of artificial UV light sources, UV polychromatic lamps, and solar light;

c) Recovery and deactivation of photocatalysts;
d) Configuration of photoreactors;
e) Photodegradation of impurities;
f) To verify the induced effects of photocatalytic treatment; such as the induction of bacterial resistance.

Apart from the abovementioned objectives, few parameters are required to scale and stipulates the use of a photoreactor. The parameters are mentioned below [83]:

a) Photocatalyst—type and particle size.
b) Dispersion of the photocatalyst (fixed or suspended).
c) Photocatalyst—type, content, and distribution.
d) Mass transfer.
e) Fluid dynamics based on laminar and turbulent flow.
f) Temperature regulation.
g) Reaction mechanism and reaction kinetics.

Furthermore, photoreactors are classified based on the types of pollutants they degrade. Few of them are described in this chapter.

1.7 Photoreactors for the Degradation of Volatile Organic Compounds

1.7.1 Annular Reactors

They consist of two or more concentric cylinders, usually made of Pyrex glass. The interior of the outer cylindrical tube surface is coated with the photocatalyst. The coating of the photocatalyst layer should be thin so that it can illuminate the radiation source. The central region of the cylindrical tube has the radiation source. In the case of a gas-phase reactor, a fluorescent black-light blue lamp is used as the light source and a P25 TiO_2 thin film was applied to the internal glass surface. The efficacy of the photoreactor could be improved by the use of P25-TiO_2 impregnated with glass fiber support amid two Pyrex glass tubes. The use of this fiberglass supports maximum exposure to UV radiation. Air flow is passed in the axial direction through the ring between the lamp and the tube. Therefore, it is called a ring reactor (Figure 1.4). In addition, the reactor is classified as a

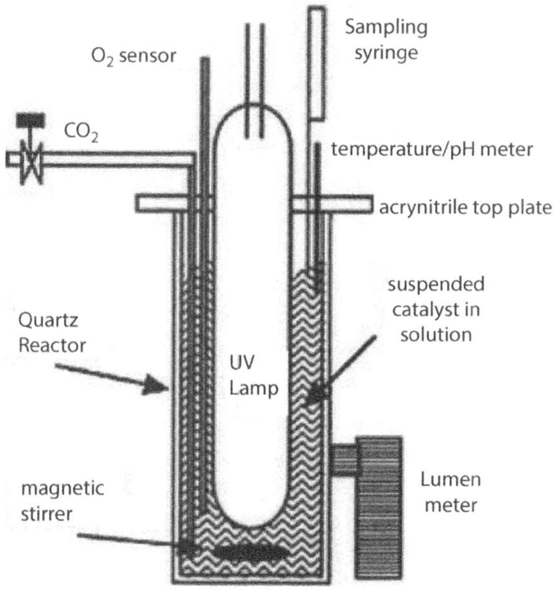

Figure 1.4 Design of annular reactor (adopted from ref [89]).

1D and 2D model on the basis of ideal flow or laminar flow conditions. Photocatalytic reaction carried out by these reactors, restricted by internal mass transfer [83–88].

1.7.2 Plate Reactor

It is considered one of the less complex types of reactors against volatile organic compounds. Plate reactors are divided into two types (internal and external source), depending on their source of radiation. It is made of stainless steel, plexiglass or polycarbonate and appears in square or rectangular box shape. Photocatalyst substances are positioned at the bottom of the device and used in powder form. In the reactor type with the internal radiation source, the lamp is attached to the upper part. In case of reactor with external source use of quartz or borosilicate window is observed (Figure 1.5). The advantageous properties of the use of these reactors are that they are less complex, have a low-pressure drop, and achieve high reaction rates. The disadvantage of this reactor, on the contrary, is that it has a small surface area for the continuation of the reaction [83].

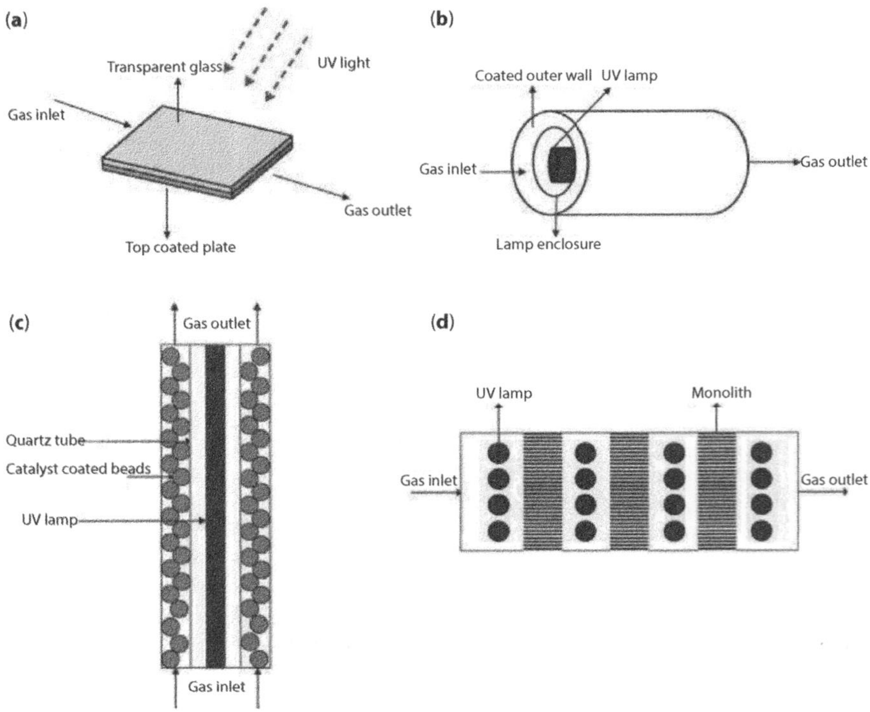

Figure 1.5 Internal structure of plate reactor (adopted from ref [90]).

1.7.3 Packed Bed Reactors

They consist of the tube-shaped usually composed of Pyrex glass or metal. Photocatalyst samples are positioned in the central area and the radiation source is located inside or outside the reactors (Figure 1.6). After numerous experiments in which material and radiation source were varied, it was concluded that the reactor could be optimized by theoretical prediction of the conversion factor, depending on volume, reaction, and molecular feed [83].

1.7.4 Honeycomb Monolith Reactors

These reactors are designed with several channels of circular or square cross-section. Inner walls of the channels coated with thin films of photocatalysts. For effective functioning, the radiation source is placed in front of the channels. Depending on the flow rate over the monolith, it was noted that an increase in the number of lamps should be followed with minimal distance between monolith and lamp, as this setup achieve an optimal configuration. Researches were also performed with computational fluid

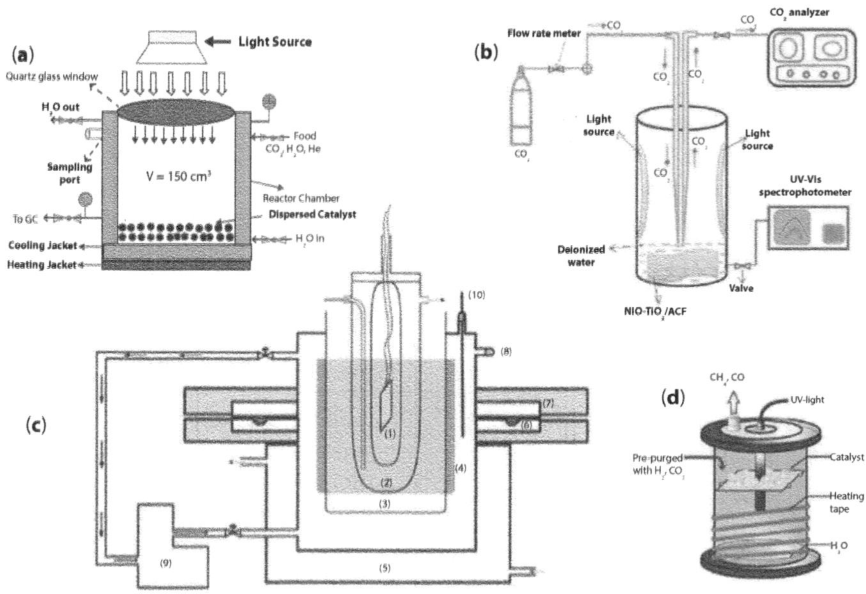

Figure 1.6 Design of packed bed photoreactor (adopted from ref [89]).

dynamics for a better outcome. Honeycomb monolith reactors are best suited for automobile combustion emission and nitrogen oxides reduction in power plant flue gases [83, 91].

1.7.5 Fluidized Bed Reactors

They consist of transparent containers. Their mechanism of action is based on the container through which air flows and is occupied with the photocatalyst bed. The light source is situated outside the apparatus. To achieve the best results, photocatalysts should be located near the high air flows used. Two-diode ultraviolet light-emitting diode (UV-LED) modules were used to intensify the photocatalytic oxidative dehydrogenation of cyclohexane in the gas phase. It was placed in front of the Pyrex windows for the emission of UV radiation [83, 92].

1.7.6 Batch Reactors

These reactors are comparatively less complex and are used specifically for the degradation of volatile organic compounds. They consist of a compartment made of Pyrex glass, and the photocatalyst is placed in the lower region of the compartment. The radiation source is aligned externally the reactor. In some studies, TiO_2 has also been coated on fiberglass fabrics using the sol-gel process with fluorescent black light lamps to achieve better performance [83, 93]. Commonly used photoreactors for organic compounds are shown in Table 1.4.

Furthermore, photolytic reactors with TiO_2 filter (500 μm) are used to inactivate bacteria (*Bacillus subtilis* and *Penicillium citrinum*). There is a fluorescent backlight lamp on the surface of the filter and glass slide. The

Table 1.4 Commonly used photoreactors for organic compounds.

S. no.	Types of chemical reactions	Types of photoreactors
1	Disinfection of water and polluted water	Annular flow reactor, packed-bed reactor, honeycomb monolithic reactor, plate reactor, wall reactor, fixed-bed reactor slurry with the immersed and external light source.
2	CO_2 conversion	Annular flow reactor, packed-bed reactor, honeycomb monolithic reactor, plate reactor, batch reactor.
3	Treatment of wastewater	Double-skin sheet reactor (DSSR), Parabolic trough reactor (PTR), Compound parabolic collecting reactor (CPCR), Wall reactor, fixed-bed reactor slurry with immersed and external light source batch reactor.
4	Treatment of polluted air	Annular flow reactor, Packed-bed reactor, Honeycomb Monolithic reactor and Plate reactor.
5	Glycerol/biomass conversion and organic synthesis	Wall reactor, Fixed-bed reactor slurry with immersed and the external light source, Batch reactor.
6	Water splitting	Twin reactor, Batch reactor.

use of photocatalytic HEPA filters (high-efficiency particle absorber filters) is also used for the disinfection of microorganisms in practice. The mechanism of action for inactivating bacteria is oxidative destruction to cell walls, membranes, enzymes, and nucleic acids by ROS [94–96].

Photocatalysis of CO_2 conversion could be carried out in types of systems

(a) Two-phase system.
(b) Three-phase system.

The former include gas photocatalysts and liquid photocatalysts. Into this gas mixture of CO_2, H_2O and methanol is added into the reactor and the CO_2 reduction takes place. Various photoreactors used for processing are sludge, fixed bed, ring, fiberglass and honeycomb monolith (Figure 1.7). The main factors for efficient photocatalysis are quite similar, i.e., the convective mass transfer rate of CO_2, the reaction rate, and the surface area of the photocatalyst [83].

In the conversion of inorganic pollutants, an oxidation process for degradation takes place. Therefore, their approach to the degradation of the

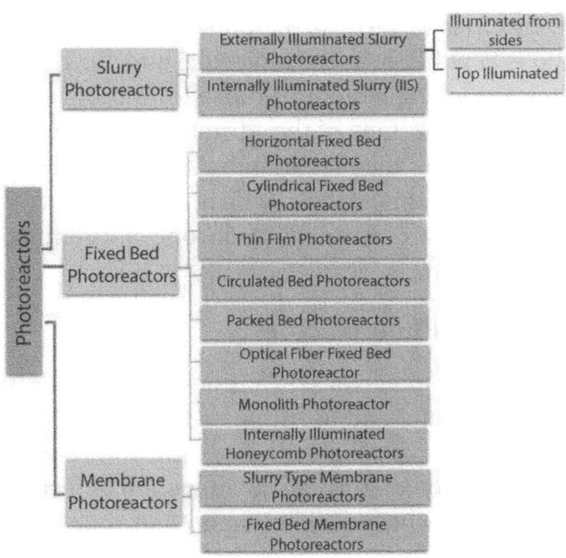

Figure 1.7 Classification of CO_2 photoreactor designs cast-off for reduction of CO_2 (adopted from ref [89]).

Table 1.5 Types of reactors used for destruction of inorganic pollutants.

Inorganic pollutants	
Type of reactors	**Mechanism of action**
Flow reactor	Oxidation of nitrogen oxide in the gas phase
Fixed bed reactor	Oxidation of nitrogen oxide, sulphur oxide, hydrogen sulphide
Plate reactor	Oxidation of nitrogen oxide in the gas phase

pollutants is summarized in Table 1.5. Some other commonly used photoreactors are parabolic trough photoreactors and inclined plane photoreactors.

1.7.7 Parabolic Trough Photoreactors

A parabolic trough photoreactor (PTP) is named because it has a parabolic reflector. This reflector accumulates the solar energy through radiation and focuses it on a photoreactor that is held on the focal line. The reflector consists of a transparent tube that contains wastewater with a suspended photocatalyst. In order to process the photocatalytic processes, the concentration ratio is kept between the range of 5 and 30. This type of photoreactor requires a tracking system in order to find out the concentrated radiation on the focal line. One of the advantages of using PTP is that this photoreactor concentrates the maximum radiations falling over it. On the contrary, due to its tracking system, which is the only axis and changes daily and seasonally, this is considered to be one of its limitations [97].

1.7.8 Inclined Flat Photoreactors

The inclined planar photoreactor (IPP) is designed as an inclined surface upon which the wastewater and the dispersed photocatalyst are allowed to flow. Exposure to the source of light. The mixture flows downward in a thin film formed by the action of gravity. The induction of turbulence depends on the inclined surface, which could be smooth or rough (step-wise) [97].

1.7.9 Gas Phase Photoreactors

This type of reactor is broadly categorized into two parts—the reactor structure and the light source. In the case of photoreactors used for

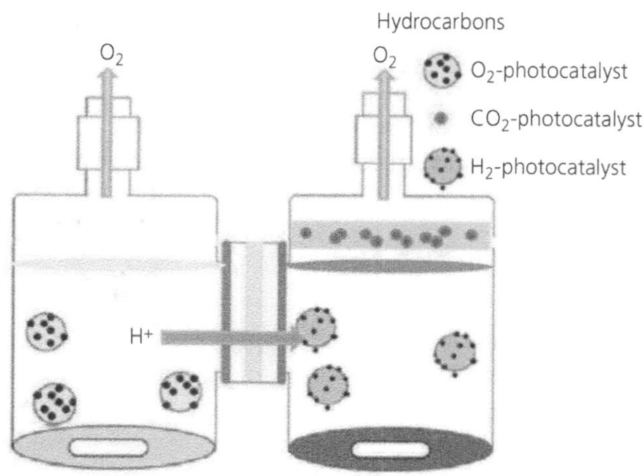

Figure 1.8 General layout of gas-phase photoreactors (adopted from ref [98]).

purification of air, source of radiation focused on photocatalyst surface, photocatalyst with large area and mass transfer, low pressure drop and long residence time are indicated (Figure 1.8). Some of the appropriately designed common photoreactors are annular, plate, slurry, honeycomb, monolith, packed bed, and fluidized bed reactors. The other types include powder layer reactor, aerosol generator, optical fibers, etc. Many of these photoreactors are limited to the use of laboratory scale. Whereas, for commercial application for environmental applications design of efficient reactors are preferred [83].

1.8 Advantages and Disadvantages of Heterogeneous Photocatalysis

Keeping in view the latest technique of degradation of industrial wastewater, a few of its advantages and disadvantages are as follows [99, 100]:

Advantages:

- stability of photocatalyst in aqueous system,
- consumable reagents are not required,
- nonselectivity of catalytic activity,

- highly active and nontoxic,
- destruction of organic and inorganic pollutants,
- low-cost stability of the photocatalyst,
- efficient recovery and reasonable recyclability,
- cost-effective in case of using solar radiation as a light source.

Disadvantages:

- high recombination rate of e^{-}–h^{+} pair,
- lack of mass transfer limitations, in augmented rate,
- difficulty in recovery from the mixture,
- expensive while using UV radiation,
- harvesting of visible light,
- photocatalyst has poor electric adsorption and treatment of high concentration of organic pollutant,
- complex O_3/UV process.

1.9 Conclusion

The persistence of pollutants in industrial wastewater is an issue of concern to contemporary society. Among varied types of advanced techniques available, structured photocatalytic systems are a great alternative in the advancement for the treatment of industrial wastewater. However, to develop these structured photocatalytic systems more efficient for scale-up and effective management of industrial waste, the efficiency of the process of photocatalysis should be focused on. Moreover, many aspects of type and effective reactor design should be structured for the proper functioning of these photocatalytic systems. Parallelly, researchers should concentrate on nanomaterial-based photocatalysis for minimal or no environmental impacts and risks. Nevertheless, continual attempts are being made in this regard, which caters to various applications.

Acknowledgment

This research was supported by Development of Marine Microplastic Pollution Response and Management Technology of Korea Institute of Marine Science & Technology Promotion (KIMST) funded by the Ministry of Oceans and Fisheries (KIMST-20220035).

References

1. Furuya, K., Hafuka, A., Kuroiwa, M., Satoh, H., Watanabe, Y., Yamamura, H., Development of novel polysulfone membranes with embedded zirconium sulfate-surfactant micelle mesostructure for phosphate recovery from water through membrane filtration. *Water Res.*, 124, 521–526, 2017.
2. Amarine, M., Lekhlif, B., Mliji, E.M., Echaabi, J., Nitrate removal from groundwater in Casablanca region (Morocco) by electrocoagulation. *Groundwater Sustainable Dev.*, 11, 100452, 2020.
3. Tang, X., Zheng, H., Teng, H., Sun, Y., Guo, J., Xie, W., Yang, Q., Chen, W., Chemical coagulation process for the removal of heavy metals from water: A review. *Desalin. Water Treat.*, 57, 1733–1748, 2014.
4. Ye, L., Chai, L., Li, Q., Yan, X., Wang, Q., Liu, H., Chemical precipitation granular sludge (CPGS) formation for copper removal from wastewater. *RSC Adv.*, 6, 115, 114405–114411, 2016.
5. Makhado, E., Pandey, S., Ramontja, J., Microwave assisted synthesis of xanthan gum-cl-poly (acrylic acid) based-reduced graphene oxide hydrogel composite for adsorption of methylene blue and methyl violet from aqueous solution. *Int. J. Biol. Macromol.*, 119, 255–269, 2018.
6. Maponya, T.C., Hato, M.J., Makhado, E., Makgopa, K., Khanuja, M., Modibane, K.D., Photocatalytic degradation of dyes in wastewater using metal organic frameworks, in: *Metal, Metal-Oxides and Metal-Organic Frameworks for Environmental Remediation. Environmental Chemistry for a Sustainable World*, vol. 64, Springer, Cham, 2021, https://doi.org/10.1007/978-3-030-68976-6_10.
7. Schiavello, M. (Ed.), *Photoelectrochemistry, photocatalysis and photo reactors, fundamentals and developments*, Dordrecht, Reidel, Springer, Holland, 1985.
8. Basile, A., Mozia, S. *et al.*, (Eds.), *Current trends and future developments on (bio-) membranes-photocatalytic membranes and photocatalytic membrane reactors*, Elesvier, Amsterdam 2018.
9. Fujishima, A., Rao, T.N. *et al.*, Titanium dioxide photocatalysis. *J. Photochem. Photobiol. C: Photochem. Rev.*, 1, 1–21, 2000.
10. Gaigneaux, E.M., D.M. *et al.*, (Eds.), Scientific bases for the preparation of heterogeneous catalysts. *Proceedings of the 10th International Symposium*, Louvain-la-Neuve, Belgium, July 11-15, 2010.
11. Khataee, A.R. and Fathinia, M. (Eds.), Recent advances in photocatalytic processes by nanomaterials, in: *New and Future Developments in Catalysis*, Elsevier, Amsterdam, 2013.
12. Bahraniab, S., Mojtaba, S. *et al.*, (Eds.), Current heterogeneous catalytic processes for environmental remediation of air, water, and soil, in: *Interface Science and Technology*, Elsevier, Amsterdam, 2021.
13. Conner, J.R. (Ed.), *Choosing the right CFS-chemical fixation and solidification of hazardous wastes*, Van. Nostrand Reinhold, New York, 1990.

14. Matthews, R.W., Photocatalytic oxidation of organic contaminants in water: An aid to environmental preservation. *Pure Appl. Chem.*, 64, 1285–1290, 1992.
15. Mills, A. and Hunte, S.L., An overview of semiconductor photocatalysis. *J. Photochem. Photobiol. A: Chem.*, 108, 1–35, 1997.
16. Chen, D., Sivakumar, M. *et al.*, Heterogeneous photocatalysis in environmental remediation. *Dev. Chem. Eng. Miner. Process.*, 8, 505–550, 2008.
17. Shukla, K., Agarwalla, S., Duraiswamy, S., Gupta, R.K., Recent advances in heterogeneous micro-photoreactors for wastewater treatment application. *Chem. Eng. Sci.*, 235, 116511, 2021.
18. Kou, J., Lu, C., Wang, J., Chen, Y., Xu, Z., Varma, R.S., Selectivity enhancement in heterogeneous photocatalytic transformations. *Chem. Rev.*, 117, 1445–1514, 2017.
19. Friedmann, D., Hakki, A., Kim, H., Choi, W., Bahnemann, D., Heterogeneous photocatalytic organic synthesis: State-of-the-art and future perspectives. *Green Chem.*, 18, 5391–5411, 2016.
20. Ahmed, S.N. and Haider, W., Heterogeneous photocatalysis and its potential applications in water and wastewater treatment: A review. *Nanotechnology*, 29, 342001, 2018.
21. Bahnemann, D., Photocatalytic water treatment: Solar energy applications. *Sol. Energy*, 7, 445, 2004.
22. Pichat, P., A brief survey of the practicality of using photocatalysis to purify the ambient air (indoors or outdoors) or air effluents. *Appl. Catal. B: Environ.*, 245, 770–776, 2019.
23. Mueses, M.A., Colina-Márquez, J., Machuca-Martínez, F., Puma, G.L., Recent advances on modelling of solar heterogeneous photocatalytic reactors applied for degradation of pharmaceuticals and emerging organic contaminants in water. *Curr. Opin. Green Sustain. Chem.*, 30, 100486, 2021.
24. Zare, E.N., Iftekhar, S., Park, Y., Joseph, J., Srivastava, V., Khan, M.A., Makvandi, P., Sillanpaa, M., Varma, R.S., An overview on non-spherical semiconductors for heterogeneous photocatalytic degradation of organic water contaminants. *Chemosphere*, 280, 130907, 2021.
25. Velempini, T., Prabakaran, E., Pillay, K., Recent developments in the use of metal oxides for photocatalytic degradation of pharmaceutical pollutants in water—A review. *Mater. Today Chem.*, 19, 100380, 2021.
26. Heng, Z.W., Chong, W.C., Pang, Y.L., Koo, C.H., An overview of the recent advances of carbon quantum dots/metal oxides in the application of heterogeneous photocatalysis in photodegradation of pollutants towards visible-light and solar energy exploitation. *J. Environ. Chem. Eng.*, 9, 105199, 2021.
27. Akerdi, A.G. and Bahrami, S.H., Application of heterogeneous nano-semiconductors for photocatalytic advanced oxidation of organic compounds: A review. *J. Environ. Chem. Eng.*, 7, 103283, 2019.

28. Chen, J., Cen, J., Xu, X., Li, X., The application of heterogeneous visible light photocatalyst in organic synthesis. *Catal. Sci. Technol.*, 6, 349, 2016.
29. Kisch, H. and Henning, H. (Eds.), *European Photochemistry Association (EPA) Newslett.*, Europe, 1983.
30. Serpone, N. and Pelizzetti, E. (Eds.), What is photocatalysis?, in: *Photocatalysis-Fundamentals and Applications*, Wiley, New York, 1989.
31. Boer, K.W. (Ed.), *Dingwang metal ions-survey of semiconductor physics*, Van Nostrand Reinhold, New York, 249, 1990.
32. Turchia, C.S. and Ollisam, D.F., Photocatalytic degradation of organic water contaminants: Mechanisms involving hydroxyl radical attack. *J. Catal.*, 122, 178–192, 1990.
33. Choi, W., Termin, A. *et al.*, The role of metal ion dopants in quantum-sized TiO_2: Correlation between photoreactivity and charge carrier recombination dynamics. *J. Phys. Chem. A*, 98, 13669–13679, 1990.
34. Ollis, D.F. and Al-Ekabi, H. (Eds.), *Photocatalytic Purification and Treatment of Water and Air: Proceedings of the 1st International Conference on TiO_2 Photocatalytic Purification and Treatment of Water and Air, - Trace Metals in the Environment 3*, London, Ontario, Canada, 8-13 November 1992, Elsevier, Amsterdam, p. 3, 1993.
35. Fujishima, A. and Honda, K., Electrochemical photolysis of water at a semiconductor electrode. *Nature*, 238, 37–38, 1972.
36. Pruden, A.L. and Ollis, D.F., Photo assisted heterogeneous catalysis: The degradation of trichloroethylene in water. *J. Catalysis*, 82, 404–417, 1983.
37. Legrini, O., Oliveros, E. *et al.*, Photochemical processes for water treatment. *Chem. Rev.*, 93, 671–698, 1993.
38. Khan, S.R., Huang, C.R. *et al.*, Oxidation of 2-chlorophenol using ozone and ultraviolet radiation. *Environ. Prog.*, 4, 229–238, 1985.
39. Zhang, T., Wang, X. *et al.*, Recent progress in TiO_2-mediated solar photocatalysis for industrial wastewater treatment. *Int. J. Photoenergy*, 2014, 12, 2014.
40. Hoffmann, M.R., Martin, S.T., Choi, W., Bahnemann, D.W., Environmental applications of semiconductor photocatalysis. *Chem. Rev.*, 95, 69–96, 1995.
41. Mishra, Y.K., Adelung, R., Kumar, G., Elbahri, M., Mohapatra, S., Singhal, R., Tripathi, A., Avasthi, D.K., Formation of self-organized silver nanocup-type structures and their plasmonic absorption. *Plasmonics*, 8, 811–815, 2013.
42. Marschall, R., Semiconductor composites: Strategies for enhancing charge carrier separation to improve photocatalytic activity. *Adv. Funct. Mater.*, 24, 2421–2440, 2014.
43. Zaleska-Medynska (Eds.), *Application of metal oxide-based photocatalysis-metal oxides, metal oxide-based photocatalysis*, Elsevier, Amsterdam, Netherlands, 2018.
44. Sonavane, S.U., Gawande, M.B., Deshpande, S.S., Venkataraman, A., Jayaram, R.V., Chemoselective transfer hydrogenation reactions over nanosized

γ-Fe_2O_3 catalyst prepared by novel combustion route. *Catal. Commun.*, 8, 1803–1806, 2007.

45. Wu, W., He, Q., Jiang, C.Z., Magnetic iron oxide nanoparticles: Synthesis and surface functionalization strategies. *Nanoscale Res. Lett.*, 3, 397–415, 2008.
46. Hasija, V., Sudhaik, A., Raizada, P., Hosseini-Bandegharaei, A., Singh, P., Carbon quantum dots supported AgI/ZnO/phosphorus doped graphitic carbon nitride as Z-scheme photocatalyst for efficient photodegradation of 2, 4-dinitrophenol. *J. Environ. Chem. Eng.*, 7, 103272, 2019.
47. Wang, Y., Wang, Q., Zhan, X., Wang, F., Safdar, M., He, J., Visible light driven type II heterostructures and their enhanced photocatalysis properties: A review. *Nanoscale*, 5, 8326–8339, 2013.
48. Sharma, S., Dutta, V., Singh, P., Raizada, P., Rahmani-Sani, A., Hosseini, A., Bandegharaei, V.K., Thakur, carbon quantum dot supported semiconductor photocatalysts for efficient degradation of organic pollutants in water: A review. *J. Cleaner Prod.*, 228, 755–769, 2019.
49. Vaiano, V., Sannino, D. *et al.*, (Eds.), *Nanomaterials for the detection and removal of wastewater pollutants-heterogeneous photocatalysis*, Elsevier, Amsterdam, Netherlands, 2020.
50. Wang, D., Wang, Y., Li, X., Luo, Q., An, J., Yue, J., Sunlight photocatalytic activity of polypyrrole–TiO_2 nanocomposites prepared by "*in situ*" method. *Catal. Commun.*, 9, 1162–1166, 2008.
51. Macák, J.M., Tsuchiya, H., Ghicov, A., Schmuki, P., Dye-sensitized anodic TiO_2 nanotubes. *Electrochem. Commun.*, 7, 1133–1137, 2005.
52. An, H., Zhou, J., Li, J., Zhu, B., Wang, S., Zhang, S., Deposition of pt on the stable nanotubular TiO_2 and its photocatalytic performance. *Catal. Commun.*, 11, 175–179, 2009.
53. Wang, X., Zhang, S., Peng, B., Wang, H., Yu, H., Peng, F., Enhancing the photocatalytic efficiency of TiO_2 nanotube arrays for H_2 production by using non-noble metal cobalt as co-catalyst. *Mater. Lett.*, 165, 37–40, 2016.
54. Yogendra, K., Naik, S., Mahadevan, K.M., Madhusudhana, N., A comparative study of photocatalytic activities of two different synthesized ZnO composites against coralene red F3BS dye in presence of natural solar light. *J. Environ. Sci.*, 1, 11–15, 2011.
55. Qiu, R., Zhang, D., Mo, Y., Song, L., Brewer, E., Huang, X., Photocatalytic activity of polymer-modified ZnO under visible light irradiation. *J. Hazard. Mater.*, 156, 80–85, 2008.
56. Huang, G., Chen, J., Wang, D., Sun, Y., Jiang, L., Yu, Y., Zhou, J., Ma, S., Kang, Y., Nb_2O_5/ZnAl-LDH composites and its calcined products for photocatalytic degradation of congo red under visible light irradiation. *Mater. Lett.*, 173, 227–230, 2016.
57. Lai, C., Wang, M.-M., Zeng, G.-M., Liu, Y.-G., Huang, D.-L., Zhang, C., Wang, R.-Z., Xu, P., Cheng, M., Huang, C., Synthesis of surface molecular imprinted TiO_2/graphene photocatalyst and its highly efficient photocatalytic

degradation of target pollutant under visible light irradiation. *Appl. Surf. Sci.*, 390, 368–376, 2016.
58. Zhang, C.F., Qiu, L.G., Ke, F., Zhu, Y.J., Yuan, Y.P., Xu, G.S., Jiang, X., A novel magnetic recyclable photocatalyst based on a core–shell metal–organic framework Fe_3O_4@ MIL-100 (Fe) for the decolorization of methylene blue dye. *J. Mater. Chem. A*, 1, 14329–14334, 2013.
59. Kadam, A., Dhabbe, R., Shin, D.S., Garadkar, K., Park, J., Sunlight driven high photocatalytic activity of Sn doped N-TiO_2 nanoparticles synthesized by a microwave assisted method. *Ceram. Int.*, 43, 5164–5172, 2017.
60. Caliskan, Y., Yatmaz, H.C., Bektas, N., Photocatalytic oxidation of high concentrated dye solutions enhanced by hydrodynamic cavitation in a pilot reactor. *Process Saf. Environ. Prot.*, 111, 428–438, 2017.
61. Hu, X., Sun, Z., Song, J., Zhang, G., Li, C., Zheng, S., Synthesis of novel ternary heterogeneous BiOCl/TiO_2/sepiolite composite with enhanced visible-light-induced photocatalytic activity towards tetracycline. *J. Colloid Interface Sci.*, 533, 238–250, 2019.
62. Jiang, H., Wang, Q., Zang, S., Li, J., Wang, Q., Enhanced photoactivity of Sm, N, Ptridoped anatase-TiO_2 nano-photocatalyst for 4-chlorophenol degradation under sunlight irradiation. *J. Hazard. Mater.*, 261, 44–54, 2013.
63. Abdullah, A.M., Al-Thani, N.J., Tawbi, K., Al-Kandari, H., Carbon/nitrogen-doped TiO_2: New synthesis route, characterization and application for phenol degradation. *Arabian J. Chem.*, 9, 229–237, 2016.
64. Crişan, M., Mardare, D., Ianculescu, A., Drăgan, N., Niţoi, I., Crişan, D., Voicescu, M., Todan, L., Oancea, P., Adomniţei, C., Dobromir, M., Iron doped TiO_2 films and their photoactivity in nitrobenzene removal from water. *Appl. Surf. Sci.*, 455, 201–215, 2018.
65. Dong, S., Cui, L., Zhang, W., Xia, L., Zhou, S., Russell, C.K., Fan, M., Feng, J., Sun, J., Double-shelled $ZnSnO_3$ hollow cubes for efficient photocatalytic degradation of antibiotic wastewater. *Chem. Eng. J.*, 384, 123279, 2020.
66. Thi, Q.V., Tamboli, M.S., Ta, Q.T.H., Kolekar, G.B., Sohna, D., A nanostructured MOF/reduced graphene oxide hybrid for enhanced photocatalytic efficiency under solar light. *Mater. Sci. Eng.: B*, 261, 114678, 2020.
67. Seftel, E.M., Niarchos, M., Vordos, N., Nolan, J.W., Mertens, M., Mitropoulos, A.Ch., Vansant, E.F., Cool, P., LDH and TiO_2/LDH-type nanocomposite systems: A systematic study on structural characteristics. *Microporous Mesoporous Mater.*, 203, 208–215, 2015.
68. Wang, L., Wang, X., Yin, J., Zhu, Y., Wang, C., Silica induced oxygen vacancies in supported mixed-phase TiO_2 for photocatalytic degradation of phenol under visible light irradiation. *Catal. Commun.*, 87, 98–101, 2016.
69. Lin, J.C.-T., Sopajaree, K., Jitjanesuwan, T., Lu, M.-C., Application of visible light on copper-doped titanium dioxide catalyzing degradation of chlorophenols. *Sep. Purif. Technol.*, 191, 233–243, 2018.

70. Venkatachalam, N., Palanichamy, M., Arabindoo, B., Murugesan, V., Enhanced photocatalytic degradation of 4-chlorophenol by Zr^{4+} doped nano TiO_2. *J. Mol. Catal. A: Chem.*, 266, 158–165, 2007.
71. Chang, N., Zhang, H., Shi, M.-S., Li, J., Shao, W., Wang, H.-T., Metal-organic framework templated synthesis of TiO_2@ MIL-101 core-shell architectures for high-efficiency adsorption and photocatalysis. *Mater. Lett.*, 200, 55–58, 2017.
72. Zhang, C.F., Qiu, L.G., Ke, F., Zhu, Y.J., Yuan, Y.P., Xu, G.S., Jiang, X., A novel magnetic recyclable photocatalyst based on a core–shell metal–organic framework Fe_3O_4@ MIL-100 (Fe) for the decolorization of methylene blue dye. *J. Mater. Chem. A*, 1, 14329–14334, 2013.
73. Kong, R.-M., Zhao, Y., Zheng, Y., Qu, F., Facile synthesis of ZnO/CdS@ZIF-8 core–shell nanocomposites and their applications in photocatalytic degradation of organic dyes. *RSC Adv.*, 7, 31365–31371, 2017.
74. Cheshme Khavar, A.H., Moussavi, G., Mahjoub, A., Yaghmaeian, K., Srivastava, V., Sillanpapa, M., Satari, M., Novel magnetic Fe_3O_4@rGO@ZnO onion-like microspheres decorated with Ag nanoparticles for the efficient photocatalytic oxidation of metformin: Toxicity evaluation and insights into the mechanisms. *Catal. Sci. Technol.*, 9, 5819–5837, 2019.
75. Wu, Q. and Zhang, Z., Visible-light-driven mitigation of antibiotic oxytetracycline and disinfection of Escherichia coli using magnetic recyclable Ag-modified zinc ferrite/diatomite ternary hybrid material. *J. Chem. Technol. Biotechnol.*, 94, 2537–2546, 2019.
76. Elhalil, A., Elmoubarki, R., Farnane, M., Machrouhi, A., Mahjoubi, F.Z., Sadiq, M., Qourzal, S., Barka, N., Synthesis, characterization and efficient photocatalytic activity of novel Ca/ZnO-Al_2O_3 nanomaterial. *Mater. Today Commun.*, 16, 194–203, 2018.
77. Abadikhah, H., Kalali, E.N., Behzadi, S., Khan, S.A., Xu, X., Shabestari, M.E., Agathopoulo, S., High flux thin film nanocomposite membrane incorporated with functionalized TiO_2@ reduced graphene oxide nanohybrids for organic solvent nanofiltration. *Chem. Eng. Sci.*, 204, 99–109, 2019.
78. Friedmann, D., Hakki, A., Kim, H., Choi, W., Bahnemann, D., Heterogeneous photocatalytic organic synthesis: State-of-the-art and future perspectives. *Green Chem.*, 18, 5391–5411, 2016.
79. Moniz, S.J.A. and Ghosh, S. (Eds.), *Visible-light-active photocatalysis: Nanostructured catalyst design, mechanisms and applications*, Wiley-VCH Verlag Gmbh & Co. KGaA, Germany, 2018.
80. Suresh, R., Rajendran, S., Hoang, T.K.A., Vo, D.-V.N., Recent progress in green and biopolymer based photocatalysts for the abatement of aquatic pollutants. Mohammad Nahid Siddiqui Lorena Cornejo-Ponce (ed.) *Environ. Res.*, 199, 111324, 2021.
81. Kowalska, E. and Rau, S., Photoreactors for wastewater treatment: A review. *Recent Pat. Eng.*, 4, 242–266, 2010.

82. Sacco, O., Vaiano, V. *et al.*, Main parameters influencing the design of photocatalytic reactors for wastewater treatment: A mini-review. *J. Chem. Technol. Biotechnol.*, 95, 2608–2618, 2020.
83. Colmenares, J.C. and Yi-Jun Xu, Y. (Eds.), *Heterogeneous photocatalysis-fundamentals to green applications*, Springer, Springer Nature Switzerland AG, 2015.
84. Imoberdorf, G., Irazoqui, H. *et al.*, Photocatalytic degradation of tetrachloroethylene in gas phase on TiO2 films: A kinetic study. *Ind. Eng. Chem. Res.*, 44, 6075–6085, 2005.
85. Imoberdorf, G.E., Cassano, A.E. *et al.*, Optimal design and modelling of annular photocatalytic wall reactors. *Catal. Today*, 129, 118–126, 2007.
86. Tomašić, V., Jović, F. *et al.*, Photocatalytic oxidation of toluene in the gas phase: Modelling an annular photocatalytic reactor. *Catal. Today*, 137, 350–356, 2008.
87. Vincent, G., Marquaire, P.-M. *et al.*, Abatement of volatile organic compounds using an annular photocatalytic reactor: Study of gaseous acetone. *J. Photochem. Photobiol. A: Chem.*, 197, 177–189, 2008.
88. Vincent, G., Marquaire, P.-M. *et al.*, Photocatalytic degradation of gaseous 1-propanol using an annular reactor: Kinetic modelling and pathways. *J. Hazard. Mater.*, 161, 1173–1181, 2009.
89. Khan, A.A. and Tahir, M., Recent advancements in engineering approach towards the design of photo-reactors for selective photocatalytic CO_2 reduction to renewable fuels. *J. CO_2 Util.*, 29, 205–239, 2019.
90. Boyjooa, Y., Sunb, H., Liu, J., Pareek, V.K., Wang, S., A review on photocatalysis for air treatment: From catalyst development to reactor design. *Chem. Eng. J.*, 310, 537–559, 2017.
91. Wang, X., Tan, X. *et al.*, Modeling of formaldehyde photocatalytic degradation in a honeycomb monolith reactor using computational fluid dynamics. *Ind. Eng. Chem. Res.*, 53, 18402–18410, 2014.
92. Palma, V., Sannino, D. *et al.*, Fluidized-bed reactor for the intensification of gas-phase photocatalytic oxidative dehydrogenation of cyclohexane. *Ind. Eng. Chem. Res.*, 49, 10279–10286, 2010.
93. Amama, P.B., Itoh, K. *et al.*, Gas-phase photocatalytic degradation of trichloroethylene on pretreated TiO_2. *Appl. Catal. B: Environ.*, 37, 321–330, 20022002.
94. Lin, C.Y. and Li, C.S., Inactivation of microorganisms on the photocatalytic surfaces in air. *Aerosol Sci. Technol.*, 37, 939–946, 2003.
95. Chotigawin, R., Sribenjalux, P. *et al.*, Airborne microorganism disinfection by photocatalytic HEPA filter. *Environ. Asia*, 3, 1–7, 2010.
96. Modesto, O., Hammer, P. *et al.*, Gas-phase photocatalytic bacteria inactivation using metal modified TiO_2 catalysts. *J. Photochem. Photobiol. A: Chem.*, 253, 38–44, 2013.

97. Bora, L.V. and Mewada, R.K., Photoreactors for heterogeneous photocatalysis for wastewater treatment. *Int. Res. J. Eng. Technol. (IRJET)*, 3, 1011–1014, 2016.
98. Ali, A., Flores, M.C. *et al.*, Gas-phase photocatalytic CO2 reduction, A brief overview for benchmarking. *Catalysts*, 9, 727, 2019.
99. Fadhel, A.Z., Pollet, P. *et al.*, Combining the benefits of homogeneous and heterogeneous catalysis with tunable solvents and nearcritical water. *Molecules*, 15, 8400–8424, 2010.
100. Singh, P., Sharma, K. *et al.*, Systematic review on applicability of magnetic iron oxides–integrated photocatalysts for degradation of organic pollutants in water. *Mater. Today Chem.*, 14, 100186, 2019.

2

Role of Heterogeneous Catalysts for Advanced Oxidation Process in Wastewater Treatment

Rupali Mishra[1*], Sadanand Pandey[2†] and Elvis Fosso-Kankeu[3]

[1]*Department of Chemistry, KUTEM Research Center, KOC University, Istanbul, Turkey*
[2]*Department of Chemistry, College of Natural Science, Yeungnam University, Daehak-Ro, Gyeongsan, Gyeongbuk, Republic of Korea*
[3]*Department of Mining Engineering, College of Science Engineering and Technology, University of South Africa, Florida Science Campus, Johannesburg, South Africa*

Abstract
Water is one of the prime substances for all plants, animals, and human for their survival. If there was no water, we cannot imagine life on earth and clean water is equally important for humans. Therefore, at present, we have to find out the way to use water from wastewater. Advanced oxidation process is excellent technique to remove organic and hybrid materials through chemical treatment procedures. In advanced oxidation process (AOP), various types of chemical materials are used. For the synthesis of chemicals, green synthesis of compounds has received great scientific attention for a decade, this is due to the economic and environmental benefits as an alternative to chemical methods. Its byproducts are nontoxic reagents that are eco-friendly. Green synthesis is used for different approaches, e.g., water splitting, renewable energy, catalytic applications, sensors, gas absorption, etc. Currently, heterogeneous catalyst is one of the most promising candidate in industrial societies due to their good thermal and chemical stability, well-defined surfaces and functionalized structure with different type of materials, which have suitable area for the reactions. Thus, heterogeneous catalysts are also known as surface catalysts. With remarkable surface properties and structural modifications, we can enhance their performances as well. In this chapter, we focus on

Corresponding author: rplmisra@gmail.com
†*Corresponding author*: sadanand.au@gmail.com

Elvis Fosso-Kankeu, Sadanand Pandey, and Suprakas Sinha Ray (eds.) Photoreactors in Advanced Oxidation Processes: The Future of Wastewater Treatment, (37–70) © 2023 Scrivener Publishing LLC

the basic concepts of different types of heterogeneous catalysts, metal oxides, perovskites, graphene, double-layered hydroxides, and metal organic frameworks (MOFs), various techniques for the synthesis with emphasis in green synthesis, scope, importance in AOPs, and in wastewater treatment.

***Keywords*:** Heterogeneous catalyst, metal organic framework, metal oxides, advanced oxidation process (AOP), water treatment

Abbreviations

Hg	Mercury
Cr	Chromium
Cd	Cadmium
As	Arsenic
Pb	Lead
SO_4	Sulfate
CO_3	Carbonate
Cl	Chlorine
NO_3	Nitrate
PO_4	Phosphate
AOPs	Advanced oxidation processes
MOF	Metal–organic framework
Zr-MOF	Zirconium-based metal–organic frameworks
$ZrCl_4$	Zirconium tetrachloride
UiO-66	Universitetet i Oslo-66
UV radiation	Ultraviolet radiation
ZnO	Zinc oxide
TiO_2	Titanium dioxide
$Zn(NO_3)_3$	Zinc nitrate
K_2CO_3	Potassium carbonate
WO_3	Tungsten trioxide
DMSO	Dimethylsulfoxide
LDH	Layered double hydroxides
rGO	Reduced graphene oxide

2.1 Introduction

World population and environmental changes that increase water demand for agricultural and industrial applications are going to be major challenges for the human and other living species. Drinking water availability is going to be a worldwide threat for all living organisms, including human health and the environment due to water contamination from various emerging sources that increase problem for scientists with each day (Figure 2.1). A diverse range of chemical contaminants, micropollutants like inorganic gases [1], carbaryls [2], coal tar [3], pesticides [4] particularly hazardous metals, such as mercury (Hg) [5], chromium (Cr) [6], cadmium (Cd) [7], arsenic (As) [8], lead (Pb) [9], and hydrocarbons [10], is discharged into the aquatic environment by transport, industry, and agriculture activities, which are not only toxic but also nonbiodegradable. Without proper controls, industrialization in different countries results in unwanted waste effluents in the soil and water. The scale of the aforementioned challenge is increasingly growing. Many of industries are not following adequate standard and rules, and because of that, they are producing large amount of wastewater.

Thus, under such scenario, we need to focus on detection of pollutant and toxic metal and their harmful effect on environment and human health. After the understanding of the science of pollutants, we can develop technologies that can help in water treatment. To explore the chemistry of toxic pollutants, first, we need to discuss surface reactivity from one surface to the next and the modification as well. Surface reactivity shows the adsorption of at least one of the reactants on the catalyst surface. Fundamental principles of surface chemistry, which are regulated by heterogeneously

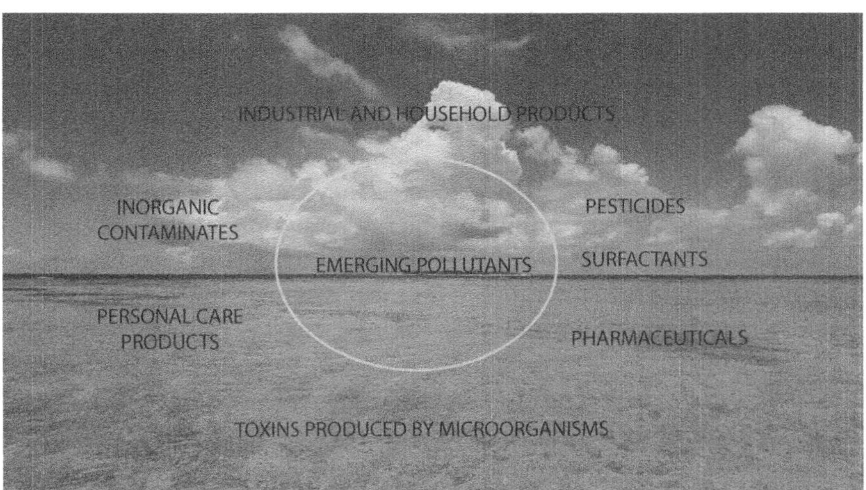

Figure 2.1 Emerging pollutants in the environment.

catalyzed reactions. In this chapter, we will explore the property, fundamental concept of heterogeneous catalysis and its synthesis.

Nowadays, heterogeneous catalysis is one of the most promising industrial processes for various fields. With the progress of technology and science, experimental methods have reached perfection, which allows us to conceive surface reactions on an atomic level to understanding and portended the action of heterogeneous catalysts. Due to their enticing capabilities, heterogeneous catalysts are more prevalent as subunits in several industries to enhance design and production through modern synthetic methods based on green concepts and the use of sustainable and recyclable materials. The surface properties of the catalyst have acidic and basic characteristics that contribute to the catalytic activity. Heterogeneous catalysts are additional and capable materials for various reactions because of simplicity and environment-friendly preparation. Many novel heterogeneous catalysts have superb activity, exceptional chemical and thermal stability, with high surface area. Based on heterogeneous nature, it can easily separate from the reaction mixture with continuous use in flow processes and products, making them attractive for recycling, and they prevent the production of inorganic salts. A group of catalysts is metallic, mixed oxide catalysts received great significant attention, in the last two decades due to their excellent potential. Many of the metallic catalysts are highly crystalline in nature and the combination of two or more metals, which allows the materials' surface properties to be coordinated, making them suitable for a particular reaction. These have a variety of advantages, including atom efficiency and green chemistry principles. Many researchers use different methods for the synthesis of heterogeneous catalyst, including green synthesis. Green synthesis is the branch of science which increase with the demands of environment-friendly processes and significant improvement by using green solvents, eco-friendly amiable techniques and renewable catalytic materials, which save energy, atom economy, and product yields. Green synthetic methods are designed so they can be conducted at room temperature, pressure or minimum energy requirement with the use and generate little or less hazardous chemical (selective catalyst or reagents) to human health and the environment for the target final product and prevent waste materials and unnecessary derivatives. Anastas and Warner defined Green Chemistry as "The invention, design and application of chemical products and processes to reduce or to eliminate the use and generation of hazardous substances" (Anastas and Warner) [11].

Chemical contaminants have been removed from contaminated water and wastewater using coagulation, filtration with coagulation, precipitation, ozonation, adsorption, ion exchange, reverse osmosis, and advanced oxidation processes. Advanced oxidation processes (AOPs) are one of these technologies that has made significant advances and achievements in the treatment of wastewater containing refractory organics. AOPs can decompose and transform toxic, hazardous, and refractory macromolecular organics into nontoxic, harmless, and biodegradable small molecular organics, such as photochemical oxidation, catalytic wet oxidation, sonochemical oxidation, electrochemical oxidation, and ozone oxidation. Carbon dioxide, water, and inorganic ions are the end products of the oxidation, without any excess sludge produced.

2.1.1 Advanced Oxidation Processes (AOPs)

In 1987, Glaze *et al.* established the concept of "advanced oxidation technologies" [12]. AOPs are widely used for overall wastewater treatment. In AOPs, free hydroxyl radicals have strong oxidizing power and electrochemical oxidant potential. Oxidizing power is one of the most important factors and ·OH is the second strongest oxidizing species with the comparison of other reactive species. Almost all recalcitrant organic contents are degraded by free hydroxyl radicals into carbon dioxide (CO_2), water, inorganic ions, or less toxic intermediate products. This technique is one of the best treatments for contaminated groundwater, surface H_2O, and non-biodegradable organic compounds presence in water. In this process, free hydroxyl radicals attack on pollutant via three possible ways: (i) dehydrogenation, (ii) hydroxylation, and (iii) redox reaction. The use of heterogeneous catalysts significantly enhances the degradation of textile sewage. In previous research found, there are very few studies, which deliver a comparative work on heterogeneous catalytic systems for wastewater treatment.

2.1.2 AOPs Classification

AOPs classification was perform based on different aspects of the process, the inclusion of light, use of precursor for HO*. Figure 2.2 shows a classification of the most common AOPs evaluated for water and wastewater based on their photochemical nature.

2.1.2.1 Catalytic Oxidation

Catalysts play an important role for minimization of waste and prevention of pollution. Commonly, we also know catalysts as "green chemicals" which can change the acceleration of the reaction process without being substantially consumed and improve the process economically. By using

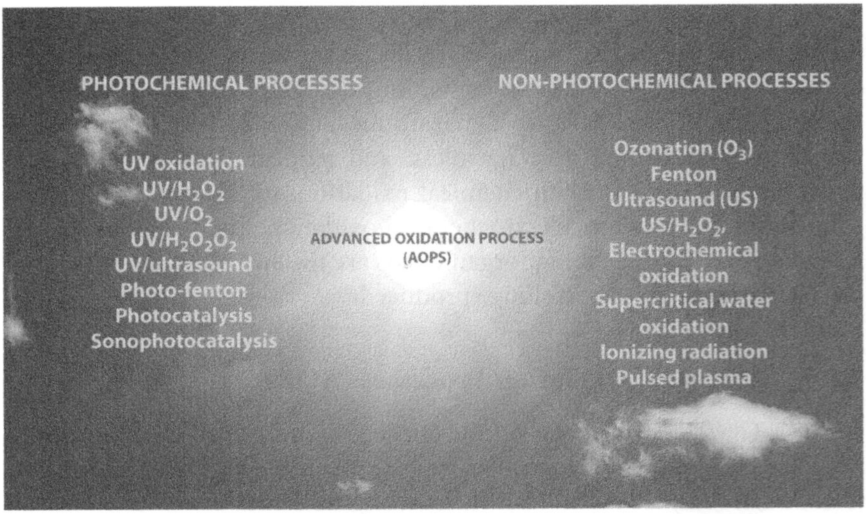

Figure 2.2 The most popular AOPs for water and wastewater that have been tested [UV, ultraviolet; US, ultrasound].

catalysts in a reaction, we can decrease the use and production of dicey material by choosing appropriate catalysts that can complete a chemical process with zero waste. According to studies, various types of catalysts can be used in conventional AOPs at a reasonable cost. Generally, two types of catalysts are used in reactions; (i) homogenous catalysts and (ii) heterogeneous catalysts.

Homogeneous reaction occurs when a homogeneous catalyst is the same phase as the reactants and is uniformly distributed in the reaction medium. Heterogeneous catalysts differ from the reactants in terms of phase. Since most reactions take place on the surface of the catalysts and in their pores, heterogeneous catalysts are also known as surface catalysts.

2.1.2.2 Heterogeneous Catalytic Oxidation

Heterogeneous catalysts have significant effects in many fields, including wastewater treatment due to high catalytic activity and high dispersion and large surface area. One of the advantages of heterogeneous catalysts is high activity under a broad range of pH due to low catalyst leaching. The porous matrices of heterogeneous catalysts play as active site. Various transition metals have been commonly used to make heterogeneous catalysts that can degrade recalcitrant organic compounds effectively. In heterogeneous AOPs for wastewater treatment, transition metal-based catalysts with different platform supports, such as Nafion, carbon aerogel, carbon nanotube (CNT), clays, polymers, alumina, and fly ashes, demonstrate excellent catalytic activity. The structure and properties of a support platform play an important role in modulating catalytic site activities.

2.2 Effect of Pollutant

Many research groups are focusing to understand and exploited the effect of pollutants, which are present in water. Pollution concentration has a major impact on AOP. With increasing pollutant concentration, wastewater significantly enhancing the collision between organic impurities and free hydroxyl ions. Increasing the concentration of pollutant showed a decrease in the treatment efficiency due to organic compounds and inorganic ions consumes on the surface of catalysts and reduce the catalytic activity. The active site of catalysts surface is deactivated after being trapped by the molecules. Many inorganic anions present in wastewater, e.g., SO_4, CO_3, CO_3, Cl, NO_3, NO_3, and PO_4, development of the Fe^{3+} complex, as well as the generation of toxic products, have an effect on the degradation of pollutants.

2.3 Type of Catalysts

The catalysts needed to have properties that could help improve reaction performance, increase the rate of a thermodynamically feasible reaction, and react at room temperature using alternative energy sources. Another important thing is that catalyst surface should be as large as possible and accessed to the reactants. In synthesis, methods are controlling the uniform structure, shape, suitable pore/particle size, and cost efficient. Both factors, cost efficiency and environment-friendly. are the important parameters to choose in the preparation method for the catalysts. There are several applications of heterogeneous photocatalysis provided in Figure 2.3. This chapter will describe the scope of varied heterogeneous catalysts, their synthesis, and contribution in the field APOs for wastewater treatment.

2.3.1 Metal Organic Frameworks

Continuous dumping of industrial waste in the river and cannel water pollution increases with each day due to toxic metal ions and dyes. To solve this problem, many research groups find the new ways, which can help

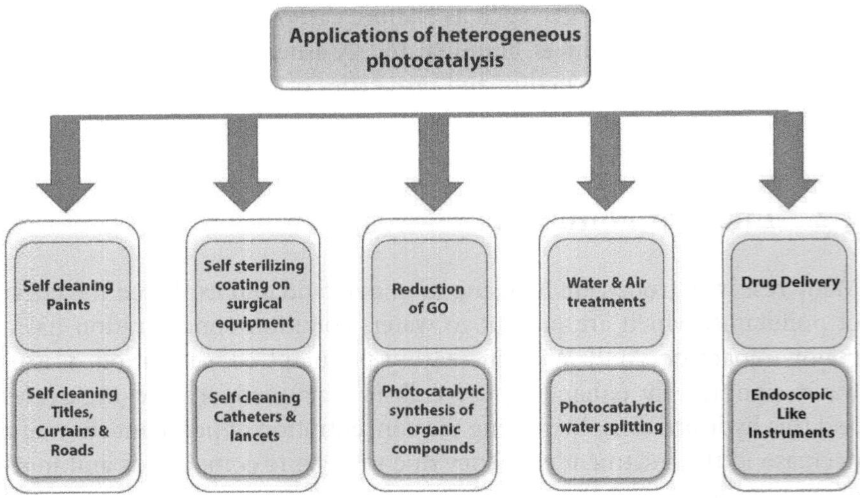

Figure 2.3 Applications of heterogeneous photocatalysis.

more to remove these contamination or developing novel materials to clean water. Metal-organic frameworks (MOFs) have garnered considerable attention in the last two decades due to their major contributions in a variety of fields, e.g., biomedical imaging [13], drug delivery [14], sensors [15], catalytic applications [16], magnetism [17], gas absorption [18], and separation [19]. In wastewater treatment, MOFs receive great interest due to some excellent physical and chemical properties, for example. In contrast to inorganic catalysts, MOFs have a lower thermal stability [20], but their surfaces are well functionalized with a variety of materials, porous in nature, high crystallinity, low density, structural diversity [21], absorption, and separation property. MOFs are hybrid materials that are constructed by of metal ions, inorganic building units, metal-oxo clusters connected by organic linkers.

MOFs have been synthesized by various method, e.g., direct mixing, in situ growth, layer-by-layer, and continuous flow synthesis. Several reasons are available to focus on the green synthesis of the promising compounds. Here, we will explore the green synthesis of MOFs, their various routes and methods. However, MOFs are mostly found as powders and crystalline in nature, processability and handling is a significant challenge as well.

To synthesize MOFs, researchers are following some of the most common techniques, such as hydrothermal, solvothermal, microwave, layered diffusion, and slow evaporation. Generally, MOF synthesis favors to high-yield product, but still, sometimes, a large amount of waste material coming out after the synthesis as organic solvent or mixed solvents are used during the synthesis or for purification. To prevent waste, we can reduce the amount of solvent and replace solvent with greener solvent, e.g., water and organic solvents produced from renewable feedstock like ethanol, N,N¢-dimethylformamide (DMF). Synthetic procedures should be designed to generate substances with no contamination. To control this parameter, we can use hydroxide or oxide of metal ions where water will form as a by-product after the combination of metal ion and acidic linker, or selective organic building units can also control unnecessary derivative production. Ligand solubility in aqueous condition is one of the important factors for the green synthesis of MOFs. Some chemists have used a modulator approach to synthesize MOF, in which single-coordination site ligands were combined with metal cations to change the rate of crystal formation [22]. MOFs have excellent architect, so synthetic route should be designed, which can preserve the product efficiency and structural functionality as well. Since energy consumption is a major factor in MOF synthesis, it should be carried out at room temperature and pressure.

2.3.1.1 Hydro (Solvo) Thermal Technique

Hydro (solvo) thermal technique has been extensively used in the synthesis of advanced materials, such as CPs, synthetic quartz, gems and complex oxides with commercial value. These techniques play an important role in the industries and in the academia as well. Hydrothermal technique depends on the solubility of minerals in a solvent or a mixture of solvents under temperature and pressure. In this reaction, precursors are dissolved in hot solvents. In this technique the solvent is water or other solvent which can provide milder and eco friendlier in reaction conditions [23]. Xu *et al.*, 2017 synthesized MIL100 (Fe) has many unsaturated metal sites, high surface area, and good hydrothermal resistance [24]. Under hydrothermal conditions at 50°C, Zr-Alg hydrogels reacted with trimesic acid, with the trimesic acid linker forming porous MOF-like structures embedded within the hydrogel [25]. Cohen group [26] synthesized Zr-MOF employed DMF and organic linker molecules in combination with $ZrCl_4$ in closed vessels at temperatures around 120°C in this processor gaseous hydrochloric acid are generated as a by-product, which is hazardous for environment. Further, Serre group reported green synthesis of UiO-66, they used water as a solvent for $ZrCl_4$ and the respective linker molecules [27].

2.3.2 Metal Oxides

Metal oxide and metal oxide nanoparticles promising class of heterogeneous catalysts and received considerable scientific emphasis due to notable electrical, optical, magnetic, and catalytic properties. Because of these excellent properties of metal oxide very useful material in industrial [28], medical [29], agricultural [30], photocatalytic oxidation processes [31], antimicrobial [32], biomedical applications [33], drug delivery [34], sensors [35], and in environmental as well. Metal oxide and metal oxide nanoparticles show good thermal stability [36] and moderate mechanical potency [37] and furnish surface area, which is very helpful to bind metal cations in mixed oxides with pollutants.

In an optimal situation, metal oxide and metal oxide nanoparticles should be able to work in a broad pH range, be cost-effective, photosensitive, and have high catalytic activity but low catalyst leaching. Various metal oxides have been used as catalysts in the surface environment to effectively degrade recalcitrant organic compounds and regulate the concentration, migration, and conversion of contaminants. Because of its low cost, high stability, and improved adsorption capability, transition metals based oxide has been widely used. It is also thermodynamically stable and has a higher performance under UV radiation. Metal oxides are crystalline in nature functional surface, well-architectural structure, including wide band gaps. The porous matrices are supporting as active side in heterogeneous reaction. Some metal oxides are also good photocatalyts. TiO_2 and ZnO are close to ideal photocatalysts due to their excellent photocatalytic properties. Metal oxide-based synthesized by several methods. Here, we will look at some of the most widely used and significant synthetic processors.

2.3.2.1 Coprecipitation Method

In this method, the preparation of metal oxide carried out at ambient temperature where precipitating medium and a metal salt solution are precipitating in an oxo-hydroxide form in a greener solvent mostly used water by adding a base solution. High pH and the nature of the base solution are favoring to establish the morphologies and properties of the metal oxides. The synthesis procedure's key advantages are its low cost and the fact that it needs only mild reaction conditions such as low temperature during the synthesis [38]. In 2014, Farahmandjou *et al.* synthesized ZnO by using $Zn(NO_3)_3$ and K_2CO_3 as precursors, whereas iron oxide nanoparticles use borohydride followed by coprecipitation method [39]. Mehmood *et al.*, have synthesized Ni Doped WO_3 nanoplates by coprecipitation shows an excellent photocatalyst activity, as a result, the band gap was narrowed, and the rate of unnecessary charge recombination processes was reduced [40].

2.3.2.2 Hydrothermal Synthesis

As previously mentioned, we addressed the hydrothermal technique and its advantages, such as reduced energy consumption, reduced time, and a solvent-free process. Basically, metal oxide complexes can have thermally decomposition by boiling the reactants in a torpid environment or in an autoclave. By adding adequate capping medium or stabilizer on proper time, agglomeration can stabilize dissolution of the particles in different solvents. In 2017, Buera *et al.* synthesized Ag-decorated ZnO-graphene nanocomposite via hydrothermal [41].

Sharma *et al.* also used a hydrothermal method to make a copper oxide/zinc oxide-tetrapods nanocomposite with high porosity, wide surface area, and low band gap that have excellent photocatalytic efficiency [42]. Yao and team reported hierarchical-structured WO3 through a facile and surfactant-free hydrothermal method, which has a composition of many nanosheets [43]. After that, in 2020, Kottam *et al.* modified tungsten oxide with carbon nanodots prepared by hydrothermal synthesis [44].

2.3.2.3 Sol-Gel Process

Metal oxides are formed by hydrolysis of metal precursors (alkoxides in an alcoholic medium) in the sol-gel process. Further, there is condensation of this hydroxide to form of a metal hydroxide polymer in dense and compact porous gel after heating and drying of the gel obtained crystalline powder of the metal oxide. Copper oxide nanoparticles were produced using the Sol-Gel Process and an ethanolic copper (II) solution [45].

2.3.2.4 Bioreduction Method

In comparison to other approaches, bioreduction requires a low amount of activation energy for reactions. Some microbes, especially bacteria, have been extensively investigated for the green synthesis of metal oxide nanoparticles [46]. Fungus has also been extensively explored for the green synthesis of metal oxide nanoparticles.

For the synthesis of metal and metal oxide or metal oxide nanoparticles, fungi are excellent biological agents. Many types of intracellular enzymes/proteins/reducing components are present in the cell or on the wall of their cell, which help in reduction process. Some members of fungal are extensively used to synthesize metal/metal oxide nanoparticles, such as gold, [47] silver, [48] zinc oxide, and titanium dioxide [49]. Many research groups have also synthesized metal oxide or metal oxide nanoparticle from yeast. Gold and silver nanoparticles by a *Saccharomyces cerevisiae*- and silver-tolerant yeast strain [50]. A wide variety of plants use the "one-pot" synthesis method to reduce and stabilize metal oxide or metallic nanoparticles. Many biomolecules contained in plants, such as carbohydrates, proteins, and coenzyme, have outstanding potential for reducing metal salts into nanoparticles. Several examples are available where metal oxide nanoparticle synthesized plant extract-assisted synthesis method, e.g., gold and silver nanoparticles extracted from aloe vera (*Aloe barbadensis* Miller) [51], oat (*Avena sativa*) [52], tulsi (*Osimum sanctum*) [53], lemon (*Citrus limon*) [54], Neem (*Azadirachta indica*) [55], coriander (*Coriandrum sativum*) [56], and mustard (*Brassica juncea*) [57] and via vivo synthesis zinc, nickel, cobalt, and copper was also extracted by mustard (*Brassica juncea*), alfalfa (*Medicago sativa*), and sunflower (*Helianthus annuus*) [58].

2.3.2.5 Solvent System-Based Green Synthesis

For any synthesis method, solvent systems are always necessary. In this process, water used as solvent have some limitations. Ionic liquids are playing an important role as a nontoxic/toxic and renewable solvent. Ionic liquids are polar and noncoordinating against polarity can be hydrophobic and hydrophilic as well, which depends on their cation or anion. Reactions can be conducted at wide range of high temperature with ionic liquids. Due to high temperature stability ionic solvents don't have evaporation issue. Every solvent have their critical point beyond that with temperature and pressure can make supercritical liquids. While an ordinary liquid become a supercritical liquid it shows drastic change in their properties such as thermal conductivity, density and in viscosity [59]. Manganese oxide (Mn_3O_4) nanoparticles were synthesized by Bussamara *et al.* in 2013 using imidazolium ionic liquids and oleylamine (a conventional solvent) [60]. Silver nanoparticles synthesis done by BmimBF4 ionic liquid and obtained isotropic spherical and large-sized anisotropic hexagonal nanoparticles in shape [61], and gold nanoparticles were synthesized in an aqueous solution using a laser ablation technique, where oxygen in the water caused partial oxidation, which increased their chemical reactivity [62]. Some other metal nanoparticles, such as aluminum, tellurium, ruthenium, iridium, and platinum, have been synthesized in ionic liquids [63].

2.3.3 Perovskites

Mixed metal oxides are used in many chemical industries. Perovskite-type oxides, in comparison to other mixed metal oxides, have a lot of scientific interest because of their low cost, adaptability, versatility, thermal stability, ideal energy bandgap value, strong absorption coefficient, spin-dependent transport, long charge diffusion length, noble magnetoresistance [64], ferroelectricity [65], and outstanding charge-carrier mobility [66].

Perovskites are excellent material for the advanced oxidation of wastewater treatment due to three reasons, (i) high temperature resistance; (ii) synthesized with wide range of elements with variable oxidation state; and (iii) high dispersion of metal particles, which are reducing the coke formation during the reaction.

In general perovskites formulated as ABX_3 with cubic structure. In formula A and B are representing the positively charged metal ions and X is negatively charged metal ions (mostly oxygen and halogen). Among all three, A is the center atom with largest atomic radius, whereas B is located on eight vertices, and X is placed at the center of the edges of the cubic cell.

Perovskite structure could be in orthorhombic by distortion of the BX and octahedral at lower temperatures. Transition metal ion-based perovskites show some interesting variation of electronic or magnetic properties on B site because of their chemical flexibility and larger certain coordination with oxygen or halides. Perovskite synthesis process requires substantial mass of energy to initiate the first step, which reduce the non-reacted reactants leave in the reaction method.

2.3.3.1 *Ultrasound-Assisted Synthesis of Perovskites*

Synthesis of perovskites via ultrasonic way is becoming popular day by day. Two component solutions were mixed in this processor by magnetic stirring at the appropriate temperature using continuous stirring at 1000 rpm or high-speed stirring at around 6000 rpm [67]. In 2015, Kesari *et al.* employed ultrasound method for the bulk synthesis of $CH_3NH_3PbI_3$ [68], and after that, Bhooshan *et al.* also announced the synthesis of perovskites in the 10 to 40 nm range by irradiating CH_3NH_3I and PbI_2 dissolved in isopropanol and processed without the use of a catalyst [69].

2.3.3.2 *Microwave-Assisted Synthesis of Perovskites*

As we know, perovskites are crystalline materials, this can be achieved by performing with a range of temperature. The microwave synthesis materials should have appropriate dipolar polarization [70] and ionic conductivity [71], which can absorb microwave energy. As we know, perovskites synthesis needs high amounts of energy to initiate the first step for that microwave radiations working as thermal initiator to get better crystallized material within a few minutes. High-temperature thermal-annealing process can change the chemical and physical properties at the atomic level and produce crystallized materials with less energy and time [72]. So, microwave synthesis is a promising route for hybrid perovskites synthesis. Solvents are one important reactant in microwave synthesis because solvents are absorbing the microwave irradiation energy and then transfers it to the perovskite powders for heating [73]. Nakamura *et al.* synthesized several perovskites with different composites of PbI_2 and MAI based on coordinating solvents, such as DMSO, DMF, and /γ-butyrolactone further, and it is stirred for 12 hours at 60°C after filtration obtained crystalline fibers [74].

2.3.3.3 Mechanosynthesis of Perovskites

The mechanochemical procedure is the most environmentally friendly and energy-saving phenomena. In the past few decades, solid-state chemical reactions are showing excellent output for the synthesis of different materials. Grinding precursors is one of the oldest methods which were followed by pharmacies, by using mortar and pestle. In grinding synthesis, the mechanical force applied on the solid materials, which accelerates the kinetic energy of the reactant and system, and this energy helps in phase transformation [75]. Kanatzidis and colleagues made $CH_3NH_3PbI_3$ by simply grinding MAI and PbI_2 in a mortar and pestle that had a lot of unreacted precursors [76]. Further, in 2015, Lewinski has reported hybrid perovskite $CH_3NH_3PbI_3$ by the mechanosynthesis processor with grinding of MAI and PbI_2, where it has taken stoichiometric ratio 1:1 in an electric ball mill for half an hour and obtained polycrystalline methylammonium lead iodide perovskite particles [77].

2.3.4 Layered Double Hydroxides

Layered double hydroxides (LDHs) are two-dimensional structural metal hydroxides materials [78] with significant properties of anion exchange abilities due to large interlayer space and high concentrations of active sites. LDHs are very applicable in the field of adsorbent [79], catalyst [80], ceramic precursor [81], and drug carrier [82], biological [83] and environmental incompatibilities [84]. Due to varied composition and micro/macroscopic morphologies of LDHs, applications are more viable. Layered double hydroxides formulated as $[M^{2+}1-xM^{3+x}(OH)_2][Ax/nn-\bullet mH_2O]$, where $M^{2+}$and M^{3+} are divalent and trivalent cations and An− anion improves their sorption potential for various pollutants. LDH could be modified by different anions according to applications. Most synthesized LDHs are found in two forms, rhombohedric [85] or hexagonal [86] unit cells. Nowadays, LDHs are getting attention because of the use in environmental remediation. Capacity to exchange ions LDHs are effective at removing heavy metal ions like Cr^{6+} and can also adsorb metal cations like Hg^{2+}, Cd^{2+}, and Pb^{2+}.

2.3.4.1 Coprecipitation by the Addition of Base

Layered double hydroxide (LDH) is prepared by mixing divalent or trivalent metal ions and base [87]. Reaction procedure used aqueous solution of metal salts and basic solution adding at constant pH or suitable pH for the precipitate. After washing precipitate LDH cake is obtained as the final product with some contaminations. Most nitrate, chloride, sulphate, and sodium or potassium presence filtrates come out. Obtained product is crystalline in nature with broad particle size. LDHs are generally in nanometer size particles and hexagonal in shape [88]. The most effective factor in this process is pH. By changing in stating duration of the reaction metal salts are using OH ions for anion exchange. After that, reaction mixture is washed and filtered to remove exchanged ions or unwanted ions.

Several LDH syntheses via green synthesis processor, e.g., in 2014, Hou group reported $Mg_2Al\text{-}NO_3$ LDH single-layer nanosheets by using $Mg(NO_3)_2/Al(NO_3)_3$ mixed solution and under magnetic stirring and nitrogen environment, an ammonia solution was applied to a beaker at the same time. The precipitate was collected by centrifugation at 12,000 rpm and washed with water after stirring for around 10 minutes [89]. Gevers and group obtained hydrocalumite LDH by using $Ca(OH)_2$ and $Al(OH)_3$ precursor [90].

2.3.5 Graphene

Graphene brought a dramatic change in the field of nanoscience and material science from carbon family. Graphene is a single atomic layer of graphite and two-dimensional structural material which has outstanding physical, chemical, optical and electronic properties. Thermal conductivity, high mobility, large specific surface area, pore structure, mechanical strength, and good stability are all desirable characteristics in catalyst supports. Because of these properties of graphene is applicable in wide

Figure 2.4 Possible applications of graphene material.

range of scientific and technical fields, such as in sensors [91], solar cells [92], field-effect transistors [93], and pharmaceutical [94] applications (Figure 2.4).

2.3.5.1 Electrochemical (EC) Processes

As we know, several types of dye molecules are contaminating water, which are coming from industries. Removal of these molecules by dye adsorption is one of the useful techniques to reuse water. Graphene and graphene family members such rGO, GO, GNS, are performing excellent output in this field. Synthesis of grapheme or grapheme sheet by electrochemical method used graphite with molten salts, such as molten NaCl [95], molten

LiCl-NaCl [96] and mixture of LiCl/KCl [97]. Molten salts have a number of advantages, including high efficiency, low cost, ease of use, and fast processing times. Increased yields and purer nanostructured graphene were obtained by tuning the synthesis process while incorporating graphite in an ionic molten salt medium. In the first step, graphite was mixed with molten salts or eutectic LiCl/KCl in a 1:10 ratio, the powders were homogenized in an agate mortar, and the mixture was then filled into a ceramic crucible and loaded into the reactor. After that system was vacuumed and flushed with nitrogen or argon gas for 5 minutes and slowly heated till desired temperature with and kept at this temperature for 5 hours. In the final step, the system was cooled to room temperature. The products are washed with water to remove the salts and dried overnight [98]. Ren and colleagues synthesized graphene oxide using an electrochemical process at room temperature and commercially available flexible graphite paper as a raw material [99].

2.3.5.2 Water Electrolytic Oxidation

This is one of the best methods because it is scalable, safe, and fast rate of reaction by electrolytic oxidation of graphite. This system is 100 times faster than any other method currently in use. Graphite preinteraction inhibits the anodic electrocatalytic oxygen evolution reaction at high voltage during the process period, allowing for rapid graphene oxidation in a matter of seconds. Water electrolytic oxidation of the graphite intercalation compound paper (GICP) in H_2SO_4 and well-dispersed graphene oxide (GO) aqueous solution sonicate in water for 5 minutes [100].

2.4 Some Recent Heterogeneous Catalysts for Advanced Oxidation Process

The synthesis of magnetic Fe_3O_4@-CD/MWCNT nanocomposites for heterogeneous Fenton-like catalysts to degrade tetrabromobisphenol A (TBBPA) was reported by Zhang *et al.*, [101]. When it came to removing TBBPA from water, Fe_3O_4@-CD/MWCNT showed good catalytic efficiency, with a removal rate of around 97 percent under ideal conditions.

Under UV–Vis light, Sun *et al.* investigated the photocatalytic efficiency of an activated carbon-supported TiO_2 catalyst (AC/TiO_2) for aflatoxin B1 (AFB1) degradation. Sun *et al.* used a simple hydrothermal synthesis to make AC/TiO_2 [102]. The higher degradation efficiency of AFB1 by AC/TiO_2 composite (98%) than bare TiO_2 (76%) was due to a higher surface area of AC/TiO_2 and increased visible-light intensity by the synergistic effect of TiO_2 and AC, according to the experiments. The mechanism of most photocatalysis processes is the same, as shown in Figure 2.5.

Chao Zhang *et al.* investigated a heterogeneous electro-Fenton (EF) catalyst for 2,4-dichlorophenol (2,4-DCP) degradation in near-neutral pH conditions using modified iron-carbon with polytetrafluoroethylene (PTFE) [103]. At a catalyst dosage of 6 g/L, an initial pH of 6.7, and a current intensity of 100 mA, the degradation efficiency of 2,4-DCP could exceed 95% after 120 minutes of treatment.

Due to the higher adsorption potential of RuO_2/AC for phenol, which had more chances of interacting with oxidizing radicals, Muhammad *et al.* reported using activated carbon (AC) as supporting materials (SMs) for the loading of RuO_2 to obtain RuO_2/AC composite that exhibited higher phenol degradation in peroxymonosulphate activation system [104].

Studies using bare Co_3O_4 nanoparticles and metal organic framework (MOF)–based composite Co_3O_4@MOF with yolk-shell configuration have verified the improved degradation of organic pollutants due to the accumulation of pollutants on catalyst surface by adsorption in the first stage and radical-induced degradation in the second stage, according to Zeng *et al.* [105].

Switchgrass BC (SB) was modified with a cationic surfactant, tetradecyltrimethyl ammonium bromide (TTAB), to produce a novel SB-TTAB

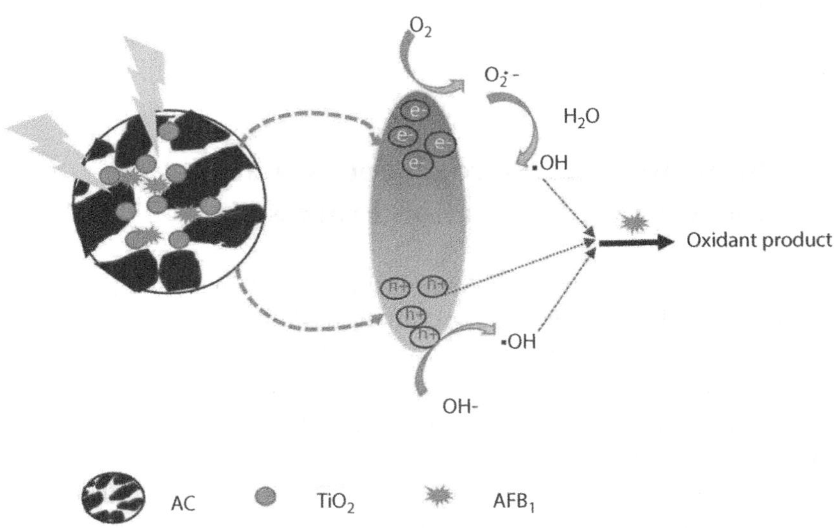

Figure 2.5 Schematic diagram for illustrating the photocatalytic mechanism of AFB1 over AC/TiO_2 photocatalyst under UV–Vis light irradiation (adopted from Ref. [102]).

adsorbent with increased adsorption potential for reactive red 195 A dye (RR-195A). When the SB-TTAB adsorbent was used in tap water, fresh water, wastewater, and seawater [106], efficient removal of the RR-195A dye was achieved with excellent recovery values in the range of 90–100%.

Hamada and colleagues demonstrated the use of aqueous ozone and ultraviolet Photolysis to treat cotton and linen fabrics in an environmentally friendly manner [107]. The authors explore the use of AOP to bleach fabrics. To achieve bleaching, these processes use electrochemically generated aqueous ozone and ultraviolet (UV) irradiation. The color difference and reflectance spectra were measured in this study after treatment to determine changes in fabric color. Photographs of the linen samples after bleaching treatment are shown in Figure 2.6. The fabric was gray until it was bleached. The fabric was bleached in the middle, where UV-irradiated aqueous ozone came into direct contact with it. The whiteness of the linen fabrics treated with AOP for 120 minutes was comparable to that of the conventional method in the central spot.

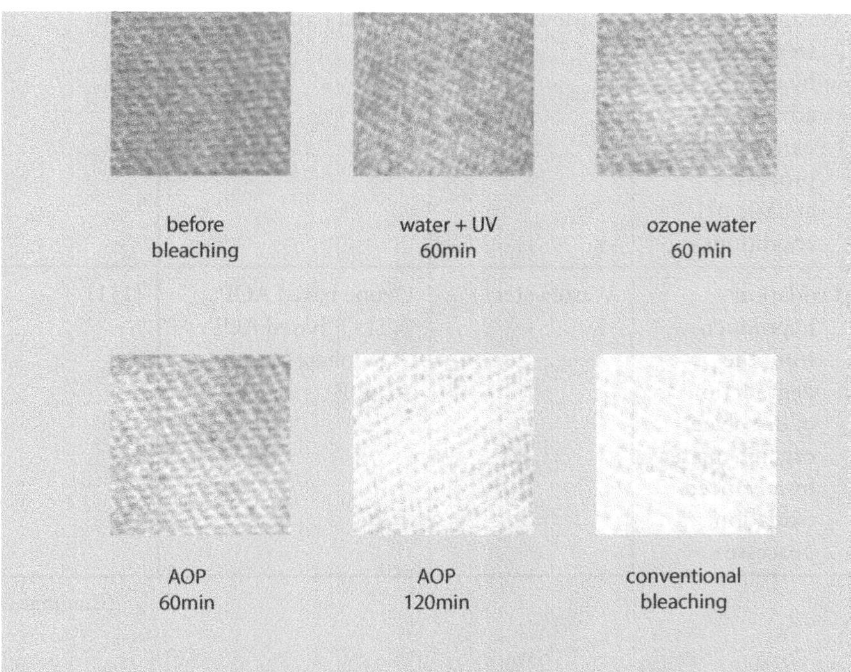

Figure 2.6 Photographs of the linen samples before bleaching, after water + ultraviolet (UV) treatment for 60 minutes, after aqueous ozone treatment for 60 minutes, after advanced oxidation processes (AOP) treatment for 60 minutes, after AOP treatment for 120 minutes, and after conventional (adopted from ref. [107]).

Table 2.1 Published articles on application of AOP.

Title of review articles	Contaminants/ wastewater	Process used	References
Nanomaterials based advanced oxidation processes for wastewater treatment	Wastewater	Nano material-based AOP	[108]
Catalytic ozonation for water and wastewater treatment: recent advances and perspective	Wastewater	Catalytic ozonation	[109]
Wastewater treatment by means of advanced oxidation processes at basic pH conditions	Wastewater	AOP at basic pH	[110]
Oxidation byproducts from the degradation of dissolved organic matter by advanced oxidation processes	Wastewater	Ozone based AOP; H_2O_2 based AOP; Sulphate based AOP	[111]

(*Continued*)

Table 2.1 Published articles on application of AOP. (*Continued*)

Title of review articles	Contaminants/ wastewater	Process used	References
Hybrid ozonation process for industrial wastewater treatment: Principles and applications	Industrial wastewater	O_3/H_2O_2; O_3/UV; O_3/ Electrocoagulation; Catalytic ozonation; Nano catalyzed ozonation	[112]
Detoxification of water and wastewater by advanced oxidation processes	Arsenic	AOP	[113]

Some of the other recent reviews published on application of AOP are shown in Table 2.1.

The synthesis of Salen–Metal Complexes (Metal=Co or Ni) Intercalated ZnCr-LDHs and their photocatalytic degradation of rhodamine B was reported by Meng *et al.* [114]. When the molar ratio of SalenM to Cr was 0.75, SalenM-intercalated ZnCr-LDHs had significantly higher photocatalytic activities than conventional LDHs, according to the authors' findings. The degradation rates of RhB are found to be close to 90%, with reasonable reusability of the products up to four cycles (see Figure 2.7).

Zhang *et al.* [115] used green leaves as biotemplates to make morph-structured TiO_2 with self-doped N. Green leaves contain N-rich chlorophylls. The absorbance intensity in the visible range increases by 103% to 258% when morph-TiO_2 is compared with TiO_2 without a template. Yang *et al.* [116] and Ma *et al.* [117] used rice husk and corn plant as biotemplates, respectively, to synthesize hierarchical porous TiO_2/SiO_2 composites. They discovered that TiO_2/SiO_2 improved not only visible-light harvesting efficiency but also dye degradation photocatalytic activity.

Liu *et al.* [118] created hierarchical artificial leaves (TiO_2–SiO_2 photocatalyst) using aquatic plant leaves (Vallisneria). The TiO_2–SiO_2 replicas degraded dye three times faster than commercial Degussa P25 TiO_2 under UV irradiation.

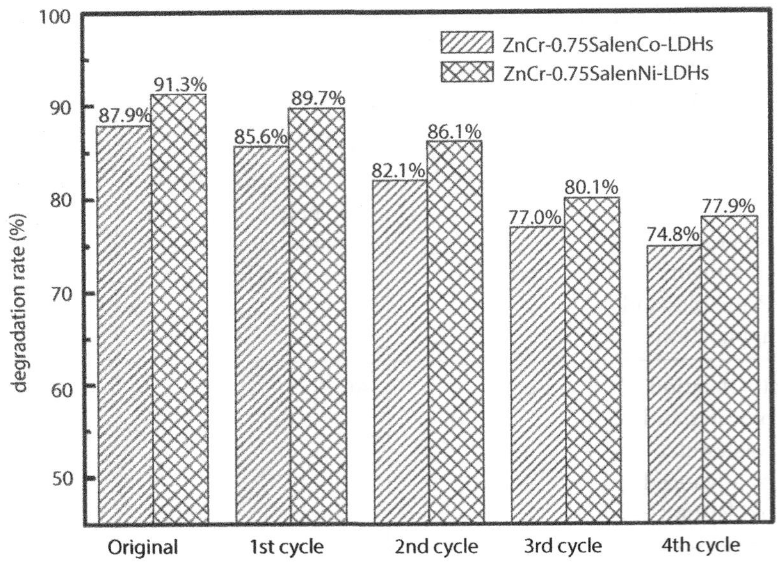

Figure 2.7 Comparison of RhB photocatalytic degradation by original ZnCr-0.75SalenM-LDHs (M = Co or Ni) and their recycled materials (adopted from ref. [114]).

There are several methods used for removal of dyes [119–125]. But for efficient dye degradation, Yan *et al.* created a biotemplated mesoporous TiO_2/SiO_2 composite derived from aquatic plant leaves [126]. This research used reed, water hyacinth, and duckweed leaves as templates and silicon precursors, resulting in a biomorphic TiO_2/SiO_2 composite with mesoporous structures. Both TiO_2/SiO_2 composites, according to the findings, are primarily composed of an anatase step with a mesoporous structure and a large specific surface area. All when compared with commercial Degussa P25 TiO_2, TiO_2/SiO_2 samples have a high light-harvesting performance, particularly in the visible light range. The TiO_2/SiO_2 samples templated by reed and water hyacinth leaves have high activity in the degradation of gentian violet, whereas the TiO_2/SiO_2 samples obtained from duckweed are inferior to P25.

2.5 Conclusions and Future Prospect

➢ Catalyst restriction

In this chapter, we have discussed about catalyst benefits, but still, there are some problems, e.g., leaching of catalyst, surface activity in whole duration of reaction, and how it can be used in practical applications. Catalysts could also be easily recoverable and reusable after reaction. Water electrolytic oxidation of the GICP in 50 wt% H_2SO_4 at 5 V for 30 seconds yielded graphite oxide (yellow area). By sonicating graphite oxide in water for 5 minutes, a well-dispersed EGO aqueous solution (5 mg/mL) was obtained.

- Control to production of by-products
  In AOP process and after that, some of by-products are generation with high polarity and H_2O solubility is compared with the final product. These by-products can be harmful for human health and aqueous species as well. There is a need to give attention on the control of the formation of by-products and long-lived intermediates during the process and after as well.
- Hydrogen peroxide
  In AOPs, processing chemicals used are mostly expensive, and during the process, there is continuous production of hydrogen peroxide. So now, researchers need to focus on the utilization of hydrogen peroxide or reduce in production.
- Ions effects
  Many ions are present in wastewater, which is degrading the process, such as NO_3^-, CO_3^-, HCO_3^-, SO_2, and CI^-. So there is a further need to study the effects of ions and absorb or remove it in the process.
- Commercial and industrial applications
  AOPs are excellent process at small scale, such as in a laboratory. More studies are required to apply this process in the industrial level. Practical or industrial use needs to focus into the increase efficiency of the AOPs process. Ozonation and UV/H_2O_2 AOPs are widely used at industrial scale for drinking and groundwater treatment but not much has been explored for wastewater treatment. Some limitation needs to explore manageability, removal efficiency, and operation safety, as well as applicability in real life.

AOPs for real wastewater treatment have to give attention on designing a required model with cost effective with high efficiency for industrial scale, which can be labile operating parameters, controlled primary and secondary pollutants, H_2O_2 and catalyst concentration, economic and environmentally viability.

Acknowledgement

The Yeungnam University supported this work.

References

1. Speight, J.G., Sources and types of inorganic pollutants. *Environ. Inorg. Chem. Eng.*, 5, 231–282, 2017.
2. Gunasekara, A.S., Rubin, A.L., Goh, K.S., Spurlock, F.C., Tjeerdema, R.S., Environmental fate and toxicology of carbaryl. *Rev. Environ. Contam. Toxicol.*, 196, 95–121, 2008.
3. Bryer, P.J., Scoggins, M., McClintock, N.L., Coal-tar based pavement sealant toxicity to freshwater macroinvertebrates. *Environ. Pollut.*, 158, 1932–1937, 2010.
4. Székács, A., Mörtl, M., Darvas, B., Monitoring pesticide residues in surface and ground water in hungary: Surveys in 1990–2015. *J. Chem.*, 2015, 1–15, 2015.
5. Driscoll, C.T., Mason, R.P., Chan, H.M., Jacob, D.J., Pirrone, N., Mercury as a global pollutant: Sources, pathways, and effects. *Environ. Sci. Technol.*, 47, 10, 4967–4983, 2013.
6. Tumolo, M., Ancona, V., Paola, D.D., Losacco, D., Campanale, C., Massarelli, C., Uricchio, V.F., Chromium pollution in european water, sources, health risk, and remediation strategies: An overview. *Int. J. Environ. Res. Public Health*, 17, 5438, 2020, 10.3390/ijerph17155438.
7. Rzętała, M.A., Cadmium contamination of sediments in the water reservoirs in Silesian Upland (Southern Poland). *J. Soils Sediments*, 16, 2458–2470, 2016, DOI 10.1007/s11368-016-1477-3.
8. Shaji, E., Santosh, M., Sarath, K.V., Prakash, P., Deepchand, V., Divy, B.V., Arsenic contamination of groundwater: A global synopsis with focus on the Indian Peninsula. *Geosci. Front.*, 12, 101079, 2021.
9. Yi, Q., Dou, X.D., Huang, Q.R., Zhao, X.Q., Pollution characteristics of Pb, Zn, As, Cd in the Bijiang River. *Proc. Environ. Sci.*, 13, 43–52, 2012.
10. Zhang, Y., Zhang, L., Huang, Z., Li, Y., Li, J., Wu, N., He, J., Zhang, Z., Liu, Y., Niu, Z., Pollution of polycyclic aromatic hydrocarbons (PAHs) in drinking water of China: Composition, distribution and influencing factors. *Ecotoxicol. Environ. Saf.*, 177, 108–116, 2019.
11. Tundo, P., Anastas, P., Black, D.S., Breen, J., Collins, T., Memoli, S., Miyamoto, J., Polyakoff, M., Tumas, W., Synthetic pathways and processes in green chemistry. Introductory overview. *Pure Appl. Chem.*, 72, 1207–1228, 2000.
12. Glaze, W.H., Kang, J.W., Chapin, D.H., The chemistry of water treatment processes involving ozone, hydrogen peroxide and ultraviolet radiation. *Ozone: Sci. Eng.*, 9, 335–352, 1987.

13. Samanta, D., Roy, S., Sasmal, R., Saha, N.D., Pradeep, K.R., Viswanatha, R., Agasta, S.S., Maji, T.K., Solvent adaptive dynamic metal-organic soft hybrid for imaging and biological delivery. *Angew. Chem. Int. Ed. Engl.*, 58, 5008–5012, 2019.
14. Wu, M.-X. and Yang, Y.-W., Metal-organic framework (MOF)-based drug/Cargo delivery and cancer therapy. *Adv. Mater.*, 29, 1606134, 2017. (b) Orellana-Tavra, C., Baxter, E.F., Tian, T., Bennett, T.D., Slater, N.K.H., Cheetham, A.K., Fairen-Jimenez, D., Amorphous metal-organic frameworks for drug delivery. *Chem. Commun.*, 51, 13878–13881, 2015.
15. Chen, B., Yang, Z., Zhu, Y., Xia, Y., Zeolitic imidazolate framework materials: Recent progress in synthesis and applications. *J. Mater. Chem. A*, 2, 16811–16831, 2014.
16. David, F., Sonia, A., Catherine, P., Metal-organic frameworks: Opportunities for catalysis. *Angew. Chem. Int. Ed.*, 48, 7502–7513, 2009. (b) Zhu, L., Liu, X.-Q., Jiang, H.-L., Sun, L.-B., Metal-organic frameworks for heterogeneous basic catalysis. *Chem. Rev.*, 117, 8129–8176, 2017. (c) Mondloch, J.E., Katz, M.J., Isley, W.C., Ghosh, P., Liao, P., Bury, W., Wagner, G.W., Hall, M.G., DeCoste, J.B., Peterson, G.W., Snurr, R.Q., Cramer, C.J., Hupp, J.T., Farha, O.K., Destruction of chemical warfare agents using metal-organic frameworks. *Nat. Mater.*, 14, 512–516, 2015.
17. Ahamad, M.N., Shahid, M., Ahmad, M., Sama, F., Cu(II) MOFs based on bipyridyls: Topology, magnetism, and exploring sensing ability toward multiple nitroaromatic explosives. *ACS Omega*, 4, 7738–7749, 2019.
18. Georgiou, Y., Perman, J.A., Bourlinos, A.B., Deligiannakis, Y., Highly cient arsenite [As(III)] adsorption by an [MIL-100(Fe)] metal–organic framework: Structural and mechanistic insights. *J. Phys. Chem. C*, 122, 9, 4859–4869, 2018.
19. Li, J.-R., Sculley, J., Zhou, H.-C., Metal-organic frameworks for separations. *Chem. Rev.*, 112, 869–932, 2012.
20. Lee, J., Farha, O.K., Roberts, J., Scheidt, K.A., Nguyen, S.T., Hupp, J.T., Metal-organic framework materials as catalysts. *Chem. Soc. Rev.*, 38, 1450–1459, 2009.
21. Davies, R.P., Less, R., Lickiss, P.D., Robertson, K., White, A.J.P., Structural diversity in metal–organic frameworks built from rigid tetrahedral [Si(p-C6H4CO2)4]4– struts. *Cryst. Growth Des.*, 10, 4571–4581, 2010.
22. Forgan, R.S., Modulated self-assembly of metal–organic frameworks. *Chem. Sci.*, 11, 4546–4562, 2020.
23. McKinstry, C., Cussen, E.J., Fletcher, A.J., Patwardhan, S.V., Sefcik, J., Effect of synthesis conditions on formation pathways of metal organic framework (MOF-5) crystals. *Cryst. Growth Des.*, 13, 5481–5486, 2013.
24. Han, L., Qi, H., Zhang, D., Ye, G., Zhou, W., Hou, C., Xu, W., Sun, Y., A facile and green synthesis of MIL-100(Fe) with high-yield and its catalytic performance. *New J. Chem.*, 41, 13504–13509, 2017.

25. Zhuang, Y., Kong, Y., Wang, X., Shi, B., Novel one step preparation of a 3D alginate based MOF hydrogel for water treatment. *New J. Chem.*, 43, 7202–7208, 2019.
26. Garibay, S.J. and Cohen, S.M., Isoreticular synthesis and modification of frameworks with the UiO-66 topology. *Chem. Commun.*, 46, 7700–7702, 2010.
27. Yang, Q., Vaesen, S., Ragon, F., Wiersum, A.D., Wu, D., Lago, A., Devic, T., Martineau, C., Taulelle, F., Llewellyn, P.L., Jobic, H., Zhong, C., Serre, C., De Weireld, G., Maurin, G., A water stable metal–organic framework with optimal features for CO2 capture. *Angew. Chem. Int. Ed.*, 52, 10316–10320, 2013.
28. Chavali, M.S. and Nikolova, M.P., Metal oxide nanoparticles and their applications in nanotechnology. *SN Appl. Sci.*, 1, 607, 2019.
29. Rai, M., Yadav, A., Gade, A., Silver nanoparticles as a new generation of antimicrobials. *Biotechnol. Adv.*, 27, 76–83, 2009, https://doi.org/10.1016/j.biotechadv.2008.09.002.
30. Younes, N.A., Hassan, H.S., Elkady, M.F., Hamed, A.M., Dawood, M.F.A., Impact of synthesized metal oxide nanomaterials on seedlings production of three Solanaceae crops. *Heliyon*, 6, e03188, 2020.
31. Yang, J., Progress of metal oxide (sulfide)-based photocatalytic materials for reducing nitrogen to ammonia. *J. Chem.*, 8, 2018, 2018. Article ID 3286782.
32. Senthilkumar, S.R. and Sivakumar, T., Green tea (Camellia sinensis) mediated synthesis of zinc oxide (ZnO) nanoparticles and studies on their antimicrobial activities. *Int. J. Pharm. Sci.*, 6, 461–465, 2014.
33. Barui, A.K., Kotcherlakota, R., Patra, C.R., Biomedical applications of zinc oxide nanoparticles, in: *Inorganic Frameworks as Smart Nanomedicines*, pp. 239–78, 2018.
34. McBrain, S.C., Yiu, H.H., Dobson, J., Magnetic nanoparticles for gene and drug delivery. *Int. J. Nanomed.*, 3, 2, 169–180, 20082008.
35. Adam, D.M. and Richard, P.V.D., Single silver nanoparticles as realtime optical sensors with zeptomole sensitivity. *Nano Lett.*, 3, 1057–1062, 2003.
36. Masoud, N., Partsch, T., de Jong, K.P., de Jongh, P.E., Thermal stability of oxide-supported gold nanoparticles. *Gold Bull.*, 52, 105–114, 2019.
37. McNamara, K. and Tofail, S.A.M., Nanoparticles in biomedical applications. *Adv. Phys. X*, 2, 54–88, 2017, https://doi.org/10.1080/23746149.2016.1254570.
38. Rodríguez, J.A. and Fernández-García, M., *Synthesis, properties, and applications of oxide nanomaterials*, John Wiley & Sons, Inc., Hoboken, NJ, USA, 2007.
39. Farahmandjou, M. and Soflaee, F., Synthesis of iron oxide nanoparticles using borohydride. *Int. J. Bio-Inorg. Hybrib Nanomater.*, 3, 203–206, 2014.
40. Mehmood, F., Iqbal, J., Ismail, M., Mehmood, A., Ni doped WO3 nanoplates: An excellent photocatalyst and novel nanomaterial for enhanced anticancer activities. *J. Alloys Compd.*, 746, 729–738, 2018.

41. Beura, R., Pachaiappan, R., Paramasivam, T. *et al.*, Photocatalytic degradation studies of organic dyes over novel Ag-loaded ZnO-graphene hybrid nanocomposites. *J. Phys. Chem. Solids*, 148, 109689, 2021.
42. Sharma, M., Poddar, M., Gupta, Y., Nigam, S., Avasthi, D.K., Adelung, R., Solar light assisted degradation of dyes and adsorption of heavy metal ions from water by CuO-ZnO tetrapodal hybrid nanocomposite. *Mater. Today Chem.*, 17, 100336, 2020.
43. Yao, S., Zhang, X., Qu, F., Umar, A., Wu, X., Hierarchical WO_3 nanostructures assembled by nanosheets and their applications in wastewater purification. *J. Alloys Compd.*, 689, 570–574, 2016.
44. Smrithi, S.P., Kottam, N., Arpitha, V., Narula, A., Anikumar, G.N., Subramanian, K.R.V., Tungsten oxide modified with carbon nanodots: Integrating adsorptive and photocatalytic functionalities for water remediation. *J. Sci. Adv. Mater. Devices*, 5, 73–83, 2020.
45. Armelao, L., Barreca, D., Bertapelle, M., Bottaro, G., Sada, C., Tondello, E., A sol-gel approach to nanophasic copper oxide thin films. *Thin Solid Films*, 442, 48–52, 2003.
46. Singh, J., Dutta, T., Kim, K.H., Rawat, M., Samddar, P., Kumar, P., Green synthesis of metals and their oxide nanoparticles: Applications for environmental remediation. *J. Nanobiotechnol.*, 16, 1–24, 2018.
47. Parveen, R. and Tremiliosi-Filho, G., A step ahead towards the green synthesis of monodisperse gold nanoparticles: The use of crude glycerol as a greener and low-cost reducing agent. *RSC Adv.*, 6, 95210–95219, 2016, https://sci-hub.tw/10.1039/C6RA14259A.
48. Bagherzade, G., Tavakoli, M.M., Namaei, M.H., Green synthesis of silver nanoparticles using aqueous extract of saffron (Crocus sativus L.) wastages and its antibacterial activity against six bacteria. *Asian Pac. J. Trop. Biomed.*, 7, 227–233, 2017, https://doi.org/10.1016/j apjtb.2016.12.014.
49. Sankar, R. and Rizwana, K., Ultra-rapid photocatalytic activity of Azadirachta indica engineered colloidal titanium dioxide nanoparticles. *Appl. Nanosci.*, 5, 731–736, 2015, https://doi.org/10.1007/ s13204-014-0369-3.
50. Mourato, A., Gadanho, M., Lino, A.R., Tenreiro, R., Biosynthesis of crystalline silver and gold nanoparticles by extremophilic yeasts. *Bioinorg. Chem. Appl.*, 1, 1, 2011, https://doi.org/10.1155/2011/54607 4.
51. Chandran, S.P., Chaudhary, M., Pasricha, R. *et al.*, Synthesis of gold nanotriangles and silver nanoparticles using aloe vera plant extract. *Biotechnol. Progr.*, 22, 577–583, 2006, https://doi.org/10.1021/bp050 1423.
52. Al-Tamimi, S.A., Biogenic green synthesis of metal oxide nanoparticles using oat biomass for ultrasensitive modified polymeric sensors. *Green Chem. Lett. Rev.*, 14, 165–178, 2021.
53. Singh, J., Mehta, A., Rawat, M., Basu, S., Green synthesis of silver nanoparticles using sun dried tulsi leaves and its catalytic application for 4-nitrophenol reduction. *J. Environ. Chem. Eng.*, 6, 1468–74, 2018, https://doi.org/10.1016/j.jece.2018.01.054.

54. Prathna, T.C., Chandrasekaran, N., Raichur, A.M., Mukherjee, A., Biomimetic synthesis of silver nanoparticles by Citrus limon (lemon) aqueous extract and theoretical prediction of particle size. *Colloids Surf. B Biointerfaces*, 82, 152–159, 2011, https://doi.org/10.1016/j.colsu rfb.2010.08.036.
55. Shankar, S.S., Rai, A., Ahmad, A., Shastry, M., Rapid synthesis of Au, Ag, and bimetallic Au core-Ag shell nanoparticles using Neem (Azadirachta indica) leaf broth. *J. Colloid Interface Sci.*, 275, 496–502, 2004.
56. Narayanan, K.B. and Sakthivel, N., Coriander leaf mediated biosynthesis of gold nanoparticles. *Mater. Lett.*, 62, 4588–90, 2008, https://doi.org/10.1016/j.matle t.2008.08.044.
57. Haverkamp, R.G. and Marshall, A.T., The mechanism of metal nanoparticle formation in plants: Limits on accumulation. *J. Nanopart. Res.*, 11, 1453–63, 2009, https://doi.org/10.1007/s1105 1-008-9533-6.
58. Marchiol, L., Synthesis of metal nanoparticles in living plants. *Ital. J. Agron.*, 7, 274–82, 2012.
59. Singh, J., Dutta, T., Kim, K.H., Rawat, M., Samddar, P., Kumar, P., Green synthesis of metals and their oxide nanoparticles: Applications for environmental remediation. *J. Nanobiotechnology*, 16, 1–24, 2018.
60. Bussamara, R., Melo, W.W.M., Scholten, J.D. *et al.*, Controlled synthesis of Mn3O4 nanoparticles in ionic liquids. *Dalton Trans.*, 42, 14473, 2013, https://doi.org/10.1039/c3dt3 2348j.
61. Lazarus, L.L., Riche, C.T., Malmstadt, N., Brutchey, R.L., Effect of ionic liquid impurities on the synthesis of silver nanoparticles. *Langmuir*, 28, 15987–15993, 2012, https://doi.org/10.1021/la303 617f.
62. Sylvestre, J., Poulin, S., Kabashin, A.V. *et al.*, Surface chemistry of gold nanoparticles produced by laser ablation in aqueous media. *J. Phys. Chem. B*, 108, 16864–9, 2004, https://doi.org/10.1021/jp047 134.
63. Chotkowski, M., Połomski, D., Czerwinski, K., Potential application of ionic liquids for electrodeposition of the material targets for production of diagnostic radioisotopes. *Materials*, 13, 5069, 2020.
64. Zhang, J., Ji, W.-J., Xu, J., Geng, X.-Y., Zhou, J., Gu, Z.-B., Yao, S.-H., Zhang, S.-T., Giant positive magnetoresistance in half-metallic double-perovskite Sr2CrWO6 thin films. *Sci. Adv.*, 3, e1701473, 2017.
65. Miyazawa, H., Natori, E., Miyashita, S., Shimoda, T., Ishii, F., Oguchi, T., Electronic states of perovskite-type oxides and ferroelectricity. *Jpn. J. Appl. Phys.*, 39, 5679–5682, 2000.
66. Kwon, S.-N., Yu, J.-H., Na, S.-I., A systematic approach to ZnO nanoparticle-assisted electron transport bilayer for high efficiency and stable perovskite solar cells. *J. Alloys Compd.*, 801, 277–284, 2019.
67. Vidyasagar, C.C., Muñoz Flores, B.M., Jiménez Pérez, V.M., Recent advances in synthesis and properties of hybrid halide perovskites for photovoltaics. *Nano-Micro Lett.*, 10, 68, 2018.

68. Kesari, Y. and Athawale, A., Ultrasound assisted bulk synthesis of CH3NH3PbI3 perovskite at room temperature. *Mater. Lett.*, 159, 87–89, 2015, https://doi.org/10.1016/j.matlet.2015.05.176.
69. Bhooshan Kumar, V., Gouda, L., Porat, Z., Gedanken, A., Sonochemical synthesis of $CH_3NH_3PbI_3$ perovskite ultrafine nanocrystal sensitizers for solar energy applications. *Ultrason. Sonochem.*, 32, 54–59, 2016, https://doi.org/10.1016/j.ultsonch.2016.02.012.
70. Guo, Y., Shoyama, K., Sato, W., Matsuo, Y., Inoue, K., Harano, K., Liu, C., Tanaka, H., Nakamura, E., Chemical pathways connecting lead(II) iodide and perovskite via polymeric plumbate(II) fiber. *J. Am. Chem. Soc.*, 137, 15907–15914, 2015, https://doi org/10.1021/jacs.5b10599.
71. Alammar, T., Hamm, I., Grasmik, V., Wark, M., Mudring, A.-V., Microwave-assisted synthesis of perovskite SrSnO3 nanocrystals in ionic liquids for photocatalytic applications. *Inorg. Chem.*, 56, 12, 6920–6932, 2017.
72. Zheng, Y.C., Yang, S., Chen, X., Chen, Y., Hou, Y., Yang, H.G., Thermal-induced volmer–weber growth behavior for planar heterojunction perovskites solar cells. *Chem. Mater.*, 27, 5116–5121, 2015, https://doi.org/10.1021/acs.chemmater. 5b01924.
73. Cao, Q., Yang, S., Gao, Q., Lei, L., Yu, Y., Shao, J., Liu, Y., Fast and controllable crystallization of perovskite films by microwave irradiation process. *ACS Appl. Mater. Interfaces*, 8, 7854–7861, 2016, https://doi.org/10.1021/acsami.6b01558.
74. Yamada, Y., Nakamura, T., Endo, M., Wakamiya, A., Kanemitsu, Y., Photocarrier recombination dynamics in perovskite CH3NH3PbI3 for solar cell applications. *J. Am. Chem. Soc.*, 136, 11610–11613, 2014, https://doi.org/10.1021/ja506624n.
75. Glisenti, A., Pacella, M., Guiotto, M., Natile, M.M., Canu, P., Largely Cu-doped LaCo1-xCuxO3 perovskites for TWC: Toward new PGM-free catalysts. *Appl. Catal. B Environ.*, 180, 94–105, 2016, https://doi.org/10.1016/j.apcatb.2015.06.017.
76. Stoumpos, C.C., Malliakas, C.D., Kanatzidis, M.G., Semiconducting tin and lead iodide perovskites with organic cations: Phase transitions, high mobilities, and near-infrared photoluminescent properties. *Inorg. Chem.*, 52, 9019–9038, 2013, https://doi.org/10.1021/ic401215x.
77. Prochowicz, D., Franckevicčius, M., Cieślak, A.M., Zakeeruddin, S.M., Grätzel, M., Lewiński, J., Mechanosynthesis of the hybrid perovskite CH3NH3PbI3: Characterization and the corresponding solar cell efficiency. *J. Mater. Chem. A*, 3, 20772–20777, 2015, https://doi.org/10.1039/C5TA04904K.
78. Meyn, M., Beneke, K., Lagaly, G., Anion-exchange reactions of layered double hydroxides. *Inorg. Chem.*, 29, 5201–5207, 1990.
79. Goh, K.H., Lim, T.T., Dong, Z.L., Enhanced arsenic removal by hydrothermally treated nanocrystalline Mg/Al layered double hydroxide with nitrate intercalation. *Environ. Sci. Technol.*, 43, 2537–2543, 2009.

80. Sipos, P. and Pálinkó, I., As-prepared and intercalated layered double hydroxides of the hydrocalumite type as efficient catalysts in various reactions. *Catal. Today*, 306, 32–41, 2018.
81. Gabrovska, M., Nikolova, D., Shopska, M., Bilyarska, L., Edreva-Kadjieva, R., Crişan, D., Stănică, N., Crişan, M., Ni–al layered double hydroxides as precursors of ceramic pigments. *Proceedings of the III Advanced Ceramics and Applications Conference*, pp. 205–220, 2016.
82. Allou, N.B., Saikia, P., Borah, A., Goswamee, R.L., Hybrid nanocomposites of layered double hydroxides: An update of their biological applications and future prospects. *Colloid Polym. Sci.*, 295, 725, 2017.
83. Rives, V., *Layered double hydroxides: Present and future*, Nova Science Publishers, Inc, New York, 2001.
84. Xu, Y., Xu, X., Hou, H., Zhang, J., Zhang, D., Qian, G., Moisture content-affected electrokinetic remediation of Cr(VI)-contaminated clay by a hydrocalumite barrier. *Environ. Sci. Pollut. Res.*, 23, 6517–6523, 2016.
85. Manohara, G.V., Prasanna, S.V., Kamath, P.V., Structure and composition of the layered double hydroxides of Mg and Fe: Implications for anion-exchange reactions. *Eur. J. Inorg. Chem.*, 2011, 16, 2624–2630, 2011.
86. Britto, S. and Kamath, P.V., Structure of bayerite-based lithium–aluminum layered double hydroxides (LDHs): Observation of monoclinic symmetry. *Inorg. Chem.*, 48, 11646–11654, 2009.
87. Ma, L., Wang, Q., Islam, S.M., Liu, Y., Ma, S., Kanatzidis, M.G., Highly selective and efficient removal of heavy metals by layered double hydroxide intercalated with the MoS_4^{2-} ion. *J. Am. Chem. Soc.*, 138, 2858–2866, 2016.
88. Rahmanian, O., Dinari, M., Neamati, S., Synthesis and characterization of citrate intercalated layered double hydroxide as a green adsorbent for Ni^{2+} and Pb^{2+} removal. *Environ. Sci. Pollut. Res.*, 25, 36267–36277, 2018, https://doi.org/10.1007/s11356-018-3584-8.
89. Pang, X., Sun, M., Ma, X., Hou, W., Synthesis of layered double hydroxide nanosheets by coprecipitation using a T-type microchannel reactor. *J. Solid State Chem.*, 210, 111–115, 2014.
90. Gevers, B.R. and Labuschagné, F.J.W.J., Green synthesis of hydrocalumite (CaAl-OH-LDH) from Ca(OH)2 and Al(OH)3 and the parameters that influence its formation and speciation. *Crystals*, 10, 672, 2020.
91. Yun, Y.J., Hong, W.G., Choi, N.J., Kim, B.H., Jun, Y., Lee, H.K., Ultrasensitive and highly selective graphene-based single yarn for use in wearable gas sensor. *Sci. Rep.*, 5, 10904, 2015.
92. Yin, Z., Wu, S., Zhou, X. *et al.*, Electrochemical deposition of ZnO nanorods on transparent reduced graphene oxide electrodes for hybrid solar cells. *Small*, 6, 307–312, 2010.
93. Zhan, B., Li, C., Yang, J., Jenkins, G., Huang, W., Dong, X., Graphene field-effect transistor and Its Application for Electronic Sensing. *Small*, 10, 4042–4065, 2014.

94. Kim, M.-G., Park, J.Y., Shon, Y., Shim, G., Oh, Y.-K., Pharmaceutical applications of graphene-based nanosheets. *Curr. Pharm. Biotechnol.*, 14, 1016–1026, 2014.
95. Kamali, A.R., Scalable fabrication of highly conductive 3D graphene by electrochemical exfoliation of graphite in molten NaCl under Ar/H_2 atmosphere. *J. Ind. Eng. Chem.*, 52, 18–27, 2017, 2017.
96. Labiadh, L. and Kamali, A.R., 3D graphene nanoedges as efficient dye adsorbents with ultra-high thermal regeneration performance. *Appl. Surf. Sci.*, 490, 383–394, 2019.
97. Abdelkader, A.M., Vallés, C., Cooper, A.J., Kinloch, I.A., Dryfe, R.A.W., Alkali reduction of graphene oxide in molten halide salts: Production of corrugated graphene derivatives for high-performance supercapacitors. *ACS Nano*, 8, 11225–11233, 2014.
98. Gürünlü, B., Yücedağ, Ç. T., Bayramoğlu, M.R., Green synthesis of graphene from graphite in molten salt medium. *J. Nanomater.*, 2020, 1–12, 2020, https://doi.org/10.1155/2020/7029601.
99. Pei, S., Weil, Q., Huang, K., Cheng, H.-M., Ren, W., Green synthesis of graphene oxide by seconds timescale water electrolytic oxidation aerogels. *Nat. Commun.*, 9, 145, 1–9, 2018.
100. Q. Wei, ., Pei, S., Wen, G., Huang, K., Wu, Z., Liu, Z., Ma, W., Cheng, H.-M., Ren, W., High yield controlled synthesis of nano-graphene oxide by water electrolytic oxidation of glassy carbon for metal-free catalysis. *ACS Nano*, 13, 9482–9490, 2019.
101. Zhang, Y., Wu, P., Chen, Z., Zhou, L., Zhao, Y., Lai, Y. *et al.*, Synergistic effect in heterogeneous fenton degradation of tetrabromobisphenol a by MWCNT and β-CD co-modified Fe_3O_4. *Mater. Res. Bull.*, 113, 14–24, 2019.
102. Sun, S., Zhao, R., Xie, Y., Liu, Y., Photocatalytic degradation of aflatoxin B1 by activated carbon supported TiO_2 catalyst. *Food Control*, 100, 183–188, 2019.
103. Zhang, C., Zhou, M., Ren, G., Yu, X., Ma, L., Yang, J. *et al.*, Heterogeneous electro-Fenton using modified iron-carbon as catalyst for 2,4-dichlorophenol degradation: Influence factors, mechanism and degradation pathway. *Water Res.*, 70, 414–424, 2015.
104. Muhammad, S., Shukla, P.R., Tadé, M.O., Wang, S., Heterogeneous activation of peroxymonosulphate by supported ruthenium catalysts for phenol degradation in water. *J. Hazard. Mater.*, 215, 183–190, 2012.
105. Zeng, T., Zhang, X., Wang, S., Niu, H., Cai, Y., Spatial confinement of Co_3O_4 catalyst in hollow metal-organic framework as nanoreactor for improved degradation of organic pollutant. *Environ. Sci. Technol.*, 49, 2350, 2014.
106. Mahmoud, M.E., Nabil, G.M., El-Mallah, N.M., Bassiouny, H.I., Kumar, S., Abdel-Fattah, T.M., Kinetics, isotherm, and thermodynamic studies of the adsorption of reactive red 195 A dye from water by modified Switchgrass Biochar adsorbent. *J. Ind. Eng. Chem.*, 37, 156–167, 2016.

107. Hamada, K., Ochiai, T., Tsuchida, Y., Miyano, K., Ishikawa, Y., Nagura, T., Kimura, N., Eco-friendly cotton/linen fabric treatment using aqueous ozone and ultraviolet photolysis. *Catalysts*, 10, 1265, 2020.
108. Bethi, B., Sonawane, S.H., Bhanvase, B.A., Gumfekar, S., Nanomaterials based advanced oxidation processes for wastewater treatment: A review. *Chem. Eng. Process.*, 109, 178–189, 2016.
109. Wang, J., Quan, X., Chen, S., Yu, H., Liu, G., Enhanced catalytic ozonation by highly dispersed CeO_2 on carbon nanotubes for mineralization of organic pollutants. *J. Hazard. Mater.*, 368, 621–629, 2019.
110. Boczkaj, G. and Fernandes, A., Wastewater treatment by means of advanced oxidation processes at basic pH conditions: A review. *Chem. Eng. J.*, 320, 608–633, 2017.
111. Ike, I.A., Karanfil, T., Cho, J., Hur, J., Oxidation byproducts from the degradation of dissolved organic matter by advanced oxidation processes - A critical review. *Water Res.*, 164, 114929, 2019.
112. Malik, S.N., Ghosha, P.C., Vaidyab, A.N., Mudliar, S.N., Hybrid ozonation process for industrial wastewater treatment: Principles and applications: A review. *J. Water Process Eng.*, 35, 101193, 2020.
113. Babu, D.S., Srivastava, V., Nidheesh, P.V., Kumar, M.S., Detoxification of water and wastewater by advanced oxidation processes. *Sci. Total Environ.*, 696, 133961, 2019.
114. Meng, Y., Luo, W., Xia, S., Ni, Z., Preparation of Salen–metal complexes (metal = Co or Ni) intercalated ZnCr-LDHs and their photocatalytic degradation of rhodamine B. *Catalysts*, 7, 143, 2017, https://doi.org/10.3390/catal7050143.
115. Li, X., Fan, T., Zhou, H., Chow, S.K., Zhang, W., Zhang, D., Guo, Q., Ogawa, H., Enhanced light-harvesting and photocatalytic properties in morph-TiO_2 from green-leaf biotemplates. *Adv. Funct. Mater.*, 19, 45–56, 2009.
116. Yang, D.L., Fan, T.X., Zhou, H., Ding, J., Zhang, D., Biogenic hierarchical TiO_2/SiO_2 derived from rice husk and enhanced photocatalytic properties for dye degradation. *PloS One*, 6, 9, e24788, 2011, https://doi.org/10.1371/journal.pone.0024788.
117. Ma, H., Liu, W.W., Zhu, S.W., Ma, Q., Fan, Y.S., Cheng, B.J., Biotemplated hierarchical TiO2 -SiO2 composites derived from Zea mays Linn. for efficient dye photodegradation. *J. Porous Mater.*, 20, 1205–1215, 2013.
118. Liu, J., Yang, Q., Yang, W., Li, M., Song, Y., Aquatic plant inspired hierarchical artificial leaves for highly efficient photocatalysis. *J. Mater. Chem.*, 1, 7760–7766, 2013.
119. Pandey, S. and Mishra, S., Catalytic reduction of p-nitrophenol by using platinum nanoparticles stabilised by guar gum. *Carbohydr. Polym.*, 113, 525–531, 2014.
120. Pandey, S. and Tiwari, S., Facile approach to synthesize chitosan based composite—Characterization and cadmium (II) ion adsorption studies. *Carbohydr. Polym.*, 134, 646–656, 2015.

121. Pandey, S., A comprehensive review on recent developments in bentonite based materials used as adsorbents for wastewater treatment. *J. Mol. Liq.*, 241, 1091–1113, 2017.
122. Kumar, S., Wani, M.Y., Koh, J., Gil, J.M., Sobral., A.J.F.N., Carbon dioxide adsorption and cycloaddition reaction of epoxides using chitosan–graphene oxide nanocomposite as a catalyst. *J. Environ. Sci.*, 69, 77–84, 2018.
123. Pandey, S., Do, J.Y., Kim, J., Kang, M., Fast and highly efficient catalytic degradation of dyes using κ-carrageenan stabilized silver nanoparticles nanocatalyst. *Carbohydr. Polym.*, 230, 115597, 2020.
124. Pandey, S., Fosso-Kankeu, E., Spiro, M.J., Waanders, F., Kumar, N., Ray, S.S., Kim, J., Kang, M., Equilibrium, kinetic, and thermodynamic studies of lead ion adsorption from mine wastewater onto MoS2-clinoptilolite composite. *Mater. Today Chem.*, 18, 100376, 2020.
125. Pandey, S., Mandari, K.K., Kim, J., Kang, M., Fosso-Kankeu, E., Recent advancement in visible-light-responsive photocatalysts in heterogeneous photocatalytic water treatment technology, in: *Photocatalysts in Advanced Oxidation Processes for Wastewater Treatment*, pp. 167–196, Scrivener Publishing LLC, Beverly, MA, United States, 2020.
126. Yan, Z., He, J., Guo, L., Li, Y., Duan, D., Chen, Y., Li, J., Yuan, F., Wang, J., Biotemplated mesoporous TiO2/SiO2 composite derived from aquatic plant leaves for efficient dye degradation. *Catalysts*, 7, 82, 2017, https://doi.org/10.3390/catal7030082.

3

Green Synthesis of Photocatalysts and its Applications in Wastewater Treatment

Premlata Kumari* **and Azazahemad A. Kureshi**

Department of Chemistry, Sardar Vallabhbhai National Institute of Technology, Surat, Gujarat, India

Abstract

The synthesis of photocatalysts using green approaches are increasing day by day over chemical or physical methods due to the elimination of undesired by-products, use of nontoxic reagents, and simple and ecofriendly methods. Nanotechnology plays an important role in the synthesis of photocatalysts called nanophotocatalysts (NPCs). In biological methods, plants, bacteria, actinomycetes, yeasts, fungi, and algae are used as reducing and capping agents. The present chapter is mainly focused on the various sources of wastewater, green synthesis of NPCs using plant extracts, biopolymers, microbes, algae, etc., their characterization using various analytical techniques, and their application in the wastewater treatment.

Keywords: Green synthesis, wastewater treatment, NPC, plant extracts

3.1 Introduction

Water is an indispensable component for all living organisms, including human beings, to survive. Several factors such as an increasing population, rapid industrial development, growth in agrochemical and petrochemicals, and domestic wastes are worsening the class and significance of water and soil globally. According to a Slovakian proverb "Pure water is the world's first and foremost medicine," and the sentence has

**Corresponding author*: pl@chem.svnit.ac.in

Elvis Fosso-Kankeu, Sadanand Pandey, and Suprakas Sinha Ray (eds.) Photoreactors in Advanced Oxidation Processes: The Future of Wastewater Treatment, (71–108) © 2023 Scrivener Publishing LLC

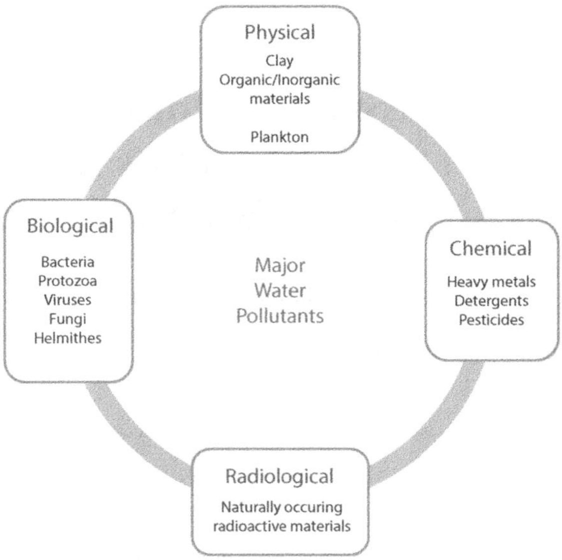

Figure 3.1 Major water-polluting sources.

a likewise important connotation in its inverted form: polluted water is the main cause of water-borne diseases consequently deaths globally. Polluted water causes various illnesses, such as cancer, gastrointestinal complications, neuronal toxicity, skin-related problems, birth defects, etc. [1–4]. According to the reports from World Health Organization, 74% (5.8 billion people) of the global population safely managed to access clean and potable water in 2020 [5]. Because of numerous reasons, such as poor sewage management, industrial wastes, marine dumping issues, radioactive waste material, some agricultural wastes, etc. [6, 7]. Water pollution has the worst effects on the environment, and it is also responsible for air contamination that affects human health directly [8]. Water pollutants can be categorized into four major types as shown in Figure 3.1.

3.2 Photocatalysts and Green Chemistry

The word "photocatalysis" comes from the Greek words "photo" and "catalysis," the breakdown of chemical substances using a light source. In recent years, Photocatalysis has attracted scientific communities vastly, due to better worldwide alertness on the protection of nonrenewable energy. Photocatalysis can be stated as a "change in the rate of a chemical reaction or its initiation under the action of ultraviolet, visible, or infrared radiation in the presence of a photocatalyst, a substance that absorbs the light and involved in the chemical transformation of the reaction partners" [9]. Nanotechnology has gained tremendous potential in evolving wastewater remediation to increase effectiveness and to the intensification of water sources by benign utilization of scarce water bodies. Nanotechnology is an emerging multidisciplinary science covering medical, physics, chemistry, biology, engineering, mechanics, toxicology, and environmental science [10–14]. Nanophotocatalysts (NPCs) have been broadly used in the wastewater treatment process in the field of environmental and ecological well-being, since having several benefits, for instance, those of cost-effective, excellent stability, great photocatalytic efficiency, nontoxicity to living organisms, etc. [15]. The word "green chemistry" was coined by Paul Anastas in the earlier 1990s of the US Environmental Protection Agency [16]. Briefly, Green chemistry is a multidisciplinary approach that includes various emerging areas, such as "green" catalysis, chemicals, synthesis, green and renewable raw materials, ecologically friendly products, and routes [17]. Green chemistry is defined as: "Green chemistry efficiently utilizes raw materials as renewable feedstocks to eliminate waste and avoids the use of toxic and/or hazardous reagents and solvents in the manufacture and application of chemical products" [18]. It is largely concentrated on pollution deterrence instead of the waste clean-up process. Moreover, Anastas has stated that the maneuvering principle was "benign by design," which means design of ecologically benign products and routes [19]. This "benign by design" concept is known as "Twelve Principles of Green Chemistry" shown in Figure 3.2. Green synthesis of nanoparticles (NPs) inspires the use of numerous microbial species for the synthesis of NPs using metals such as Au, Ag, Pt, Pd, and several semiconducting oxides *viz.* TiO_2, ZnO_2, etc. The green synthesis of highly stable metal NPs is carried out utilizing microbial species including algae, bacteria, and fungi. Currently, crude extracts/phytochemicals from plants are gaining acceptance due to their simple process and low-risk hold. NPCs produced by green routes showed promising photocatalytic efficiency, lower consumption of costly and harmful reagents [20]. These are the main bases of choosing green synthesis for the NPCs over the other available routes. NPCs can be prepared by physical, chemical, or biological methods. Physical methods (top-down approach) include mechanical milling, etching, sputtering, laser ablation, electro-explosion, etc. NPCs can also be fabricated using chemical methods (Bottom-up approach),

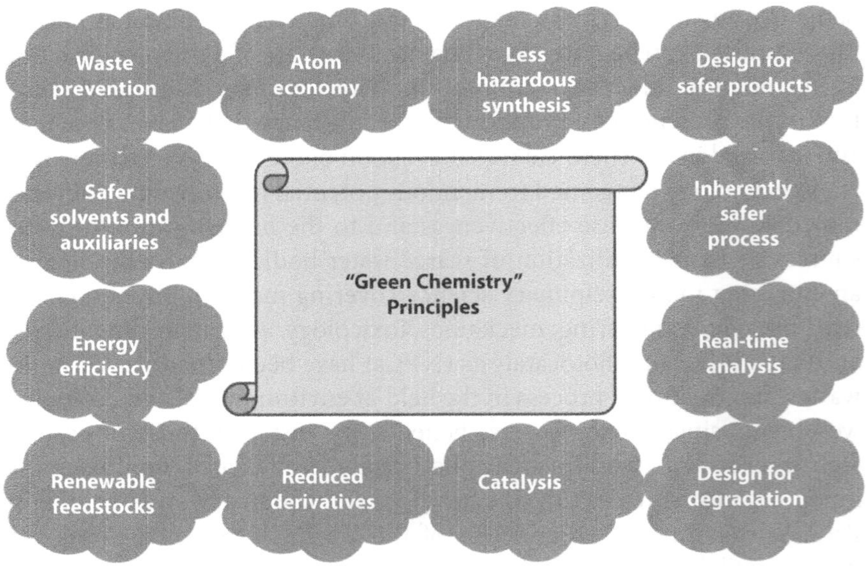

Figure 3.2 The 12 principles of green chemistry.

such as sol-gel, colloidal method, chemical vapor deposition, spray pyrolysis, supercritical fluid synthesis, etc. However, these methods require high pressures, temperatures, longer reaction times, and more energy consumption. Moreover, for the reduction and stabilization step, hazardous chemical reagents such as sodium borohydride, hydrazine hydrate, sodium hypophosphite, polyol, and isopropanol with cetyltrimethylammonium bromide. Due to the use of such hazardous chemical reagents, these methods have become less favorable for environmental applications [21]. To overcome these problems NPCs can also be synthesized using biological methods (Bottom-up approach) utilizing plant extracts, bacteria, actinomycetes, yeasts, fungi, algae, etc. In the present chapter, we have covered the synthesis (*via* green route) of NPCs using plant extracts, biopolymers, microbes, algae, and their characterization using various analytical techniques.

3.2.1 Nanophotocatalysts (NPCs)

The definition of "photocatalysis" can be given as "change in the rate of a chemical reaction or its initiation under the action of ultraviolet, visible, or infrared radiation in the presence of a photocatalyst, a substance that absorbs the light and involved in the chemical transformation of the reaction partners" [9]. Nanomaterial-based catalysts are popularly known as NPCs. Nanomaterials have been in consideration for several decades. The mechanism behind photocatalysis is shown in Figure 3.3. Two types of steps take place in the photocatalysis process, *viz.* mineralization and degradation of organic pollutants [22]. In the degradation process, the organic pollutants (for example, methylene blue dye) are splitting or decomposed into several products whereas, in the mineralization step, the ample degradation of organic pollutants takes place into H_2O, CO_2, and several inorganic ions. Once pollutant degradation is done by using NPCs in the presence of artificial or direct sunlight, chromatographic techniques, such as high-performance liquid chromatography or gas chromatography coupled with mass spectroscopy are applied to analyze the formed by-products. Additionally, it is required to analyze the produced degraded products, if these degradation products rather have or less toxicity as relative to the parent molecule. As per the above discussions, the mineralization step is essentially tantamount to the complete photodegradation process. Simultaneously, with the photodegradation of pollutants into carbon dioxide and water, occasionally others minerals also formed during the mineralization of inorganic ions such as SO_4^-, NH_4^+, S^{2-}, F^-, SO_3^{2-}, Cl^-, PO_4^{3-}, NO_2^-, etc. [23]. NPCs namely titanium oxide, zinc oxide, zinc sulfide, zirconium oxide, tungsten oxide, and cadmium sulfide, etc. absorb photon energy (hυ) which is similar to the energy that of bandgap (Eg) of

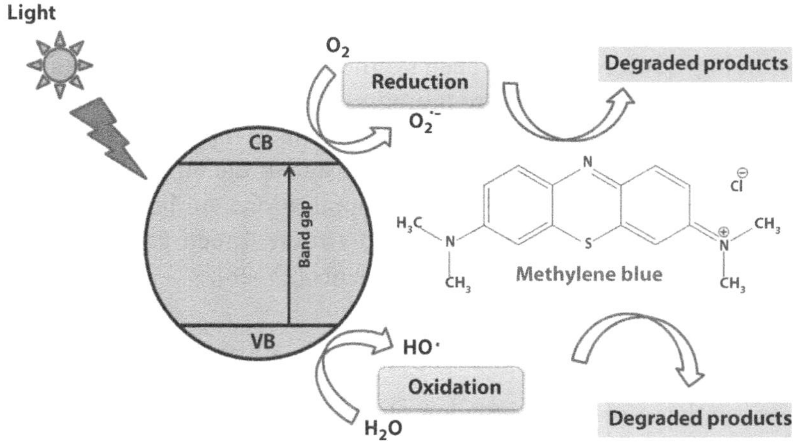

Figure 3.3 Mechanism of photocatalysis with methylene blue dye as an organic pollutant.

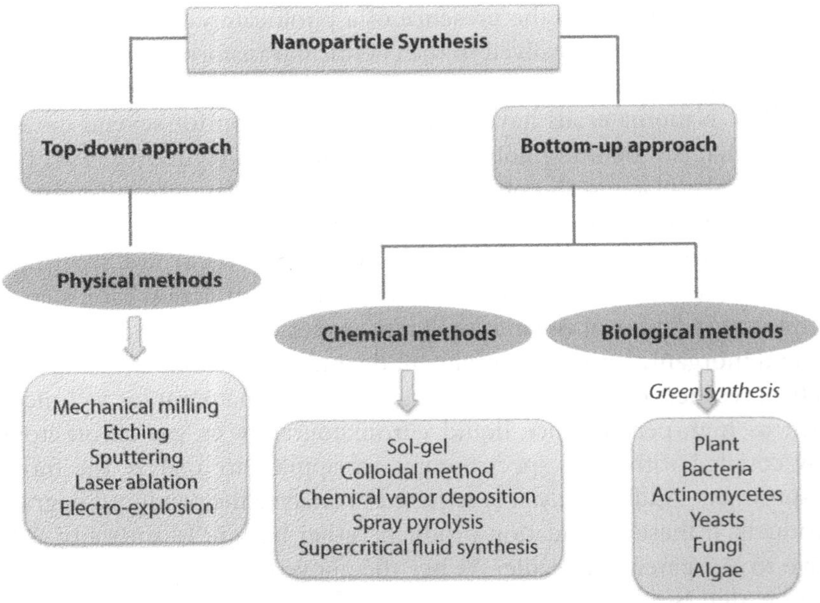

Figure 3.4 The classification of methods used for NP synthesis.

the NPC from the light source, and form an electron-hole pair. The electron (e^-_{CB}) present in the conduction band, the NPCs can be utilized in the reduction of any compound, whereas the hole ($hþ_{VB}$) present in the valence band could be exploited in the oxidation of numerous substrates.

An imperative feature of nanotechnology is the synthesis of nanomaterial having a size less than 100 nm and the control over synthesized NPCs' morphology [24]. NPCs have been synthesized using existing physical and chemical since a time, but current progress demonstrates the crucial role of utilizing microorganisms and biological systems in the preparation of metal NPs (Figure 3.4). The utilization of organisms in the biogenic preparation of NPs is evolving area due to their vast applications. Additionally, biogenic synthesis of metal NPs is an eco-friendly route (green pathway) without using unfavorable deadly and costly reagents [25–28].

3.2.2 Plant-Mediated Green Synthesis of NPCs

Green synthesis of NPCs utilizing plants or their crude extracts or phytochemicals is deliberated economically and thus could be used as a

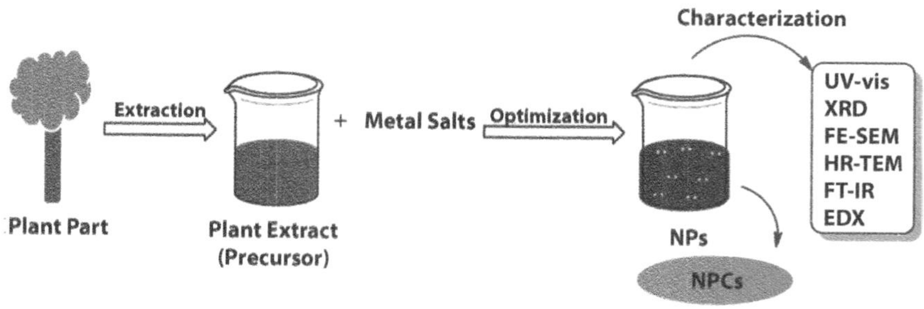

Figure 3.5 Schematic representation for the synthesis of plant-based NPCs.

commercially viable and valid substitute for large-scale production [28]. Plants remain an attractive platform for NPs synthesis having a broad spectrum of its usage [29]. Due to cost-effective production, ease of scaling up the production volume, shorter preparation and work-up time, and safety, phyto-synthesized NPs have distinct benefits over all other biological systems. Plant-mediated biosynthesized NPs offer advantages in terms of lower toxic effects and higher biocompatibility as reduction/capping, and successive formation of NPs are facilitated by various metabolites produced by plant parts [30]. The overall reaction scheme is embodied in Figure 3.5. In the synthesis of NPCs, metals/metal oxides, such as Ag, Au, Pd, ZnO, CuO, Fe_2O_3, TiO_2, CeO_2, etc., could be utilized. Plant metabolites present in the extracts may act as reducing/stabilizing agents (precursor) in NPCs green synthesis. The bioreduction of metal is facilitated by combinations of phytomolecules present in plant extracts such as proteins, amino acids, enzymes, vitamins, polyphenols, flavonoids, alkaloids, terpenes, and organic acids are plant metabolites present in different plant parts. This method is ecologically viable, however, chemically complex.

Several researchers have previously reported the green synthetic approach for the formation of metal NPCs from different plant extracts prepared using different parts such as fruits, leaves, root, flowers, bark, seeds having a wide variety of prospective uses [21, 26, 29, 31].

3.2.3 Biopolymer-Mediated Synthesis of NPCs

According to the International Union of Pure and Applied Chemistry (IUPAC) biopolymers are the macromolecules (including proteins, nucleic acids, and polysaccharides) formed by living organisms [32]. Natural

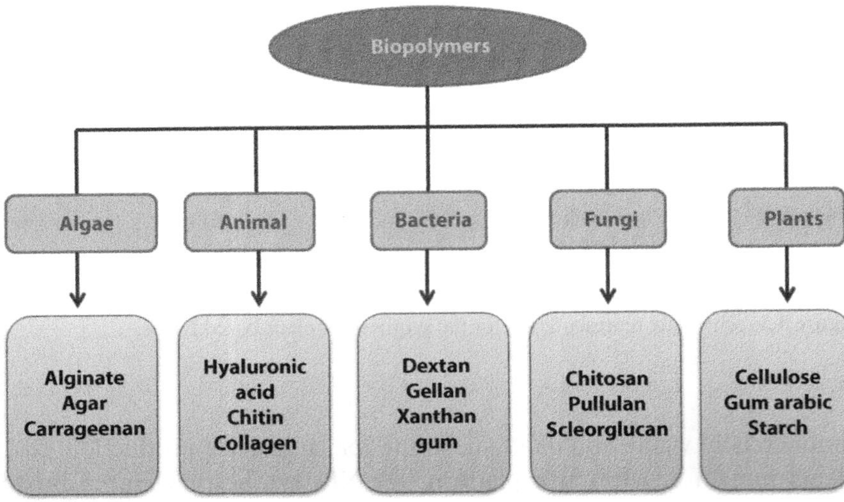

Figure 3.6 Biopolymer and their sources.

polymers and their products including wood, cotton, silk, animal skin, natural rubber, etc. have been used since ancient times by all living beings for survival [33]. Biopolymers are unique in nature and can be derived from algae, animals, bacteria, fungi, and plants (Figure 3.6). Biopolymeric NPCs have been widely used in wastewater remediation. Biopolymers assist the NPC synthesis since biomass morphology is usually structured at the nanolevel that ranges between 10 nm and 1 μm. Biopolymers supported NPCs can be used in the photocatalytic degradation of organic dyes, phenol, and natural organic matter. Recently, numerous reports have appeared in the field of biopolymer-mediated nanomaterial synthesis for photocatalytic applications.

3.2.3.1 Alginic Acid

Alginic acid is one of the polysaccharides biosynthesized abundantly in the cell walls of brown algae. Alginic acid comprises (1,4)-linked β-D-mannuronic acid and α-L-guluronic acid and has wide application in the textile, paper, food, and pharmaceutical industries [34, 35]. Kolya *et al.* [36] reported the synthesis of AuNP–alginate beads as an efficient heterogeneous catalyst, which offers significant degradation efficiency for azo dye [36].

3.2.3.2 *Carrageenan*

Carrageenan is a sulfated linear polysaccharide and found in marine red algae and mainly comprises D-galactose residues linked alternately by (1–3)-linked β-D-galactopyranose (unit G) and (1–4)-linked α-D-galactopyranose (unit D). Pandey *et al.* [37] recently reported the synthesis of AgNPs by using high-molecular weight κ-carrageenan. The biopolymeric AgNPs successfully employed for the degradation of rhodamine B and methylene blue, the degradation efficiency was ~ 100% in a very short exposure time [37].

3.2.3.3 *Chitin and Chitosan*

Chitin is one of the core structural components of an invertebrate's exoskeleton and the cell walls of fungi [38, 39]. Chitin is made up of a linear copolymer with glucosamine and *N*-acetyl glucosamine units [38–40]. Chitosan which is extracted from Chitin is one of the most abundant natural polymers and used for nanotechnology allied applications due to its wide compatibility [41, 42]. Chitosan-mediated metal (Ag, Au, Cu, Fe, Pd, La), metal oxide (Al_2O_3, Fe_3O_4, MnO_2, TiO_2, ZnO), and metal sulfide (ZnS) NPCs have been synthesized for the wastewater remediation. Sargin *et al.* [43] reported Chitosan c-Nanotube supported PdNPs for the degradation of Congo red, methylene blue, methyl orange, and methyl red dyes [43]. Kasiri [44] reported Chitosan-mediated ZnO NPCs which efficiently used in the photocatalytic degradation of Cr complex dye, direct blue 78, and acid black 26 [44]. Mansur and Mansur [45] reported ZnS-Chitosan synthesis and its application in the photocatalytic degradation of methylene blue dye under UV light with approximately 90% efficiency [45].

3.2.3.4 *Guar Gum*

Guar gum is a polysaccharide found abundantly in nature with unique properties including nontoxicity, eco-friendliness, biodegradability, and cost-effectiveness. It is a renewable and biopolymer mainly consists of a backbone of β-D-(1–4) mannopyranosyl units with α-D-galactopyranosyl units as side chains [46]. Several reports have appeared in the past decade stating that guar gum can be utilized in the fabrication of NPs with numerous nanotechnological applications. Pandey *et al.* [47] investigated the application of guar gum as a reducing and capping agent in biosynthesis of AuNPs, under basic conditions and at 80°C [47]. Vanaamudan *et al.* [41] investigated the catalytic activity of a binary dye using $AgNO_3$NPs mediated by Chitosan and Guar gum [41].

3.2.3.5 Cellulose

Cellulose is a renewable natural biopolymer found abundantly, comprises a linear chain of β (1,4)-linked D-glucose [48, 49]. Li *et al.* [49] reported a green synthesis of PdNPs facilitated by a cellulose derivative called carboxymethyl cellulose. The PdNPs displayed outstanding catalytic activity for azo dyes, namely p-aminoazobenzene, acid red 66, acid orange 7, scarlet 3G, and reactive yellow 179 [49]. Cellulose film containing MnO_2 NPC is reported which successfully degraded Indigo carmine dye solution in 25 minutes under ambient light irradiation [50].

3.2.3.6 Xanthan Gum

It is naturally occurring extracellular polysaccharides comprises repeating pentasaccharide units of two glucose units, two mannose units, and one glucuronic acid unit [51, 52]. Xanthan gum has been broadly used in the food and pharmaceutical industries because of its outstanding nontoxic and biocompatible properties [53]. Inamuddin [54] reported xanthan/TiO_2 nanocomposite with photocatalytic ability to degrade methyl orange dye [54].

3.2.4 Green Synthesis of NPCs Using Bacteria, Algae, and Fungus

Researchers have also utilized various bacteria, algae, and fungus as precursors to synthesize NPCs. Microbes such as bacteria can also be used as reducing and stabilizing agents in the green synthesis of NPCs. Various functional groups such as an alkane, alkene, alkyne, amine, carbonyl, amide, ether, halide, phosphate, nitro, nitrile, etc. present in the cells of bacteria, algae, and fungus may take part in the synthesis of NPCs. *Bacillus licheniformis* is a thermophilic nonpathogenic bacteria, which is successfully utilized for the biosynthesis of ZnONPs. The synthesized ZnONPs showed excellent photocatalytic activity against methylene blue dye with 83% efficiency [55]. Rajkumar *et al.* [56] utilized freshwater green alga, *Chlorella vulgaris* for the synthesis of AgNPs. The synthesized AgNPs were successfully employed for the degradation of methylene blue dye with 96.51% efficiency [56]. There are numerous reports available in literature in which fungus has also been used for the fabrication of NPCs. Kalpana *et al.* [57] reported biosynthesis of ZnONPs using *Aspergillus niger* fungus. The biosynthesized ZnONPs showed 90% degradation of Bismarck brown dye [57]. The mechanism behind the extracellular production of NPs using bacteria and fungi is not yet completely understood. Though scientists have discovered that NADH and NADH-dependent nitrite reductase enzymes play a vital role in the bio-reduction of metal ions to metal NPs [58]. Saif Hasan *et al.* [59] proposed that NPs are formed by intracellular mechanisms followed by the death of the cells [59].

3.2.5 Characterization of NPCs Using Various Analytical Techniques

Analytical techniques are very important to characterize synthesized NPCs and to determine their physical and chemical properties. The following table shows the usefulness of each analytical technique used for the characterization (Table 3.1).

3.2.5.1 UV-Visible Spectroscopy

This is the most important and primary technique to observe reduction and progress of NPC synthesis. This is due to the existence of surface plasmon resonance (SPR) band, which occurs because of the shared electron oscillation around the surface mode of the particles. Rajan *et al.* [60] reported a one-pot green synthesis of gold nanocrystals using *G. cambogia* fruit extract. The formation and nucleation of gold nanocrystals are examined for intervals of time and the UV–Vis spectrum of gold nanocrystals for reaction time is shown in Figure 3.7 [60]. Table 3.2 represents UV-vis range and band gap energy for different metal-based NPCs synthesized from various plants.

Table 3.1 Analytical techniques used for the characterization of NPCs.

Interpretation	Analytical techniques
Distinct surface plasmon resonance (SPR) band for different metallic NPCs	UV–visible spectroscopy
Phase identification and crystallinity	X-ray diffraction (XRD)
Surface morphology	Scanning Electron Microscopy (SEM)
Particle size	High Resolution – Transmission Electron Microscopy (HR-TEM)
Identification of involving functional groups	Fourier Transform – Infrared spectroscopy (FT-IR)
Elemental composition in metallic NPCs	Energy-dispersive spectroscopy (EDX)
Determination of surface charge and particle size distribution	Dynamic Light Scattering (DLS)
Specific particle size, texture, and morphology	Atomic Force Microscopy (AFM)
Surface area of porous materials	Brunauer-Emmett-Teller (BET)
Pore size and volume	Barrett-Joyner-Halenda (BJH)

3.2.5.2 XRD

XRD analysis is a useful tool to know the crystalline nature and average particle size of the biosynthesized NPCs. XRD is a powerful nondestructive analytical technique for crystalline materials [68]. For example, in the studies carried out by Khan and co-workers [69], the XRD pattern displays that the green synthesized ZnONPs are purely crystalline. Diffraction bands were observed at (101), (110), (111), (200), (220), (311), and (222) planes (Figure 3.8) [69].

3.2.5.3 SEM, HR-TEM, EDX, and AFM

SEM (Figure 3.9) and HR-TEM (Figure 3.10) are adapted to study morphology including shape and size and distribution of the synthesized NPCs

Table 3.2 UV-vis range (nm) and band gap energies of reported green synthesized NPCs.

No.	NPCs	Plant used	UV-vis range (nm)	Band gap energy (eV)	Ref.
1	AgNPs	*Kalanchoe pinnata*	420	-	[61]
		Solanum tuberosum	430	-	[62]
2	Fe_2O_3NPs	*Emblica officinalis*	530	~1.97 – 2.05	[63]
3	PdNPs	*Sapium sebiferum*	274	-	[64]
4	ZnONPs	*Solanum nigrum*	400	~3.38	[31]
5	CuONPs	*Carica papaya*	250-300	~2.17	[65]
6	CeO_2NPs	*Citrullus lanatus*	341	~5.57	[66]
7	TiO_2NPs	*Parthenium hysterophorus*	420	~4.1 – 5.1	[67]

at the nanomicro scale [70]. Yet, the resolution of HR-TEM is a thousand times greater compared to SEM [71]. In a study carried out by Tahir and co-workers [64], the surface morphology of PdNPs has been demonstrated using SEM, it shows uniform-sized spherical NPs have been formed with no aggregation. EDX is an important tool to know the composition of elements present in NPCs. The EDX analysis showed a strong peak corresponding to palladium metal. However, several small peaks were also observed corresponding to carbon and oxygen (Figure 3.9) [64]. In Figure 3.10, HR-TEM and selected area electron diffraction (SAED) images of AuNPs mediated by *Garcinia indica* (Kokum) fruit pericarp extracts (authors' own) [29]. The particle sizes of the synthesized AuNPs were ranging from 2 to 10 nm measured by HR-TEM. AFM is an essential tool to determine the specific particle size, to analyze texture and morphology [72, 73]. Figure 3.11 shows the AFM image of AgNPs synthesized using

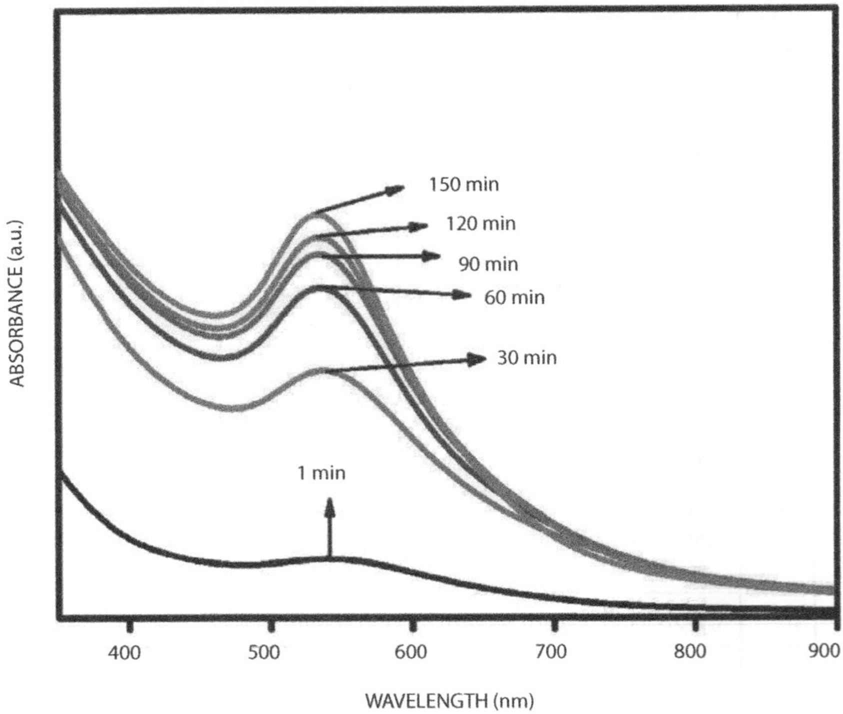

Figure 3.7 UV-visible absorbance of R6 colloid measured as a function of time (reproduced with permission from Rajan *et al.* [60] copyright Elsevier).

poly(*N,N*-dimethylamino)ethylmethacrylate (PDMAEMA)/sodium carboxymethyl cellulose (CMC) multilayer films. The presented AFM shows that AgNPs are spherical and not aggregated after 120 minutes of reaction time [74].

3.2.5.4 Fourier Transform Infrared Spectroscopy

Fourier Transform Infrared (FTIR) technique is employed to acquire an infrared spectrum of absorption/emission of solid, liquid, or gas. FTIR is an important technique to identify involving functional groups from the biomolecules responsible for the capping and stabilization of NPCs. It is known that biomolecules such as phenols, flavonoids, amino acids, terpenes, alkaloids, etc. present in plants can reduce the metal ions used for the NPCs synthesis. Figure 3.12 represents the FTIR spectra of extract, CeO_2NPs mediated by *M. oleifera* peel extract, and Cerric Ammonium Nitrate (CAN).

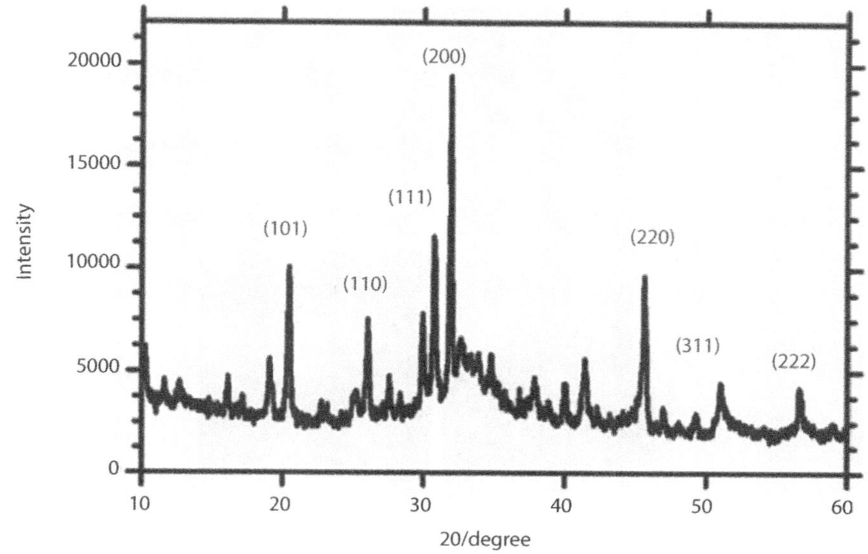

Figure 3.8 XRD pattern of synthesized ZnONPs (reproduced with permission from Khan *et al.* [69] copyright Elsevier).

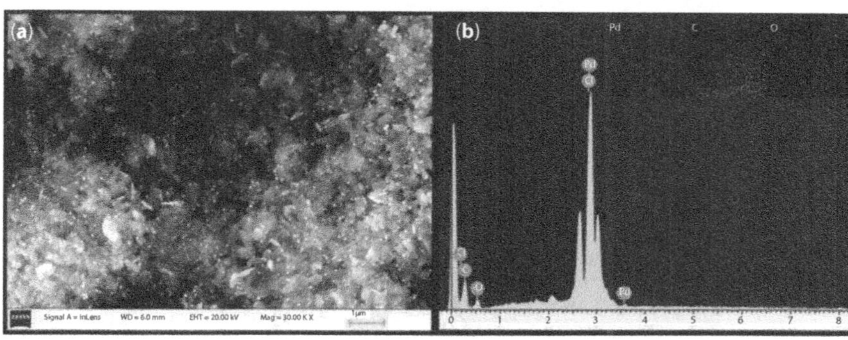

Figure 3.9 (a) SEM analysis of PdNPs synthesized at optimized conditions (10 mL extract concentration and 60°C temperature) (b) EDX profile of *Sapium sebiferum*-mediated PdNPs and mapping for C, Pd, and O (reproduced with permission from Tahir *et al.* [64] copyright Elsevier).

3.2.5.5 *Dynamic Light Scattering*

Dynamic light scattering (DLS), known as photon correlation spectroscopy (PCS), is a very useful tool to measure the fluctuations in light intensity over a period of time. This method is accurate, reproducible, and rapid. DLS has been proven as one of the routine essential techniques to

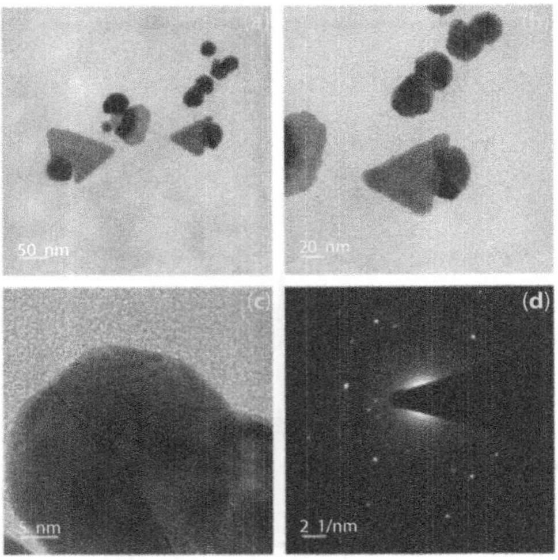

Figure 3.10 (a), (b), (c) are HR-TEM images and (d) SAED image of AuNPs/GI (reproduced with permission from our work Kureshi *et al.* [29] copyright TUOMS press).

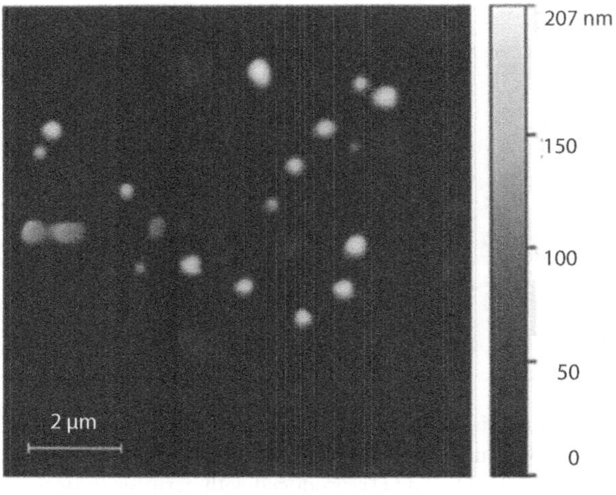

Figure 3.11 AFM height image (10×10 µm) of AgNPs obtained using the $C_{4.5}$ composite (CAgNO3=5×10−3 mol/L; reaction time =120 min) (reproduced with permission from Ghiorghita *et al.* [74] copyright Elsevier).

characterize the NPCs. It measures charge on the surface and distribution of the varying sizes of the NPs (Figure 3.13) [75, 76].

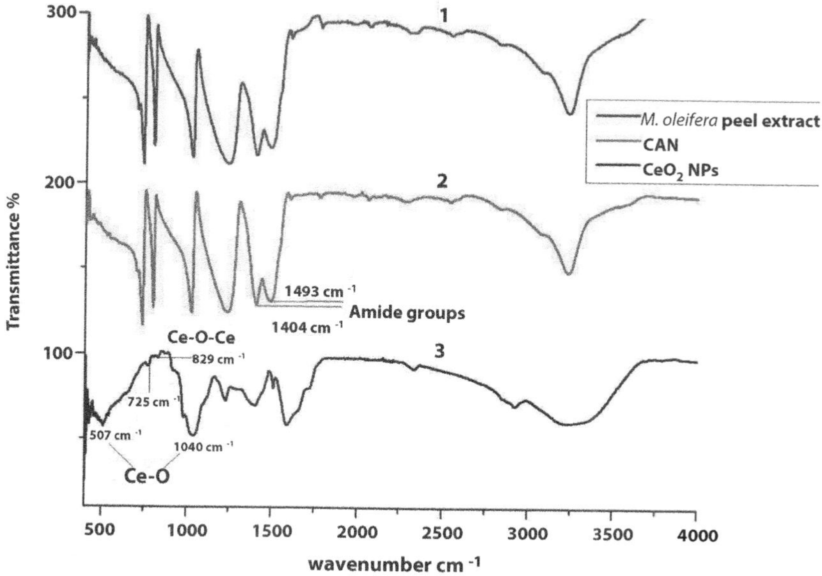

Figure 3.12 FTIR spectra of CeO2 NPs, (1) *M. oleifera* peel extract, (2) CAN (3) Synthesized CeO_2NPs (reproduced with permission from Surendra *et al.* [77] copyright Elsevier).

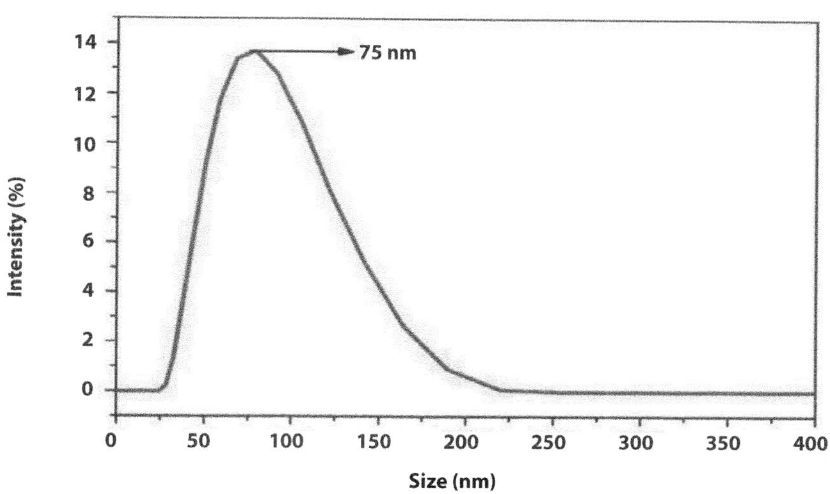

Figure 3.13 Particle size distribution curve of green synthesized TiO_2 NPs by DLS (reproduced with permission from Goutam *et al.* [76] copyright Elsevier).

3.2.5.6 *Brunauer-Emmett-Teller (BET)*

BET method is an essential technique to study the surface areas of porous materials such as metal-organic frameworks (MOFs). Though it has been revealed that the BET method does not give precise surface area always, specifically for high surface area MOFs [78].

3.2.5.7 *Barrett-Joyner-Halenda*

Barrett-Joyner-Halenda (BJH) analysis is useful to calculate the surface area, pore size distribution, and pore volume of the biosynthesized NPCs. For example, the specific surface areas and pore volume of Cu NPs/perlite were determined by this technique. The N_2 adsorption-desorption isotherm (NADI) and BJH pore size distribution plot of Cu NPs/perlite shown in Figure 3.14.

3.2.6 Application of Green Synthesized NPCs in Wastewater Treatment

Green chemistry and nanotechnology collaboratively have a broad spectrum of applications in various areas. Nevertheless, NPCs synthesized with the principles of green chemistry have diverse applications in phytoremediation (Table 3.3).

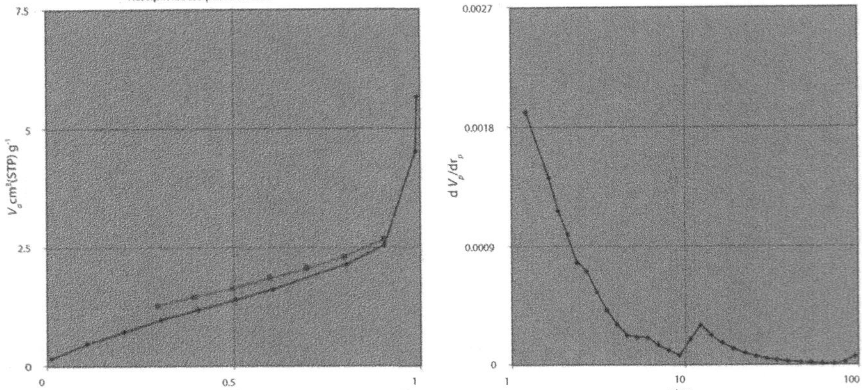

Figure 3.14 The N2 adsorption-desorption isotherm and Barrett–Joyner–Halenda (BJH) pore size distribution plot of catalyst (reproduced with permission from Nasrollahzadeh *et al.* [79] copyright Elsevier).

Table 3.3 Accomplishment of various green synthesized NPCs in wastewater treatment.

No.	Plant name	NPC	Size (nm)	Pollutant	Source	Degradation efficiency (%)	Ref.
1	*Costus speciosus*	(i) ZnO (ii) ZnS	50–100	RR 120	UV	82 (ZnO) 96 (ZnS)	[80]
					Sunlight	59 (ZnO) 80 (ZnS)	
2	*Passiflora edulis*	Ag	7	Methyl Orange Methylene Blue	Visible	100	[81]
3	*Artocarpus Heterophyllus*	ZnO	15-25	Rose Bengal dye	UV	> 80	[82]
4	*Centella asiatica*	Ag	30-50	Eosin Y Phenol Red Methyl Orange Methyl Red	Solar	76.2–97.88	[83]
		CuO	20-30				
5	*Pyrus pyrifolia*	ZnO	~45	Methylene Blue	Solar	> 90	[84]

(Continued)

Table 3.3 Accomplishment of various green synthesized NPCs in wastewater treatment. (*Continued*)

No.	Plant name	NPC	Size (nm)	Pollutant	Source	Degradation efficiency (%)	Ref.
6	*Cucumis sativus*	Ag	8-10	Methylene Blue	Sunlight	> 90	[62]
7	*Vaccinium floribundum* Kunth	Ag	~20.5	Methylene Blue	Sunlight	29.09	[85]
8	*Coccinia grandis*	Ag	20-30	Coomassie Brilliant Blue G-250	UV	Not mentioned	[86]
9	*Carissa edulis*	ZnO	50-55	Congo red	UV-Vis light	97	[87]
10	*Jatropha curcas* L.	TiO_2	75	Tannery wastewater	Sunlight	82.26	[76]
11	*Corymbia citriodora*	ZnO	64	Methylene Blue	Visible	83.45	[88]
12	*Brassica oleracea*	CuO	~22.20	Methylene Blue	UV	96.28	[89]
	Solanum tuberosum		~31.60			87.37	
	Pisum sativum		~24.70			79.11	

(*Continued*)

Table 3.3 Accomplishment of various green synthesized NPCs in wastewater treatment. (*Continued*)

No.	Plant name	NPC	Size (nm)	Pollutant	Source	Degradation efficiency (%)	Ref.
13	*Nephelium lappaceum* L.	ZnO	25-40	Methyl Orange	UV	83.99	[90]
14	*Camellia japonica*	Ag	12-25	Eosin-Y	Visible	> 97	[91]
15	*Trichodesma indicum*	Ag	20-35	Methylene Blue	Sunlight	82	[92]
16	*Trianthema portulacastrum*	ZnO	25-90	Synozol Navy Blue-KBF	Sunlight	91	[69]
17	*Dimocarpus longan*	Au	25	Methylene Blue	Visible	76	[93]
18	*Ulva lactuca*	Ag	48.59	Methyl Orange	Visible	not mentioned	[111]
19	*Cordia dichotoma*	Ag	~10	Methyl Orange Congo Red	Sunlight	not mentioned	[94]
20	*Tabernaemontana divaricata*	ZnO	20-50	Methylene Blue	Sunlight	~100	[95]

(*Continued*)

Table 3.3 Accomplishment of various green synthesized NPCs in wastewater treatment. (*Continued*)

No.	Plant name	NPC	Size (nm)	Pollutant	Source	Degradation efficiency (%)	Ref.
21	*Gynostemma pentaphyllum*	ZnO	35.41	Malachite Green	UV	89	[96]
22	*Eriobutria japonica*	ZnO	18-27	Methylene Blue	UV	not mentioned	[97]
23	*Withania coagulans*	Iron oxide	~16	Safranin	Sunlight	not mentioned	[98]
24	*Carica papaya*	Iron oxide	21.59	Remazol yellow RR dye	Sunlight	76.6	[99]
25	*Brassica oleracea*	Ag	5-50	Methylene Blue	Sunlight	97.57	[100]
26	*Calotropis procera*	CeO_2	7	Methyl Orange	Sunlight	98	[101]
27	*Citrus sinensis*	SnO_2	4.5	Methylene Blue	UV	94.4	[102]

(*Continued*)

Table 3.3 Accomplishment of various green synthesized NPCs in wastewater treatment. *(Continued)*

No.	Plant name	NPC	Size (nm)	Pollutant	Source	Degradation efficiency (%)	Ref.
28	*Amomum longiligulare*	ZnO	50	Methylene Blue	UV	66	[103]
				Malachite Green		38.1	
29	*Aegle marmelos*	TiO_2	18.83	Ornidazole	UV	66.15	[104]
30	*Ficus benghalensis*	Se	45-95	Methylene Blue	UV	57.63	[105]
31	*Trigonella foenum-graecum*	Ag	82.53	Rhodamine B	UV	93	[106]
32	*Ceropegia bulbosa* Roxb	Se	55.90	Methylene Blue	Halogen	96	[107]
33	*Ludwigia octovalvis*	Ag	28-50	Alizarin Red	Sunlight	92.3	[108]
				Congo Red		76	
				Methylene Blue		94.5	
				Rhodamine B		91.1	

(Continued)

Table 3.3 Accomplishment of various green synthesized NPCs in wastewater treatment. (*Continued*)

No.	Plant name	NPC	Size (nm)	Pollutant	Source	Degradation efficiency (%)	Ref.
34	*Moringa oleifera*	Ag	04	Methylene Blue	Sunlight	81	[109]
				Orange Red		82	
				4-Nitrophenol		75	
				Pb		> 80	
35	*Ruellia tuberosa*	Ag	55.65	Crystal Violet	Sunlight	87	[110]
				Coomassie Brilliant Blue		74	

Ravikumar *et al.* [80] studied photocatalytic degradation of reactive red 120 dye utilizing *Costus speciosus* Koen leaf extract. The degradation efficiency was ranging from 59% to 96% for both the prepared NPCs [80]. Methyl orange and methylene blue are two major dyes found in the effluent of dye and textile industries. These dyes were efficiently degraded using the biologically synthesized AuNPs mediated by *Passiflora edulis f. flavicarpa* aqueous leaf extract [81]. Rose bengal is an organic anionic, a water-soluble and photosensitive dye that belongs to the xanthene derivative and is widely used in various industries. Vidya and co-workers [82] reported the synthesis of ZnONPs mediated by jackfruit leaf extract with > 80% dye degradation efficiency under the presence of UV light [82]. Raina *et al.* [83] reported the green synthesis of AgNPs and CuONPs using *Centella asiatica* leaves. The biologically synthesized NPCs were employed to degrade methyl red (98.49%), methyl orange (98.84%), and phenyl red (99.62%) dyes [83]. Sundaramurthy and Parthiban [84] biogenically synthesized ZnONPs using *Pyrus Pyrifolia* leaves and successfully degraded Methylene Blue with catalytic efficiency > 90% upon solar irradiation [84]. Roy *et al.* [62] reported a one-step biogenic synthesis of AgNPs using cucumber fruit extract, the biosynthesized AgNPs were able to degrade > 90% of methylene blue dye on sunlight exposure of 6 hours. Furthermore, the synthesized AgNPs showed effective bactericidal properties against the *S. aureus*, *K. pneumonia*, and *E. coli* bacterial strains [62]. Kumar *et al.* revealed that the AgNPs biosynthesized using Mortiño (*Vaccinium floribundum* Kunth) berry were capable of degrading around 29.09% of methylene blue dye on sunlight exposure of 60 minutes [85]. *Coccinia grandis* L. (Cucurbitaceae) is a herb that is found around the globe. The leaves of this herb were utilized for the biosynthesis of AgNPs having 20 to 30 nm particle size. The results revealed that these AgNPs have suitable promising photocatalytic efficiency for the remediation of Coomassie Brilliant Blue G-250 dye [86]. *Carissa edulis* fruits were utilized for the biosynthesis of ZnONPs, which showed 97% degradation efficiency of Congo red dye [87]. Goutam and co-workers [76] synthesized TiO_2NPs mediated by *Jatropha curcas* L. leaves, which was effectively employed for the phytoremediation of tannery wastewater with 82.26% efficiency [76]. Zheng *et al.* [88] also reported the synthesis of ZnONPs but utilizing *Corymbia citriodora* leaf and hydrothermal methods. The green synthesized ZnONPs showed 83.45% photocatalytic efficiency higher than the hydrothermal synthesized ZnONPs having 59.47% efficiency using visible light [88]. Ullah *et al.* [89] biosynthesized CuONPs using Cauliflower, Potato, and Peas peel which is often being considered waste. These CuONPs showed significant photocatalytic efficiency (%) for methylene blue dye in the following order: cauliflower peels (96.28%) > potato peels (87.37%) > peas peels (79.11%). *Nephelium lappaceum* L. (Rambutan) peel extract was successfully employed for the biosynthesis of ZnONPs followed by photocatalytic degradation of methyl orange with 83.99% efficiency [90]). Karthik *et al.* [91] reported a green synthetic route of AgNPs using *Camellia japonica* leaf extract which was utilized for the photocatalytic degradation of eosin-Y dye with above 97% degradation [91]. Kathiravan biogenically synthesized AgNPs utilizing *Trichodesma indicum* leaf extract. The green synthesized AgNPs showed significant photocatalytic activity against methylene blue with 82% efficiency under solar irradiation. Moreover, the AgNPs also showed antibacterial activity against two bacterial strains *viz. B. cereus* and *E. coli* [92]. ZnONPs were biosynthesized using *Trianthema portulacastrum* weed and showed 91% degradation capacity for synozol navy blue-KBF textile dye up on sunlight [69]. Khan *et al.* [69] biosynthesized AuNPs using *Dimocarpus longan* (Longan) fruit which was further characterized and used as NPC for the treatment of methylene blue upon visible light illumination, the green synthesized AuNPs showed 76% photocatalytic degradation efficiency [69]. *Ulva lactuca* is seaweed that was utilized for the biosynthesis of AgNPs for the photocatalytic application. It demonstrated significant efficiency against the photocatalytic degradation of methyl orange [111]. Biosynthesized AgNPs using *Cordia dichotoma* leaf extract was utilized for the treatment of methylene blue and Congo red, which showed significant photocatalytic as well as antibacterial activity [94]. Raja *et al.* [95] explored ZnONPs biosynthesized using *Tabernaemontana divaricate* and investigated the degradation of methylene blue dye upon sunlight illumination. Nearly ample photocatalytic degradation of the dye was observed in 90 minutes. Furthermore, these ZnONPs displayed antibacterial activity [95]. Recently, Park *et al.* [96] investigated *Gynostemma pentaphyllum* extract for the green synthesis of ZnONPs. The biosynthesized ZnONPs were successfully employed for the treatment of malachite green under UV light [96]. Shabaani *et al.* [97] synthesized ZnONPs using *Eriobutria japonica* (loquat) seed aqueous extract; it was observed that the highest bioremediation against methylene blue was attained with an initial ZnONP concentration of 12 mg/mL upon ultraviolet irradiation. Furthermore, higher antibacterial activity and antioxidant activity were also observed for NPCs by ferric reducing antioxidant power (FRAP) assay [97]. Qasim and co-workers [98] reported *Withania coagulans* extract mediated biosynthesis of iron nanorods. The synthesized iron nanorods were capable of degrading safranin dye, and they also exhibited antibacterial activity against *S. aureus* and *P. aeuroginosa* [98]. Bhuiyan *et al.* [99] biosynthesized Iron oxide NPs using Papaya leaves extract which was further used for the treatment of remazol blue RR dye, the results revealed that the synthesized NPs were successfully degraded 76.6% of the dye. Additionally, the biosynthesized NPs exhibited significant antibacterial. Additionally, the cytotoxic properties of green synthesized NPs against HeLa (Cervical), BHK-21 (baby hamster kidney), and Vero (kidney) cell line was found to be toxic at maximum doses, however, it could be reflected for tumor cell damage since its promising cytotoxic activity against the studied cell lines [99] Kadam *et al.* [100] investigated *Brassica oleracea* (cauliflower) leaf for the biosynthesis of AgNPs, these green synthesized AgNPs were successfully employed for the treatment of 98% methylene blue upon solar irradiation. Mercury is one of the toxic heavy metals, which is harmful to human and other living organisms. Therefore, the AgNPs were utilized for the biosensing of Hg^{2+} ions [100]. CeO_2NPs were synthesized using flower extract of *Calotropis procera*. The green synthesized CeO_2NPs were successfully employed for the treatment of Methyl orange up on solar illumination. CeO_2NPs were fabricated using the green chemistry aspect; significant antibacterial activity [101]. Luque *et al.* [102] reported an improved biosynthetic method for the synthesis of SnO_2NPs using the *Citrus sinensis* (orange) peels. Photocatalytic degradation of methylene blue was efficient (94.4%) upon UV irradiation [102]. Liu *et al.* [103] reported a simple biogenic synthesis of ZnONPs utilizing *Amomum longiligulare* fruit extract and studied their photocatalytic efficiency. The biosynthesized ZnONPs demonstrated photocatalytic degradation of two dyes namely methylene blue (66%) and malachite green (38.1%) upon UV light irradiation [103]. Ornidazole is a third-generation 5-nitroimidazole derivative that is used to treat several infections caused by bacterial and protozoa. Ornidazole is a harmful chemical compound found in almost all kinds of water sources. Ahmad *et al.* [112] synthesized TiO_2NPs which were able to photodegrade ornidazole up to 66.15% upon UV light irradiation [112]. Tripathi *et al.* [105] reported a biosynthesis of highly stable fluorescent SeNPs mediated by fresh leaves of *Ficus benghalensis*. Green synthesized SeNPs were attempted for the treatment of methylene blue dye. It was observed that there was 57.63% degradation of methylene blue dye upon ultraviolet illumination [105]. Awad *et al.* [106] synthesized fenugreek (*Trigonella foenum-graecum*) seeds mediated AgNPs with 82.53 nm particle size. The synthesized AgNPs were capable of degrading nearly 93% rhodamine B dye upon UV irradiation [106]. Cittrarasu *et al.* [107] have reported the synthesis of SeNPs with spherically shaped having 55.90 nm particle size. The SeNPs successfully degraded the 96% methylene blue dye under halogen irradiation [107]. Kannan *et al.* [108] reported AgNPs having particle sizes ranging from 28 to 50 nm, the synthesized AgNPs had excellent photocatalytic activity against four harmful dyes such as alizarin red, Congo red, methylene blue, and rhodamine B [108]. Mehwish *et al.* [109] reported the biogenic synthesis of AgNPs utilizing seed extract of *Moringa oleifera,* the results revealed that the AgNPs showed significant photocatalytic activity against methylene blue (> 81%), orange red (> 82%), and 4-nitrophenol (> 75%)) upon sunlight irradiation. Moreover, the synthesized AgNPs also showed potential degradation activity against toxic metal ions with above 80% efficiency [109]. *Ruellia tuberosa* leaf extract was utilized for the biosynthesis of AgNPs, the measured average sizes of the AgNPs were 55.65 nm. The green fabricated AgNPs were successfully employed for the degradation of crystal violet and Coomassie brilliant blue dye with degradation efficiency of 87% and 74%, respectively [110].

3.3 Limitations and Future Aspects

In plant-mediated synthesis NPCs, plant extract carrying various plant metabolites, which can act as reducing/capping agents. Green synthesized NPCs have no toxicity and are eco-friendly in nature. These green materials are successfully applied for environmental remediation for wastewater treatment. However, there are several limitations and future projections that should be considered [113].

- Toxicity studies of the green synthesized NPCs need to be carried out before its commercialization.
- Several physico-chemical parameters such as solution extracts volume, temperature, solvent type, pH, the strength of precursor, and functional groups from plant metabolites are required to be heightened since it can change magnetic behavior and saturation magnetization value of the NPCs.
- More efforts to be made on NPCs morphology and saturation magnetization value by optimizing various parameters for better and efficient treatment of wastewater.

- New NPCs using various plant metabolites and synthesis methods for multiple applications in wastewater treatment is further needed to be explored.
- To commercialize NPCs cost-management analysis to be performed due to lack of data.
- Moreover, biocompetency studies are essential for biomedical applications.

3.4 Conclusion

Green chemistry and nanotechnology are rapidly emerging sciences for the remediation of toxic dye molecules, heavy metals, and other pollutants from wastewater. Researchers shifting toward the greener route for the NPs are elaborated here. Green synthesized NPCs utilizing plant extracts have greater benefits over other chemical and physical methods of NPCs. Moreover, general protocol and characterization techniques for the development of novel NPCs have been discussed thoroughly in this chapter. Wastewater treatment technologies used currently can remove organic, inorganic, heavy metals, etc. pollutants from the wastewater, nevertheless, these technologies are costly and energy demanding because of their incompetence to decontaminate wastewater thoroughly. The green synthesized NPCs are easy to synthesize and separate. Also, the NPCs can be reused owing to their unambiguous nature and higher stability. Furthermore, NPCs synthesized utilizing green chemistry principles require low cost and energy and can be promising for the low-economy countries for water remediation. Hence, green synthesized NPCs would be commercialized in near future due to their vast benefits in environmental protection.

References

1. Mazumder, D.N.G., Haque, R., Ghosh, N., De, B.K., Santra, A., Chakraborty, D., Smith, A.H., Arsenic levels in drinking water and the prevalence of skin lesions in West Bengal, India. *Int. J. Epidemiol.*, *27*, 5, 871–877, 1998.
2. Morales, K.H., Ryan, L., Kuo, T.L., Wu, M.M., Chen, C.J., Risk of internal cancers from arsenic in drinking water. *Environ. Health Perspect.*, *108*, 7, 655–661, 2000.
3. Waller, K., Swan, S.H., DeLorenze, G., Hopkins, B., Trihalomethanes in drinking water and spontaneous abortion. *Epidemiology*, *9*, 2, 134–140, 1998.

4. Wright, J., Gundry, S., Conroy, R., Household drinking water in developing countries: A systematic review of microbiological contamination between source and point-of-use. *Trop. Med. Int. Health*, *9*, 1, 106–117, 2004.
5. World Health Organization (WHO), https://www.who.int/news-room/fact-sheets/detail/drinking-water. [Internet], 2022.
6. Ahmad, A., Mohd-Setapar, S.H., Chuong, C.S., Khatoon, A., Wani, W.A., Kumar, R., Rafatullah, M., Recent advances in new generation dye removal technologies: novel search for approaches to reprocess wastewater. *RSC adv.*, *5*, 39, 30801–30818, 2015.
7. Zhang, Y., Wu, B., Xu, H., Liu, H., Wang, M., He, Y., Pan, B., Nanomaterials-enabled water and wastewater treatment. *NanoImpact*, *3*, 22–39, 2016.
8. Yaqoob, A.A., Parveen, T., Ahmad, K., Mohamad Ibrahim, M.N., Role of nanomaterials in the treatment of wastewater: A review. *Water*, *12*, 2, 495, 2020.
9. McNaught, A.D. and Wilkinson, A., *IUPAC gold book*, Blackwell Scientific Publications, Oxford, 1997.
10. Atrak, K., Ramazani, A., Fardood, S.T., Green synthesis of amorphous and gamma aluminum oxide nanoparticles by tragacanth gel and comparison of their photocatalytic activity for the degradation of organic dyes. *J. Mater. Sci.: Mater. Electron.*, *29*, 10, 8347–8353, 2018.
11. Zhang, W., Zhao, W., Zhou, Z., Yang, Z., Facile synthesis of α-MnO_2 micronests composed of nanowires and their enhanced adsorption to Congo red. *Front. Chem. Sci. Eng.*, *8*, 1, 64–72, 2014.
12. Song, G., Wu, X., Xin, F., Yin, X., $ZnFe_2O_4$ deposited on BiOCl with exposed (001) and (010) facets for photocatalytic reduction of CO_2 in cyclohexanol. *Front. Chem. Sci. Eng.*, *11*, 2, 197–204, 2017.
13. Saeidian, H. and Moradnia, F., Benign synthesis of N-aryl-3, 10-dihydroacridin-1 (2H)-one derivatives via ZnO nanoparticle-catalyzed Knoevenagel condensation/intramolecular enamination reaction. *Q. J. Iran. Chem. Commun.*, *5*, 252–261, 2017.
14. Soo, J.Z., Ang, B.C., Ong, B.H., Influence of calcination on the morphology and crystallinity of titanium dioxide nanofibers towards enhancing photocatalytic dye degradation. *Mater. Res. Express*, *6*, 2, 025039, 2018.
15. Tahir, M.B., Kiran, H., Iqbal, T., The detoxification of heavy metals from aqueous environment using nano-photocatalysis approach: A review. *Environ. Sci. Pollut. Res.*, *26*, 11, 10515–10528, 2019.
16. Anastas, P.T. and Kirchhoff, M.M., Origins, current status, and future challenges of green chemistry. *Acc. Chem. Res.*, 35, (2002) 686–694, 2002.
17. Sheldon, R.A., Engineering a more sustainable world through catalysis and green chemistry. *J. R Soc. Interface*, *13*, 116, 20160087, 2016.
18. Sheldon, R.A., Selective catalytic synthesis of fine chemicals: Opportunities and trends. *J. Mol. Catal. A: Chem.*, *107*, 1-3, 75–83, 1996.
19. Anastas, P.T., Heine, L.G., Williamson, T.C., Green chemical syntheses and processes: Introduction. *ACS Symp. Ser.*, 767, 1–6, 2000.

20. Agarwal, H., Kumar, S.V., Rajeshkumar, S., A review on green synthesis of zinc oxide nanoparticles–An eco-friendly approach. *Resour.-Effic. Technol.*, *3*, 4, 406–413, 2017.
21. Nagajyothi, P.C., Cha, S.J., Yang, I.J., Sreekanth, T.V.M., Kim, K.J., Shin, H.M., Antioxidant and anti-inflammatory activities of zinc oxide nanoparticles synthesized using *Polygala tenuifolia* root extract. *J. Photochem. Photobiol. B: Biol.*, *146*, 10–17, 2015.
22. Umar, K., Haque, M.M., Mir, N.A., Muneer, M., Farooqi, I.H., Titanium dioxide-mediated photocatalysed mineralization of two selected organic pollutants in aqueous suspensions. *J. Adv. Oxid. Technol.*, *16*, 2, 252–260, 2013.
23. Rasolevandi, T., Naseri, S., Azarpira, H., Mahvi, A.H., Photo-degradation of dexamethasone phosphate using UV/Iodide process: Kinetics, intermediates, and transformation pathways. *J. Mol. Liq.*, *295*, 111703, 2019.
24. Fawcett, D., Verduin, J.J., Shah, M., Sharma, S.B., Poinern, G.E.J., A review of current research into the biogenic synthesis of metal and metal oxide nanoparticles via marine algae and seagrasses. *J. Nanosci.*, *2017*, 1–15, 2017.
25. Ahmad, A., Mukherjee, P., Senapati, S., Mandal, D., Khan, M.I., Kumar, R., Sastry, M., Extracellular biosynthesis of silver nanoparticles using the fungus *Fusarium oxysporum*. *Colloids Surf. B: Biointerfaces*, *28*, 4, 313–318, 2003.
26. Shankar, S.S., Rai, A., Ankamwar, B., Singh, A., Ahmad, A., Sastry, M., Biological synthesis of triangular gold nanoprisms. *Nat. Mater.*, *3*, 7, 482–488, 2004.
27. Ankamwar, B., Damle, C., Ahmad, A., Sastry, M., Biosynthesis of gold and silver nanoparticles using *Emblica officinals* fruit extract, their phase transfer and transmetallation in an organic solution. *J. Nanosci. Nanotechnol.*, *5*, 10, 1665–1671, 2005.
28. Huang, J., Li, Q., Sun, D., Lu, Y., Su, Y., Yang, X., Wang, H., Wang, Y., Shao, W., Chen, C., Biosynthesis of silver and gold nanoparticles by novel sundried *Cinnamomum camphora* leaf. *Nanotechnology*, *18*, 10, 105–104, 2007.
29. Kureshi, A.A., Vaghela, H.M., Kumar, S., Singh, R., Kumari, P., Green synthesis of gold nanoparticles mediated by *Garcinia* fruits and their biological applications. *Pharm. Sci.*, *27*, 2, 238–250, 2020.
30. Kumar, V. and Yadav, S.K., Plant-mediated synthesis of silver and gold nanoparticles and their applications. *J. Chem. Technol. Biotechnol.: International Research Process, Environmental & Clean Technology*, *84*, 2, 151–157, 2009.
31. Ramesh, M., Anbuvannan, M., Viruthagiri, G., Green synthesis of ZnO nanoparticles using *Solanum nigrum* leaf extract and their antibacterial activity. *Spectrochim. Acta Part A: Mol. Biomol. Spectrosc.*, *136*, 864–870, 2015.
32. Vert, M., Doi, Y., Hellwich, K.H., Hess, M., Hodge, P., Kubisa, P., Rinaudo, M., Schué, F., Terminology for biorelated polymers and applications (IUPAC Recommendations 2012). *Pure Appl. Chem.*, *84*, 2, 377–410, 2012.

33. Cheremisinoff, P., *Handbook of Engineering Polymeric Materials*, CRC Press, Boca Raton, 1997.
34. Zhou, W., He, W., Zhong, S., Wang, Y., Zhao, H., Li, Z., Yan, S., Biosynthesis and magnetic properties of mesoporous Fe_3O_4 composites. *J. Magn. Magn. Mater.*, *321*, 8, 1025–1028, 2009.
35. Chen, W.P., Chen, J.Y., Chang, S.C., Su, C.L., Bacterial alginate produced by a mutant of *Azotobacter vinelandii*. *Appl. Environ. Microbiol.*, *49*, 3, 543–546, 1985.
36. Kolya, H., Pal, S., Pandey, A., Tripathy, T., Preparation of gold nanoparticles by a novel biodegradable graft copolymer sodium alginate-g-poly (*N*, *N*-dimethylacrylamide-co-acrylic acid) with anti-micro bacterial application. *Eur. Polym. J.*, *66*, 139–148, 2015.
37. Pandey, S., Do, J.Y., Kim, J., Kang, M., Fast and highly efficient catalytic degradation of dyes using κ-carrageenan stabilized silver nanoparticles nanocatalyst. *Carbohydr. Polym.*, *230*, 115597, 2020.
38. Ruel-Gariepy, E. and Leroux, J.-C., Chitosan: A natural polycation with multiple applications. *ACS Symposium Series*, Oxford University Press, pp. 243–259, 2006.
39. Jain, A., Gulbake, A., Shilpi, S., Jain, A., Hurkat, P., Jain, S.K., A new horizon in modifications of chitosan: Syntheses and applications. *Crit. Rev. Ther. Drug Carrier Syst.*, *30*, 2, 91–181, 2013.
40. Kumar, M.N.R., A review of chitin and chitosan applications. *React. Funct. Polym.*, *46*, 1, 1–27, 2000.
41. Vanaamudan, A., Sadhu, M., Pamidimukkala, P., Chitosan-Guar gum blend silver nanoparticle bionanocomposite with potential for catalytic degradation of dyes and catalytic reduction of nitrophenol. *J. Mol. Liq.*, *271*, 202–208, 2018.
42. Morin-Crini, N., Lichtfouse, E., Torri, G., Crini, G., Applications of chitosan in food, pharmaceuticals, medicine, cosmetics, agriculture, textiles, pulp and paper, biotechnology, and environmental chemistry. *Environ. Chem. Lett.*, *17*, 4, 1667–1692, 2019.
43. Sargin, I., Baran, T., Arslan, G., Environmental remediation by chitosan-carbon nanotube supported palladium nanoparticles: Conversion of toxic nitroarenes into aromatic amines, degradation of dye pollutants and green synthesis of biaryls. *Sep. Purif. Technol.*, *247*, 116987, 2020.
44. Kasiri, M.B., Application of chitosan derivatives as promising adsorbents for treatment of textile wastewater, in: *The Impact and Prospects of Green Chemistry for Textile Technology*, pp. 417–469, 2019.
45. Mansur, H.S. and Mansur, A.A.P., Nano-photocatalysts based on ZnS quantum dots/chitosan for the photodegradation of dye pollutants, in: *IOP Conference Series: Materials Science and Engineering*, vol. 76, No. 1, p. 012003, IOP Publishing, March, 2015.

46. Ahmed, H.B., Recruitment of various biological macromolecules in fabrication of gold nanoparticles: Overview for preparation and applications. *Int. J. Biol Macromol., 140*, 265–277, 2019.
47. Pandey, S., Goswami, G.K., Nanda, K.K., Green synthesis of polysaccharide/gold nanoparticle nanocomposite: An efficient ammonia sensor. *Carbohydr. Polym., 94*, 1, 229–234, 2013.
48. Lindhorst, T.K., *Essentials of carbohydrate chemistry and biochemistry*, vol. 3, Wiley-Vch, Weinheim, 2007.
49. Li, G., Li, Y., Wang, Z., Liu, H., Green synthesis of palladium nanoparticles with carboxymethyl cellulose for degradation of azo-dyes. *Mater. Chem. Phys., 187*, 133–140, 2017.
50. Oliveira, L.V., Bennici, S., Josien, L., Limousy, L., Bizeto, M.A., Camilo, F.F., Free-standing cellulose film containing manganese dioxide nanoparticles and its use in discoloration of indigo carmine dye. *Carbohydr. Polym., 230*, 115621, 2020.
51. Garcıa-Ochoa, F., Santos, V.E., Casas, J.A., Gomez, E., Xanthan gum: Production, recovery, and properties. *Biotechnol. Adv., 18*, 7, 549–579, 2000.
52. Katzbauer, B., Properties and applications of xanthan gum. *Polym. Degrad. Stab., 59*, 1-3, 81–84, 1998.
53. Ayodhya, D., Venkatesham, M., Santoshi Kumari, A., Reddy, G.B., Ramakrishna, D., Veerabhadram, G., Photocatalytic degradation of dye pollutants under solar, visible and UV lights using green synthesised CuS nanoparticles. *J. Exp. Nanosci., 11*, 6, 418–432, 2016.
54 Inamuddin, Xanthan gum/titanium dioxide nanocomposite for photocatalytic degradation of methyl orange dye. *Int. J. Biol. Macromol.*, 121, 1046–1053, 2019.
55. Tripathi, R.M., Bhadwal, A.S., Gupta, R.K., Singh, P., Shrivastav, A., Shrivastav, B.R., ZnO nanoflowers: novel biogenic synthesis and enhanced photocatalytic activity. *J. Photochem. Photobiol. B, 141*, 288–295. 2014.
56. Rajkumar, R., Ezhumalai, G., Gnanadesigan, M., A green approach for the synthesis of silver nanoparticles by *Chlorella vulgaris* and its application in photo- catalytic dye degradation activity. *Environ. Technol. Innovation, 21*, 101282, 2021.
57. Kalpana, V.N., Kataru, B.A.S., Sravani, N., Vigneshwari, T., Panneerselvam, A., Rajeswari, V.D., Biosynthesis of zinc oxide nanoparticles using culture filtrates of *Aspergillus niger*: Antimicrobial textiles and dye degradation studies. *OpenNano, 3*, 48–55, 2018.
58. Hulkoti, N.I., and Taranath, T.C., Biosynthesis of nanoparticles using microbes—A review. *Colloids Surf. B: Biointerfaces, 121*, 474–483, 2014.
59. Saif Hasan, S., Singh, S., Parikh, R.Y., Dharne, M.S., Patole, M.S., Prasad, B.L.V., Shouche, Y.S., Bacterial synthesis of copper/copper oxide nanoparticles. *J. Nanosci. Nanotechnol., 8*, 6, 3191–3196, 2008.

60. Rajan, A., MeenaKumari, M., Philip, D., Shape tailored green synthesis and catalytic properties of gold nanocrystals. *Spectrochim. Acta Part A: Mol. Biomol. Spectrosc.*, *118*, 793–799, 2014.
61. Phatak, R.S. and Hendre, A.S., Sunlight induced green synthesis of silver nanoparticles using sundried leaves extract of *Kalanchoe pinnata* and evaluation of its photocatalytic potential. *Der Pharm. Lett.*, *7*, 5, 313–324, 2015.
62. Roy, K., Sarkar, C.K., Ghosh, C.K., Photocatalytic activity of biogenic silver nanoparticles synthesized using potato (*Solanum tuberosum*) infusion *Spectrochim. Acta Part A: Mol. Biomol. Spectrosc.*, *146*, 286–291, 2015.
63. Malarkodi, C., Malik, V., Uma, S., Synthesis of Fe2O3 using *Emblica officinalis* extract and its photocatalytic efficiency. *Mater. Sci.: Indian J.*, *16*, 1–10, 2018.
64. Tahir, K., Nazir, S., Li, B., Ahmad, A., Nasir, T., Khan, A.U., Ali Shah, S.A., Haq Khan, Z.U., Yasin, G., Hameed, M.U., *Sapium sebiferum* leaf extract medicated synthesis of palladium nanoparticles and *in vitro* investigation of their bacterial and photocatalytic activities. *J. Photochem. Photobiol. B: Biol.*, *164*, 164–173, 2016.
65. Sankar, R., Manikandan, P., Malarvizhi, V., Fathima, T., Shivashangari, K.S., Ravikumar, V., Green synthesis of colloidal copper oxide nanoparticles using *Carica papaya* and its application in photocatalytic dye degradation. *Spectrochim. Acta Part A: Mol. Biomol. Spectrosc.*, *121*, 746–750, 2014.
66. Yadav, L.R., Manjunath, K., Archana, B., Madhu, C., Naika, H.R., Nagabhushana, H., Kavitha, C., Nagaraju, G., Fruit juice extract mediated synthesis of CeO_2 nanoparticles for antibacterial and photocatalytic activities. *Eur. Phys. J. Plus*, *131*, 5, 1–10, 2016.
67. Thandapani, K., Kathiravan, M., Namasivayam, E., Padiksan, I.A., Natesan, G., Tiwari, M., Giovanni, B., Perumal, V., Enhanced larvicidal, antibacterial, and photocatalytic efficacy of TiO_2 nanohybrids green synthesized using the aqueous leaf extract of *Parthenium hysterophorus*. *Environ. Sci. Pollut. Res.*, *25*, 11, 10328–10339, 2018.
68. Kohli, R., Methods for monitoring and measuring cleanliness of surface, in: *Developments in surface contamination and cleaning*, vol. 12, pp. 23–105, 2012.
69. Khan, I., Saeed, K., Khan, I., Nanoparticles: Properties, applications and toxicities. *Arabian J. Chem.*, *12*, 7, 908–931, 2019.
70. Schaffer, B., Hohenester, U., Trugler, A., Hofer, F., High-resolution surface plasmon imaging of gold nanoparticles by energy-filtered transmission electron microscopy. *Phys. Rev. B*, *79*, 4, 041401, 2009.
71. Eppler, A.S., Rupprechter, G., Anderson, E.A., Somorjai, G.A., Thermal and chemical stability and adhesion strength of Pt nanoparticle arrays supported on silica studied by transmission electron microscopy and atomic force microscopy. *J. Phys. Chem. B*, *104*, 31, 7286–7292, 2000.

72. Kumar, B.R. and Rao, T.S., AFM studies on surface morphology, topography and texture of nanostructured zinc aluminum oxide thin films. *Dig. J. Nanomater. Biostructures*, *7*, 4, 1881–1889, 2012.
73. Soonmin, H., Atomic force microscopy studies on sulfur-, selenium- and tellurium-based metal chalcogenide thin films: A review. *Afr. J. Pure Appl. Chem.*, *11*, 5, 42–49, 2017.
74. Ghiorghita, C.A., Dragan, E.S., Bucatariu, F., Schwarz, D., Blegescu, C., Mihai, M., Green synthesis of Ag nanoparticles with uncommon behaviour towards $NaBH_4$ in presence of Congo red using polyelectrolyte multilayers containing sodium carboxymethyl cellulose. *Colloids Surf. A: Physicochem. Eng. Asp.*, *585*, 124157, 2020.
75. Jiang, J., Oberdorster, G., Biswas, P., Characterization of size, surface charge, and agglomeration state of nanoparticle dispersions for toxicological studies. *J. Nanopart. Res.*, *11*, 1, 77–89, 2009.
76. Goutam, S.P., Saxena, G., Singh, V., Yadav, A.K., Bharagava, R.N., Thapa, K.B., Green synthesis of TiO_2 nanoparticles using leaf extract of *Jatropha curcas* L. for photocatalytic degradation of tannery wastewater. *Chem. Eng. J.*, *336*, 386–396, 2018.
77. Surendra, T.V. and Roopan, S.M., Photocatalytic and antibacterial properties of phytosynthesized CeO2 NPs using *Moringa oleifera* peel extract. *J. Photochem. Photobiol. B: Biol.*, *161*, 122–128, 2016.
78. Datar, A., Chung, Y.G., Lin, L.C., Beyond the BET analysis: The surface area prediction of nanoporous materials using a machine learning method. *J. Phys. Chem. Lett.*, *11*, 14, 5412–5417, 2020.
79. Nasrollahzadeh, M., Sajadi, S.M., Rostami-Vartooni, A., Bagherzadeh, M., Safari, R., Immobilization of copper nanoparticles on perlite: Green synthesis, characterization and catalytic activity on aqueous reduction of 4-nitrophenol. *J. Mol. Catal. A: Chem.*, *400*, 22–30, 2015.
80. Ravikumar, S., Pandiyan, V., Alam, M., Ahmad, N., Nithya, V., Krishnakumar, B., Sobral, A.J., *Costus speciosus* koen leaf extract assisted cs-znx (X= O or S) nanomaterials: Synthesis, characterization and photocatalytic degradation of rr 120 dye under uv and direct sunlight. *J. Mol. Struct.*, *1225*, 129176, 2021.
81. Thomas, B., Vithiya, B., Prasad, T., Mohamed, S.B., Magdalane, C.M., Kaviyarasu, K., Maaza, M., Antioxidant and photocatalytic activity of aqueous leaf extract mediated green synthesis of silver nanoparticles using *Passiflora edulis f. flavicarpa J. Nanosci. Nanotechnol.*, *19*, 5, 2640–2648, 2019.
82. Vidya, C., Prabha, M.C., Raj, M.A., Green mediated synthesis of zinc oxide nanoparticles for the photocatalytic degradation of Rose Bengal dye. *Environ. Nanotechnol. Monit. Manage.*, *6*, 134–138, 2016.
83. Raina, S., Roy, A., Bharadvaja, N., Degradation of dyes using biologically synthesized silver and copper nanoparticles. *Environ. Nanotechnol. Monit. Manage.*, *13*, 100278, 2020.

84. Sundaramurthy, N. and Parthiban, C., Biosynthesis of copper oxide nanoparticles using *Pyrus pyrifolia* leaf extract and evolve the catalytic activity. *Int. Res. J. Eng. Technol.*, *2*, 6, 332–338, 2015.
85. Kumar, B., Vizuete, K.S., Sharma, V., Debut, A., Cumbal, L., Ecofriendly synthesis of monodispersed silver nanoparticles using Andean Mortino berry as reductant and its photocatalytic activity. *Vacuum*, *160*, 272–278, 2019.
86. Arunachalam, R., Dhanasingh, S., Kalimuthu, B., Uthirappan, M., Rose, C., Mandal, A.B., Phytosynthesis of silver nanoparticles using *Coccinia grandis* leaf extract and its application in the photocatalytic degradation. *Colloids Surf. B: Biointerfaces*, *94*, 226–230, 2012.
87. Fowsiya, J., Madhumitha, G., Al-Dhabi, N.A., Arasu, M.V., Photocatalytic degradation of Congo red using *Carissa edulis* extract capped zinc oxide nanoparticles *J. Photochem. Photobiol. B: Biol.*, *162*, 395–401, 2016.
88. Zheng, Y., Fu, L., Han, F., Wang, A., Cai, W., Yu, J., Peng, F., Green biosynthesis and characterization of zinc oxide nanoparticles using *Corymbia citriodora* leaf extract and their photocatalytic activity. *Green Chem. Lett. Rev.*, *8*, 2, 59–63, 2015.
89. Ullah, H., Ullah, Z., Fazal, A., Irfan, M., Use of vegetable waste extracts for controlling microstructure of CuO nanoparticles: Green synthesis, characterization and photocatalytic applications. *J. Chem.*, *2017*, 1–15, 2017.
90. Karnan, T. and Selvakumar, S.A.S., Biosynthesis of ZnO nanoparticles using rambutan (*Nephelium lappaceum* L.) peel extract and their photocatalytic activity on methyl orange dye. *J. Mol. Struct.*, *1125*, 358–365, 2016.
91. Karthik, R., Govindasamy, M., Chen, S.M., Cheng, Y.H., Muthukrishnan, P., Padmavathy, S., Elangovan, A., Biosynthesis of silver nanoparticles by using *Camellia japonica* leaf extract for the electrocatalytic reduction of nitrobenzene and photocatalytic degradation of Eosin-Y. *J. Photochem. Photobiol. B: Biol.*, *170*, 164–172, 2017.
92. Kathiravan, V., Green synthesis of silver nanoparticles using different volumes of *Trichodesma indicum* leaf extract and their antibacterial and photocatalytic activities. *Res. Chem. Intermed.*, *44*, 9, 4999–5012, 2018.
93. Khan, A.U., Yuan, Q., Wei, Y., Khan, G.M., Khan, Z.U.H., Khan, S., Ali, F., Tahir, K., Ahmad, A., Khan, F.U., Photocatalytic and antibacterial response of biosynthesized gold nanoparticles. *J. Photochem. Photobiol. B: Biol.*, *162*, 273–277, 2016.
94. Kumari, R.M., Thapa, N., Gupta, N., Kumar, A., Nimesh, S., Antibacterial and photocatalytic degradation efficacy of silver nanoparticles biosynthesized using *Cordia dichotoma* leaf extract. *Adv. Nat. Sci.: Nanosci. Nanotechnol.*, *7*, 4, 045009, 2016.
95. Raja, A., Ashokkumar, S., Marthandam, R.P., Jayachandiran, J., Khatiwada, C.P., Kaviyarasu, K., Raman, R.G., Swaminathan, M., Eco-friendly preparation of zinc oxide nanoparticles using *Tabernaemontana divaricata* and its photocatalytic and antimicrobial activity. *J. Photochem. Photobiol. B: Biol.*, *181*, 53–58, 2018.

96. Park, J.K., Rupa, E.J., Arif, M.H., Li, J.F., Anandapadmanaban, G., Kang, J.P., Kim, M., Ahn, J.C., Akter, R., Yang, D.C., Kang, S.C., Synthesis of zinc oxide nanoparticles from *Gynostemma pentaphyllum* extracts and assessment of photocatalytic properties through malachite green dye decolorization under UV illumination-A green approach. *Optik*, *239*, 166249, 2021.
97. Shabaani, M., Rahaiee, S., Zare, M., Jafari, S.M., Green synthesis of ZnO nanoparticles using loquat seed extract; Biological functions and photocatalytic degradation properties. *LWT*, *134*, 110133, 2020.
98. Qasim, S., Zafar, A., Saif, M.S., Ali, Z., Nazar, M., Waqas, M., Iqbal, F.F., Shu, X.G., Hasan, M., Green synthesis of iron oxide nanorods using *Withania coagulans* extract improved photocatalytic degradation and antimicrobial activity. *J. Photochem. Photobiol. B: Biol.*, *204*, 111784, 2020.
99. Bhuiyan, M.S.H., Miah, M.Y., Paul, S.C., Aka, T.D., Saha, O., Rahaman, M.M., Islam Sharif, H.J., Habiba, O., Ashaduzzaman, M., Green synthesis of iron oxide nanoparticle using *Carica papaya* leaf extract: Application for photocatalytic degradation of remazol yellow RR dye and antibacterial activity. *Heliyon*, *6*, 8, e04603, 2020.
100. Kadam, J., Dhawal, P., Barve, S., Kakodkar, S., Green synthesis of silver nanoparticles using cauliflower waste and their multifaceted applications in photocatalytic degradation of methylene blue dye and Hg2+ biosensing. *SN Appl. Sci.*, *2*, 4, 1–16, 2020.
101. Muthuvel, A., Jothibas, M., Mohana, V., Manoharan, C., Green synthesis of cerium oxide nanoparticles using *Calotropis procera* flower extract and their photocatalytic degradation and antibacterial activity. *Inorg. Chem. Commun.*, *119*, 108086, 2020.
102. Luque, P.A., Nava, O., Soto-Robles, C.A., Chinchillas-Chinchillas, M.J., Garrafa-Galvez, H.E., Baez-Lopez, Y.A., Valdez-Nunez, K.P., Vilchis-Nestor, A.R., Castro-Beltran, A., Improved photocatalytic efficiency of SnO_2 nanoparticles through green synthesis. *Optik*, *206*, 164299, 2020.
103. Liu, Y.C., Li, J., Ahn, J., Pu, J., Rupa, E.J., Huo, Y., Yang, D.C., Biosynthesis of zinc oxide nanoparticles by one-pot green synthesis using fruit extract of *Amomum longiligulare* and its activity as a photocatalyst. *Optik*, *218*, 165245, 2020.
104. Ahmad, W., Singh, A., Jaiswal, K.K., Gupta, P., Green synthesis of photocatalytic TiO_2 nanoparticles for potential application in photochemical degradation of ornidazole. *J. Inorg. Organomet. Polym. Mater.*, *31*, 2, 614-623. 2021.
105. Tripathi, R.M., Hameed, P., Rao, R.P., Shrivastava, N., Mittal, J., Mohapatra, S., Biosynthesis of highly stable fluorescent selenium nanoparticles and the evaluation of their photocatalytic degradation of dye. *Bionanoscience*, *10*, 2, 389–396, 2020.
106. Awad, M.A., Hendi, A.A., Ortashi, K.M., Alzahrani, B., Soliman, D., Alana A., Alenazi, W., Taha, R.M., Ramadan, R., El-Tohamy, M., Almasoud, N., Alomar, T.S., Biogenic synthesis of silver nanoparticles using *Trigonella*

foenum- graecum seed extract: Characterization, photocatalytic and antibacterial activities. *Sens. Actuators A: Phys.*, *323*, 112670, 2021.
107. Cittrarasu, V., Kaliannan, D., Dharman, K., Maluventhen, V., Easwaran, M., Liu, W.C., Arumugam, M., Green synthesis of selenium nanoparticles mediated *from Ceropegia bulbosa* Roxb extract and its cytotoxicity, antimicrobial, mosquitocidal and photocatalytic activities. *Sci. Rep.*, *11*, 1, 1–15, 2021.
108. Kannan, D.S., Mahboob, S., Al-Ghanim, K.A., Venkatachalam, P., Antibacterial, antibiofilm and photocatalytic activities of biogenic silver nanoparticles from *Ludwigia octovalvis*. *J. Clust. Sci.*, *32*, 2, 255–264, 2021.
109. Mehwish, H.M., Rajoka, M.S.R., Xiong, Y., Cai, H., Aadil, R.M., Mahmood, Q., Zhu, Q., Green synthesis of a silver nanoparticle using *Moringa oleifera* seed and its applications for antimicrobial and sun-light mediated photocatalytic water detoxification. *J. Environ. Chem. Eng.*, *9*, 4, 105290, 2021.
110. Seerangaraj, V., Sathiyavimal, S., Shankar, S.N., Nandagopal, J.G.T., Balashanmugam, P., Al-Misned, F.A., Pugazhendhi, A., Cytotoxic effects of silver nanoparticles on *Ruellia tuberosa*: Photocatalytic degradation properties against crystal violet and coomassie brilliant blue. *J. Environ. Chem. Eng.*, *9*, 2, 105088, 2021.
111. Kumar, P., Govindaraju, M., Senthamilselvi, S., Premkumar, K., Photocatalytic degradation of methyl orange dye using silver (Ag) nanoparticles synthesized from *Ulva lactuca*. *Colloids Surf. B: Biointerfaces*, *103*, 658–661, 2013.
112. Ahmad, W., Singh, A., Jaiswal, K.K., Gupta, P., Green synthesis of photocatalytic TiO_2 nanoparticles for potential application in photochemical degradation of ornidazole. *J. Inorg. Organomet. Polym. Mater.*, 1–10, 2020.
113. Ali, I., Peng, C., Naz, I., Khan, Z.M., Sultan, M., Islam, T., Abbasi, I.A., Phytogenic magnetic nanoparticles for wastewater treatment: A review. *RSC Adv.*, *7*, 64, 40158–40178, 2017.

4

Green Synthesis of Metal Ferrite Nanoparticles for the Photocatalytic Degradation of Dyes in Wastewater

Aubrey Makofane[1,2], David E Motaung[3] and Nomso C. Hintsho-Mbita[1,2*]

[1]Department of Chemistry, University of Limpopo, Sovenga, Polokwane, South Africa
[2]DSI-NRF Centre of Excellence in Strong Materials, South Africa
[3]Department of Physics, University of Limpopo, Private Bag, Sovenga, South Africa

Abstract
Organic dyes can be classified as a large group of contaminants produced mostly from the cosmetic, textile, and rubber industry. These dyes are highly toxic and can have carcinogenic effects on humans. Photocatalytic degradation of these contaminants has become a method of choice as it is able to degrade these harmful pollutants into environmentally safe materials. Metal ferrites with the formula (MFe_2O_4) have been widely used in photocatalysis. One of their limitations has been their synthesis approach, which uses harsh solvents as they are not environmentally safe. Green chemistry through the use of plant extracts, yeast, fungus, and bacteria has recently assisted in overcoming these shortcomings. This chapter will highlight the synthesis of green derived metal ferrite nanoparticles and their use as photocatalysts in the degradation of organic dye pollutants. The effect of low bandgap semiconductor metal oxide materials and carbon nanomaterials on the photodegradation efficiency of metal ferrites is also explored.

***Keywords*:** Organic dyes, photodegradation, metal ferrites nanoparticles and metal ferrites-based composites, green technology

**Corresponding author*: nomso.hintsho-mbita@ul.ac.za

Elvis Fosso-Kankeu, Sadanand Pandey, and Suprakas Sinha Ray (eds.) Photoreactors in Advanced Oxidation Processes: The Future of Wastewater Treatment, (109–150) © 2023 Scrivener Publishing LLC

Abbreviations

AOPs	Advanced oxidation processes
BBV	Basic blue violet
BOD	Biological oxygen demand
$CoFe_2O_4$	Cobalt Ferrite
$CuFe_2O_4$	Copper Ferrite
CR	Congo Red
CVD	Chemical vapor deposition
DBPs	Disinfection by products
EO	Electrochemical oxidation
MB	Methylene blue
MFe_2O_4	Metal Ferrite
MG	Malachite Green
MO	Methyl orange
RhB	Rhodamine B
ROS	Reactive oxygen species
SEM	Scanning electron microscope
WHO	World Health Organization
$ZnFe_2O_4$	Zinc Ferrite
ZnS	Zinc Sulphide

4.1 Introduction

The lack of water supply owing to the climatic conditions and poor conservation of water resources has become a global problem in the last few decades [1]. Access to clean water has become an enormous problem in a growing global economy and populated nations [2].

In the textile, food, and leather industry, synthesized dyes (i.e., methylene blue [MB], methyl orange [MO], Congo red [CR], malachite green [MG], rhodamine B [RhB], etc.) are the most used around the world, and the by-products from these operations are washed straight through natural water supplies. These dyes tend to spread around the bodies of freshwater, which leads to a serious impact on aquatic life as they are harmful to plants and animals [3]. They have a complex structure and a strong chemical persistence that persists over large areas in freshwater bodies [1]. This results in slowing down of the photosynthetic mechanisms, inhibits marine or aquatic life due to the prevention of sunshine, and the use of dissolved oxygen, thus reducing the recreation value of the stream. Due to their enormous production volume, slow biodegradation, low decolouration and high toxicity, the degradation of dyes in industrial wastewater has created a significant interest [4].

The World Health Organisation (WHO) estimates that almost 780 million citizens have inadequate access to sufficient drinking water [5]. Organic dyes are indeed the major contaminants that impact the marine, as well as land life [6]. The conventional chemical, biological, and physical techniques, which include osmosis, adsorption, and coagulation are constantly utilized to reduce or eliminate the organic dye's harmful implications [5]. However, these methods have several limitations, they are usually time-consuming and produce intermediate products that are more toxic than the dye itself. Consequently, more efficient methods of degrading such toxins from drinking water sources have been established without producing secondary harsh pollutants [7].

Interest in heterogeneous photocatalysis, utilizing metal ferrites has been on the upsurge in recent years because of its useful applications for both environmental and organic synthesis. Metal ferrite nanoparticles are characterized by their unique physical and chemical properties. They are cost-effective, require low manufacturing costs, and can be used for the degradation of various pollutants [8].

Methods that have been used for metal ferrite synthesis include microwave combustion, coprecipitation, hydrothermal treatment accompanied by sol-gel phase, and combustion reaction [9]. These techniques require long-term reactions, sophisticated machinery utilizing chemicals, and high temperatures, thereby leading to some adverse environmental effects. Hence, researchers in the last decade have geared toward investigating other methods of synthesis that are environmentally friendly, easier and affordable.

Thus, among the synthesis methods, green chemistry has become an innovative way of preparing environmentally friendly products because it utilizes nonhazardous resources, such as organic materials and several other environmentally benign materials. A productive method to green technology has become the hydrothermal method, which utilizes water as fuel as well as plant extracts, fungi, bacteria, yeast, etc as replacement solvents. These are used as reducing and capping agents that facilitate the formation of nanoparticles using bioactive molecules.

In this chapter, the focus is on the synthesis of green derived metal ferrites using bioactive molecules from plant extracts for their use as photocatalysts in the degradation of organic dye pollutants. The effect of depositing noble and transitional metals (e.g. Ag, Au, Cu) and carbon nanomaterials (carbon spheres, carbon nanotubes, carbon dots) on their photodegradation efficiency of metal ferrites will be highlighted. Lastly, their future perspectives and use in other biological applications will be explored.

4.2 Metal Ferrite Nanoparticles

Metal ferrites with the general formula MFe_2O_4 (M = Cu, Mn, Zn, Co, Ni, and other metals) can be defined as the cubic structure of a closely packed arrangement of M^{2+} and Fe^{3+} ions. Their oxygen atoms occupy either the octahedral (B) or tetrahedral (A) positions as illustrated in Figure 4.1. The nanoferrites' physicochemical characteristics are extremely reliant on the kinds, quantities and sites of the metal cations in the crystal structure [10, 11].

These materials have gained a lot of interest due to their unique catalytic, electronic, and magnetic properties in particular for wastewater treatment. They have been investigated intensively in recent years as they possess unique magnetic, electrical and optical properties [12]. In addition to the above features, spinel ferrite nanoparticles are important among the magnetic nanoparticles because of their thermal and chemical firmness [13].

Several synthesis methods, such as sol-gel, combustion, ball milling, and coprecipitation have been used to synthesize these nanoparticles [14]. Nanomaterial synthesis through green synthesis is preferable over numerous methods because it is safe, environmentally friendly, has a lower reaction temperature and is clear of unwanted toxic by-products. The green

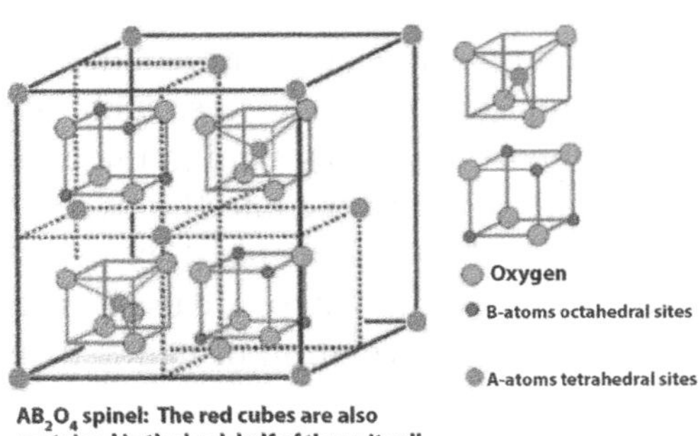

Figure 4.1 Schematic diagram displaying the spinel structure of MFe_2O_4, indicating oxygen atom (sky blue), Fe^{+3}/M^{+2} octahedral(red), Fe^{+3} tetrahedral (green) [14].

production of metal nanoparticles has been extensively implemented utilizing natural materials, such as plants, fungi, yeast, and bacteria. Though that has been the case, before green synthesis, conventional synthesis methods have been explored whereby through the previous decades, magnetic spinel ferrite nanoparticles have drawn enormous attention toward their enhanced properties relative to those of "normal" grain-sized materials > 10 μm. The synthesis of their special formulation and microstructure has resulted in a high potential for various applications, which includes magnetic processing, conservation of magnetic energy, catalysis, bioengineering, and disposal of wastewater.

4.3 General Synthesis Methods of Metal Ferrites and Their Limitations

Metal ferrites are known to be ferromagnetic due to their special magnetic, optical, and electrical activities. These materials have been utilized in numerous applications, such as catalysis, drug delivery, pigments, sensors, fluids, magnetic recording, and so forth. For ecological remediation, ferrites are interesting candidates because of their narrow bandgap (1.1–2.3 eV) as they are nonharmful, require minimal cost, and have shown chemical and thermal stability. Numerous strategies have been implemented for the synthesis of spinel ferrite, for example, mechanical alloying, microwave heating, combustion, ultrasonication, coprecipitation, ball milling, sol-gel, etc. [15].

Nanoparticle synthesis approaches are generally classified under two categories: the physical as well as the chemical strategies [16]. During the physical process, the grinding mechanism is the most realistic representation of the physical processes. Highly efficient machines are used to isolate the nanometric particles, whereby size-reducing devices are also used to grind the samples to macroscale level [17]. The particles arising afterward are air graded to retrieve oxidized nanoparticles. The conditions which have a crucial effect on the characteristics of the resulting nanoparticles include materials and period of milling as well as the atmospheric environment. In another physical method, pyrolysis (see Figure 4.2), an organic intermediate (be it a fluid or even a gas) is pushed under higher pressure within an orifice and burned. The resultant ash is air treated to extract nanoparticles that have oxidized.

During the chemical synthesis process, a reduction of metal ions or decomposition of precursors to form atoms, followed by aggregation takes place. An average methodology of these includes the formation of

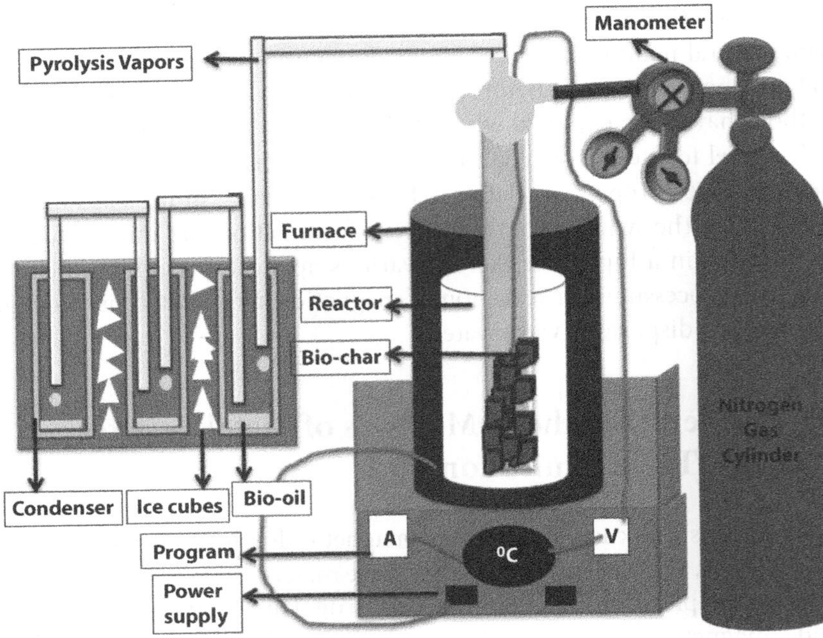

Figure 4.2 Reactor utilized for the pyrolysis of biomass feedstock [18].

nanoparticles in a fluid medium containing different reactants, specifically reducing agents, such as sodium borohydride or methoxy polyethylene glycol or potassium bitartrate or hydrazine. Furthermore, to avoid agglomeration of nanostructured materials, the solution mixture is also combined with a stabilizing agent, which includes polyvinyl pyrrolidone or either sodium dodecyl benzyl sulphate. The nanoparticles produced through these chemical techniques have been proven to have a small size distribution.

Naseri *et al.*, [19] demonstrated the synthesis of cobalt ferrite nanoparticles via thermal treatment method utilizing polyvinyl pyrrolidone as an agglomeration capping agent and aqueous solutions of metal nitrates. The mean particle size of the as-prepared nanoparticles at 623 K and 923 K varied from 12.5 to 39 nm. It was observed that magnetization, remanent magnetization, and magnetic saturation of the cobalt ferrite nanoparticles increased with an increase in temperature, while both the remanence ratio and coercivity field increased until reaching the maximum value and then decreased.

From the methods listed in the physical and chemical synthesis, many of these techniques used during nanoparticle production were shown to be too costly and required the use of harmful and toxic substances that are responsible for different biological hazards, thus now the focus has geared toward biological synthesis.

4.4 Biological Synthesis of Metal Ferrite Nanostructures

The advancement of green techniques to produce nanoparticles has developed into a significant field in nanotechnology, whereby green nanotechnology interacts with the safety and environmentally friendly approaches to produce nanomaterials as a substitute for conventional physical and chemical approaches [20].

Biosynthesis of nanomaterial is an effective method of forming nanoparticles through the use of plants and microorganism having both catalytic and biomedical applications. This method is cost-effective, ecologically sustainable, biocompatible, and safe [21].

There are two classifications to which preparation techniques for nanoparticles are divided, for example, "top-down," as well as "bottom-up." In "top-down" approach, the particles that are generated as chemicals are smashed to create smaller particles, whereas in the "bottom-up" approach, the particles get generated as ions and are chemically mixed [22]. There are numerous synthetic methods in "bottom-up" approach, such as microwave heating, hydrothermal, flame spray pyrolysis, co-precipitation, sol-gel, thermal decomposition, solvothermal, microemulsion, chemical vapor decomposition, and sonochemical methods. While the "top-down" approach only includes the pulsed laser ablation as well as the mechanical milling, which are the most widely recognized methods [14].

The synthesis of nanomaterials utilizing plant extracts has been the most widely adopted process since they are environmentally sustainable. The green processing of nanomaterials can serve as a source of certain active ingredients, since they are also simpler to treat and easier to access. Fungi, as well as bacteria throughout the development media, need a substantially longer implantation time to reduce a metallic ion and water-soluble phytochemicals [23]. As a result, plants are among the potential candidates for the production of metal nanostructures relative to bacteria or fungi. The synthesis of nanomaterials focused on plants operates at lower temperatures and only includes moderate and environmentally friendly materials.

Though plants have been used extensively for the synthesis of metal ferrites, fungi and bacteria have also been of significance.

4.4.1 Synthesis of Metal Ferrite Nanostructures Using Bacteria

The subsurface Fe (III)-reducing bacteria have been an underexplored tool for magnetic nanomaterial growth since these pathogens can generate significant amounts of nanoscale magnetite (Fe_3O_4) at room temperatures. Metal decreasing bacteria live in oxygen-deficient habitats and generate power for development by oxidizing hydrogen or natural electron donors, combined with the lowering of oxidized metals, e.g., minerals containing Fe (III) [24]. These can lead to magnetite formulation through the extracellular alleviation of amorphous Fe (III)-oxyhydroxides leading to a release of soluble Fe (II), as well as the completed recrystallization of amorphous minerals toward a new stage [25].

Several studies have shown changes in the formation of biogenic magnetite generated by Fe(III)-degrading bacteria for industrial production and environmental operating systems. Scientific studies on the economic utilization of bacteria to generate magnetic materials have focused entirely on magnetotactic bacteria that structurally forms magnetosome magnetite nanoparticles [26]. Magnetotactic bacteria are located at the surface water interfaces using an inner nanomagnet to direct them through the Earth's magnetic field to their desired environmental niche. Given that magnetotactic bacteria typically develop efficiently under highly controlled microaerobic circumstances, the cultivation systems for such organisms are difficult but mostly result in minimal nanomagnetite returns. Considering these drawbacks, *in vitro* magnetotactic bacteria change the magnetic properties of the established material [27].

Synthesis of nanoparticles utilizing bacteria (as shown in Table 4.1), is a sustainable solution but has numerous drawbacks. These include microbial imaging as it is a time-consuming procedure and proper control of culture broth is needed to prevent contaminations. Loss of control over the size of nanoparticles, morphology, and costs of the medium that is used to cultivate bacteria is also quite high [28]. Figure 4.3 displays the synthetic approach used for the formation of cobalt ferrite supported onto the bacterial cellulose.

Menchaca-Nal *et al.*, [25] reported on the green synthesized cobalt ferrite ($CoFe_2O_4$) nanoparticles utilizing the bacterial cellulose nanoribbons, which were used as a template. These nanoparticles with a diameter in the range of 9 to 13 nm within the bacterial nanocellulose aerogel

Table 4.1 Green-derived metal ferrite nanoparticles using bacterial extracts.

Name of nanoparticles	Bacterial species	Size	SEM/TEM morphology	Refs.
$CoFe_2O_4$	bacterial cellulose nanoribbons	9–13 nm	Nanoribbons	[25]
$CoFe_2O_4$	bacterial cellulose	2.0 mm	Fibers	[26]
$ZnFe_2O_4$	Thermoanaerobacter	7, 49, and 25 nm	Crystals	[29]
$CoFe_2O_4$	*Magnetospirillum magneticum* strain AMB-1	40 –100 nm	Crystals	[27]
$CoFe_2O_4$	*Geobacter sulfurreducens*	8–16 nm	Crystals	[31]
(MFe_2O4, M = Mn, Co, Ni, and Cu)	bacterial cellulose	~100 nm	Fibers	[32]
$CoFe_2O_4$	Bacterial cellulose	6–10 nm	Fibers	[30]

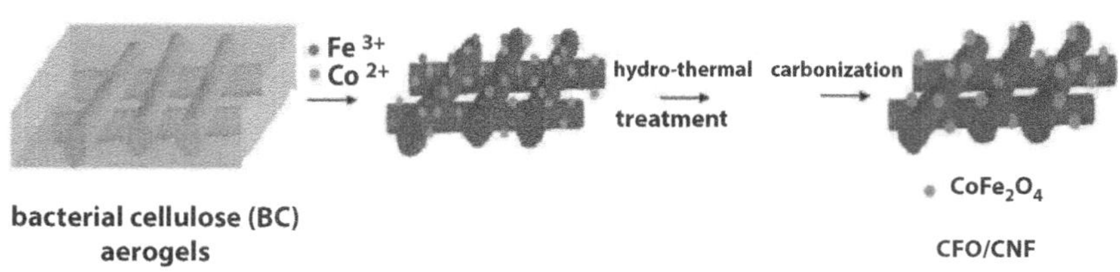

Figure 4.3 Synthesis of cobalt ferrite supported onto the bacterial cellulose [25].

were homogeneously distributed. The synthesized nanoparticles exhibited some crystalline defects that can be contributed to a lesser extent on the microstrains contained in the crystalline lattice and, to a greater extent, on the small crystalline size. The as-prepared $CoFe_2O_4$ nanoparticles displayed magnetic activity at lower temperatures. It was further confirmed that the magnetic strength was determined by the magnitude of the nanoparticles in a superparamagnetic environment.

Ren *et al.*, [26] synthesized cobalt ferrite nanocrystals utilizing carbonized bacterial cellulose through the solvothermal technique. It was observed that the implementation of bacterial cellulose conductive network onto the cobalt ferrite nanoparticle improved the electrical conductivity and also played a crucial part as a substitute for the non-aggregated growth of magnetic particles.

Yeary *et al.*, [29] prepared zinc ferrites (Zn-substituted magnetite, $Zn_yFe_{1-y}Fe_2O_4$) nanoparticles using a metal-reducing bacterium (*Thermoanaerobacter*, strain TOR-39) through a microbial technique. The mean crystallite particle sizes obtained were 67, 49, and 25 nm respectively. The average crystallite size was considerably decreased due to the increase in Zn insertion, and both the unit cell volume and the lattice parameter displayed a steady increase.

Liu *et al.*, [30] prepared cobalt ferrite nanoparticles supported onto the bacterial cellulose-derived carbon nanofiber through a facile and green approach. The average particle size of the as-prepared cobalt ferrite composite (nanofibers) varied from 6.1 to 10.3 nm which was also confirmed using XRD (6.1 nm). The synthesized composites were found to have a high electrocatalytic property and high stability toward the oxygen reduction reaction as well as oxygen evolution reaction.

Coker *et al.*, [31], worked on the preparation of cobalt ferrite nanoparticles using the bacterial strain of *Geobacter sulfurreducens* via the biotechnological route. The crystalline structured particles with an average size in the range of 8 to 16 nm were obtained. It was confirmed that cobalt ferrite nanocomposites with lower temperature coercivity of approximately 8 kOe, as well as an effective anisotropy constant of approximately 10^6 cm^{-3} can always be produced through this synthetic method.

As a result of the several drawbacks associated with the formation of nanoparticles through the use of bacteria as capping or reducing agents, the fungal strains were found to be the potential candidates over the bacterial strains due to their greater tolerance and metal bioaccumulation properties.

4.4.2 Synthesis of Metal Ferrites Nanostructures Using Fungi

Fungi are a kingdom of typically multicellular eukaryotic organisms that are heterotrophic and play an important role in the cycling of nutrients in the environment. These species are ideally adapted to the intracellular aggregation of metallic materials due to their capacity to accumulate and withstand high amounts of metals [33]. Fungi can be used as potential candidate biomaterials toward the green preparation of nanoparticles with a broad range of benefits. These include strong metal resistance rates, the secretion of significant quantities of extracellular matrix proteins, and better handling in the research laboratories [34]. These species are commonly used as reducing and capping agents because of their heavy metal resistance and capability to rationalize and bioaccumulate chemicals. Also, fungus species can be effectively produced over a wide-scale ("nanofactories"), which can generate, regulate the size and morphology of nanoparticles [23]. Figure 4.4 shows the biosynthesis of cobalt ferrite nanoparticles utilizing the fungal extract of *Saccharomyces cerevisiae* [35].

Fungi benefits over other microorganisms are in such a way that they contain greater amounts of enzymes and proteins that can also be utilized to synthesize nanoparticles rapidly and sustainably. This is due to fungi producing proteins and enzymes as a capping agent that can be used from their salts to synthesize metal nanoparticles [36]. Since some fungi are pathogenic, one must be vigilant when operating with them [37]. Under the same circumstances, fungal biomass usually grows faster than bacteria. Although the bacterial synthesis of metallic nanocomposites is widespread, the fungal synthesis is much more beneficial since their mycelia have a wide range of interactions [38]. Fungi can also secrete quite a large amount of protein unlike bacteria; thus, it is very easy to transform metal salts into metal nanoparticles. The development of biogenic metal nanoparticles

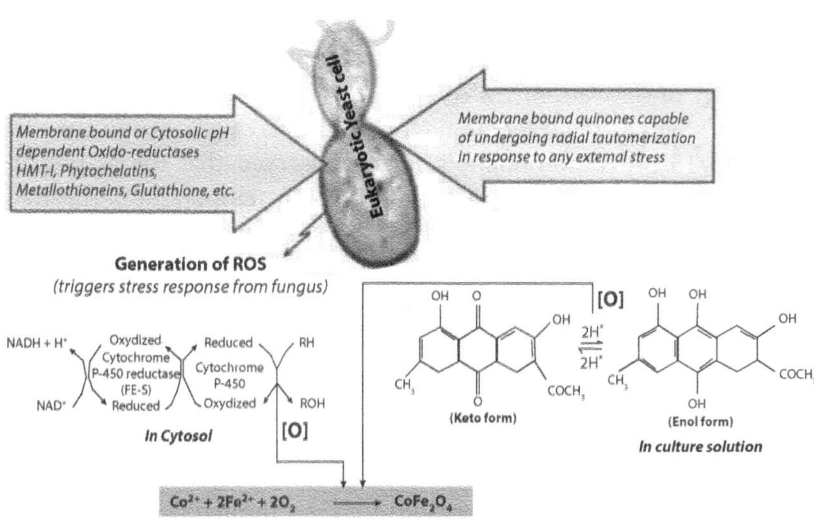

Figure 4.4 Biosynthesis of cobalt ferrite nanoparticles utilizing the fungal extract of *Saccharomyces cerevisiae* [35].

requires the bioreduction of metal salts into metallic ions that can be stabilized by biological compounds found in microbial species, such as bacteria and fungi [39].

Another way of producing metal nanoparticles is through biosorption, where the metal ions are fused to the surface of the organisms' cell wall in the aqueous medium. Fungi and yeasts are favoured over other species for greater development of nanoparticles. If fungi are introduced to metallic salts such as silver nitrate or chloroauric acid, it develops metabolites and enzymes to defend itself against harmful foreign materials [33].

Also, the fungi develop naphthoquinones and anthraquinones that serve as reduction agents. A unique enzyme can, therefore, operate on a metal. For example, to reduce ferric ion to iron nanoparticles, a nitrate reductase is essential [40]. It has been stated that not only are enzymes required but also the electron shuttle is needed for the reduction of the metal ion.

El-Eayed *et al.*, [41] synthesized cobalt ferrite nanoparticles utilizing the fungus extract of *Manascus purpureus* through the extracellular biosynthetic approach. Spherically shaped cobalt ferrite nanoparticles with an average size of 6.50 nm were obtained. The as-prepared nanoparticles exhibited superparamagnetic properties, as well as the antifungal and antibacterial properties against the human microbial pathogens such as cancer cells (liver and breast) with the MIC in the range of 250 to 500 μg mL^{-1}.

Anal *et al.* [35], also prepared cobalt ferrite nanoparticles using the yeast extract of *Saccharomyces cerevisiae* via the biosynthetic technique. Spherically shaped cobalt ferrite nanoparticles having a size in the range of 3 nm to 15 nm were obtained. It was observed that the as-prepared nanoparticles were superparamagnetic at room temperature. The maximum value of coercivity was found to be 130 Oe which was higher than that of its bulk counterparts.

Through these studies (Table 4.2), it is clear that there is not a lot of research that has been done using fungi, though it possesses a lot of advantages over its counterparts, such as bacteria. This could be due to some of the limitations mentioned earlier on. Though this material has been mentioned to operate through a wide range of synthesis strategies during nanoparticles synthesis, some of the fungi have been found to be toxic which then could be one of the reasons for their limited usage. In the next section, the synthesis of these ferrites will be explored using plant extracts.

Table 4.2 Green derived metal ferrite nanoparticles using fungal extracts and their application.

Name of nanoparticles	Fungal species	Shape	Size	Application	Refs.
$CoFe_2O_4$	*Monascus purpureus*	Spherical	6.50 nm	Antioxidant, anticancer & antimicrobial	[41]
$CoFe_2O_4$	*Saccharomyces cerevisiae*	Spherical	3–15 nm	Magnetic	[35]
$CuFe_2O_4$@Ag	*Chlorella Vulgaris*	Spherical	20 nm	Antibacterial	[42]

4.4.3 Synthesis of Metal Ferrites Nanostructures Using Plant Extracts

Plant materials have been found to possess various metabolites, which include phenols, vitamins, polysaccharides, amino acids, and carbohydrates, for which they can be used as a stabilizing agent, capping agents, chelating agents, and reducing agents for rendering a co-ordination reaction in capturing the metal ions [43]. These plant extracts also play a vital role as fuel to dissolve various salts. The utilization of plant materials in various synthetic pathways can significantly influence the morphological form, the shape, and the size of the synthesized nanoparticles.

Throughout the biosynthesis of metal nanomaterials utilizing natural plants, three essential parameters are required; metallic salt, stabilizing agent, and a reducing agent to regulate the scale of nanomaterials and to avoid their aggregate [44]. Several organic compounds in plant species which include enzymes, phenolic compounds, amino acids, vitamins, saponins, sugars, alkaloids and tannins, may lead to metallic nanoparticle bioreduction, formulation and optimization [45]. The reduction ability for ions and degradation of plant efficiency, which relies mostly on the availability of enzymes, polyphenols, as well as capping agents found within plant materials, has a significant value toward the output quantities of nanoparticles [46]. Table 4.3 shows the effect of the different plant extracts on the morphology and particle sizes of the synthesized materials.

Table 4.3 Green-derived metal ferrite nanoparticles using plants extracts.

Name of nanoparticles	Plant species	Size	Morphology TEM/SEM	Refs.
$ZnFe_2O_4$	*Moringa Oleifera*	5–10 nm and 10–25 nm	Spherical	[47]
$NiFe_2O_4$	*Hydrangea paniculata* flower	10–45 nm	spherical, oval, and irregular	[50]
$CuFe_2O_4$	*Hibiscus rosa-Sinensis* leaf	17 nm	Spherical	[51]
$ZnFe_2O_4$	*Petroselinum crispum*	~ 50 nm	Granules	[52]
$NiFe_2O_4$, $ZnFe_2O_4$, $CuFe_2O_4$	*Aloe vera*	15–70 nm	Spherical	[48]
$ZnFe_2O_4$	*Citrus aurantium*	9–20 nm	Spherical	[53]
$ZnFe_2O_4$	*Lawsonia inermis*	17.12 nm	Nearly spherical and rectangular	[54]
$NiFe_2O_4$	*Persa Americano seeds*	15–20 nm	Cubic and irregular	[49]

Matinise *et al.* [47], investigated the physical and electrochemical properties of zinc ferrite ($ZnFe_2O_4$) nanoparticles via an inexpensive and ecologically friendly route. $ZnFe_2O_4$ nanocomposites were synthesized using the plant extract of *Moringa Oleifera* as a fuel. The synthesized materials ($ZnFe_2O_4$) were found to exhibit a good voltammetric activity, high electro-activity, and excellent performance in potential electrochemical applications.

Laokul *et al.*, [48] worked on the synthesis of metal ferrite (MFe_2O_4, M=Zn, Ni, and Cu) nanoparticles through a modified sol-gel technique utilizing the metal nitrates with high purity as well as the plant extract of *Aloe vera* as a fuel. Spherically shaped nanoparticles with average crystallite size in the range of 15 to 70 nm were obtained. It was observed that the crystallite size and saturation-specific magnetization (for both nickel and copper ferrites) increased with an increase in calcination temperatures 600°C to 900°C, while zinc ferrite exhibited paramagnetic properties at temperatures which are above the critical point (T_N = 10 K).

Bashir *et al.*, [49] worked on the green synthesis of Nickel Ferrite ($NiFe_2O_4$) nanoparticles using the seeds extract of *Persa Americano*. The morphological analysis of the nanoparticles indicated the cubic shapes with an average particle size of 15 nm to 20 nm as well as the irregular shapes. The magnetic analysis revealed a decrease in saturation magnetization (Ms) of $NiFe_2O_4$ nanoparticles as compared with their bulk system which was due to the surface spin disorder. Electrochemical analysis indicated that the electrochemical properties of the as-prepared nanoparticles were extremely influenced by the charge transfer, as well as the diffusion process. The high electronic conductivity and electrochemical stability of these nanoparticles led to these particles being suitable candidates toward the electrochemical applications.

Plants have become the most desirable source of synthesizing metal ferrite nanoparticles since they contribute to large-scale production, as well as the production of stable, variable shape, and size nanoparticles. Fungus-based synthesis of metal ferrite nanoparticles has gained significant interest due to their widespread use in various disciplines. As compared to bacteria, fungi can produce greater quantities of nanoparticles since they can secrete larger amounts of proteins that directly result in increased production of nanoparticles. The plant-derived nanoparticle could have a significant role in the field of textile, pharmaceutical, and cosmetic industries and thus become a major field of study.

4.5 Plant-Derived Metal Ferrites as Photocatalysts for Dye Degradation

Photocatalyst semiconductors, generally their function is to initiate or accelerate the oxidation/reduction reactions in the presence of the irradiated semiconductor. These species become irradiated with charged particles (*i.e.*, photons) whose energy should be equal or higher than the energy bandgap of the semiconductor catalysts in order to accomplish the oxidation and reduction reactions. The hole and electron pairs recombine and distribute their energy input in the form of heat or light. These recombinations are, however, promoted by scavengers or crystalline defects. Improved crystalline structure can, therefore, minimize trapping positions, as well as the recombination sites, leading to an improvement in the efficiency of photo-generated carriers toward the desired reaction mechanism [55].

Semiconductor photocatalysis has stimulated the science community throughout its broad variety of applications to minimize water contaminants. Water contamination is a complicated issue around the globe with increasing population growth. Organic dyes play a crucial part in water pollution and are extremely toxic, mostly utilized across multiple industries, which include textiles, fiber, foods, cosmetics, polymers and the textile industry recently ranked first with dyes [56]. They are challenging to degrade due to their complex structure [4]. Over the years, many semiconductor photocatalysts were produced to degrade organic pollutants in the contaminated water.

Solar energy effective photocatalysts are an encouraging source for purification of wastewater. Attempts have been made to synthesize substances that could use solar radiation to degrade industrial waste contaminants and dyes. A large application of organic dyes is in their use as chemical stains, redox indicators as well as pharmaceutical products. The decomposition of chemical contaminants and dyes (see Figure 4.5) such as bromophenol blue (BB), methylene blue (MB), methylene orange (MO), bromophenol blue (BB) and Chicago sky blue (CSB) from the textile wastes remains a major challenge owing to the poor visible light catalytic performance of sulphides and metal oxide nanoparticles. The broadband gaps of TiO_2, Ag_3VO_4, ZnS, and $SrTiO_3$ restrict them from working as an active catalyst for visible light photodegradation.

Transition-metal and alkaline earth metal ferrites with the generic form MFe_2O_4 are potentially valuable compounds that display unique chemical and physical properties. Apart from the broad bandgap semiconductors, these metal ferrite nanoparticles have a bandgap of about 2.0 eV or less and can efficiently absorb a large fraction of visible light radiation resulting in enhanced photocatalytic activity [5].

However, the most advantageous benefit about the metal ferrites as a photocatalytic species is their intrinsic absorption abilities over visible light radiation. The larger extent of catalytic sites and the lower bandgaps makes the ferrites a potential candidate for adsorption of organic dyes and noxious industrial waste. The magnetic properties of these ferrites gives them an alternative recovery after the photocatalytic processes. Therefore, as shown in Table 4.4, the alternative approach to photocatalytic activity needs the absorption of solar irradiation or any other form of energy source [57].

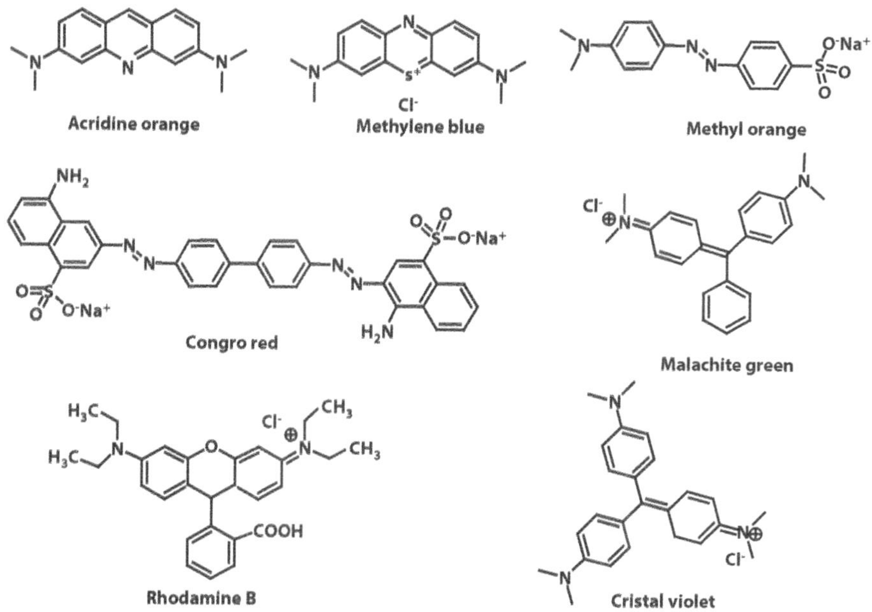

Figure 4.5 Various types of organic dyes [1].

Madhukara *et al.*, [15] synthesized $ZnFe_2O_4$ nanoparticles utilizing *Limonia acidissima* plant extract via the microwave-assisted routes. Spherically shaped $ZnFe_2O_4$ nanoparticles with a crystalline size of about 27.34 ± 0.56 nm were produced. The photocatalytic ability of this material was found to be significantly active against dyes, such as Evans blue (89%) and methylene blue (99.66%).

Prasad *et al.*, [58] formed spherically shaped Ni/Fe_3O_4 nanostructures using *Moringa Oleifera* plant extract, with a crystalline particle size of approximately 16 to 20 nm. The synthesized materials were found to have strong catalytic activity against MG dye.

Surendra *et al.*, [46] synthesized zinc ferrite ($ZnFe_2O_4$) nanoparticles using the plant extract of *Jatropha* through the combustion route and also studied the photocatalytic activity of this ferrite against MG. These nanoparticles formed materials that flaked in shape with an average size of 18 nm. Upon testing their photoactivity, they were found to be active with a 98% removal of the dye upon ultraviolet radiation exposure.

Patil *et al.*, [59] prepared $ZnFe_2O_4$ nanoparticles using the combustion method whereby the sugar cane juice extract was used as a solvent.

Table 4.4 Green-derived metal ferrite nanoparticles and their photocatalytic activity.

Name of nanoparticles	Plant species	Shape	Size	Pollutants	Percentage degradation (%)	Refs.
$ZnFe_2O_4$	*Limonia acidissima* (wood apple)	Spherical	20 nm	EB and MB	89% and 99.66%	[15]
Ni/Fe_3O_4	*Moringa oleifera* (MO)	Spherical	16–20 nm	MG	91.6%	[58]
$ZnFe_2O_4$	Sugar cane juice	Spherical	22.13 nm	MB and RB	98% 99%	[59]
$CoFe_2O_4$-SiO_2	*Salix alba* bark	irregular sphere	15–51 nm	MG	-	[3]
$ZnFe_2O_4$	*Nyctanthes arbor-tristis*	-	4.5–12 nm	RhB	-	[63]
$ZnFe_2O_4$	*Jatropha curcas*	Flakes	~ 18 nm	MG	98%	[45]

(Continued)

Table 4.4 Green-derived metal ferrite nanoparticles and their photocatalytic activity. *(Continued)*

Name of nanoparticles	Plant species	Shape	Size	Pollutants	Percentage degradation (%)	Refs.
$NiFe_2O_4$	*Hibiscus rosa-Sinensis* leaf	Rod-like and spherical	13, 15 & 21 nm	Batik waste and DY	84% 95%	[60]
$ZnFe_2O_4$	*Opuntia dilenii haw*	Spherical	32 nm and 73 nm	Glycerol	96% 85%	[12]
$ZnO\text{-}CoFe_2O_4$	*Nephelium lappaceum*	Rice-like grains & granular	24 and 27 nm	DR- 81	99.6%	[60]
$CuFe_2O_4$	*Hibiscus rosa-Sinensis*	Spherical	15 nm- 25 nm	Benzylic alcohols	97.89% 98.56%	[51]
$MgFe_2O_4$	Tragacanth gum	Spherical	11 nm	MG	98%	[61]

*MB- Methylene Blue *MG- Malachite Green EB- Evans Blue
*RB- Rose Bengal *DR81- Direct Red 81 DY- Direct Yellow
*Rh-B- Rhodamine B dye

The spinel like cubic structured $ZnFe_2O_4$ nanoparticles with the average size of 22.13 nm was formed. The photocatalytic activity of the $ZnFe_2O_4$ nanoparticles against the mixed organic dyes, such as Rose Bengal (RB) and Methylene blue (MB) gave a percentage removal of 98% and 99%, respectively, under visible light.

Amiri *et al.*, [3] synthesized cobalt ferrite ($CoFe_2O_4$)-silica nanocomposites through a self-propagating sol-gel route using a bark extract of *Salix alba* as the fuel. The morphology, size, and magnetic hysteresis of the as-prepared samples were investigated. It was observed that the size of the prepared $CoFe_2O_4$ silica nanoparticles was influenced by the calcination temperature as it was between 15 and 51 nm. Furthermore, the magnetic property was affected by the size and the structure of the nanoparticles. It was further suggested that these materials could be used as an effective magnetic adsorbent toward the removal of Malachite green (MG) from wastewater.

Rahmayeni *et al.*, [60] prepared ZnO-$CoFe_2O_4$ via the hydrothermal method utilizing the peel extract of *Nephelium lappaceum* as a capping agent. Small granules and rice-like morphology nanoparticles with an average crystalline size of 27.88 nm and 24.02 nm were observed. The ZnO-$CoFe_2O_4$ nanocomposites were shown to have a 99.6% photocatalytic activity toward direct red 81 organic dye.

Fardood *et al.*, [61] prepared $MgFe_2O_4$ nanoparticles through the sol-gel route utilizing the tragacanth gel as a solvent. A spherically shaped nanoparticle with an average size of 11 nm was formed. Upon testing their photocatalytic activity against Malachite Green (MG), 98% of the dye was degraded. Moreover, this material showed stability as it was reused up to six times and could be easily removed from the reaction mixture using an external magnet.

Though very high photocatalytic activity has been reported in the above analysis, especially with the formation of the spherical nanoparticles and low particle size distributions, one of the key drawbacks for the applications of metal ferrites nanoparticles is their poor electron and hole separation efficiency, which contributes to a much lower photocatalytic performance compared to other semiconductors. They also possess various limitations, such as agglomeration when particles are at a higher concentration and leaching which results when particles are at the lower concentration. Photocorrosion, a high rate of holes and electrons recombination are also some of the shortfalls. This tends to reduce the photocatalytic activity due to alkaline and strong acidic conditions [45]. Studies show that the deposition of a novel metal at the nanoparticle surface enhances the separation rate of electrons and holes and promotes the transition to the interfacial surface [62].

Metal nanoparticles, such as Ag, Au, and Pt, deposited onto semiconductor materials, such as MFe_2O_4, play a crucial role because they serve as electron trappers, promote electron-hole separation and enhance the interfacial electron transfer mechanism. Hence, these metal nanoparticles could improve the photocatalytic performance of semiconductor materials under both visible light and UV light [44].

4.5.1 Effect of Depositing Noble and Transition Metal on Metal Ferrites for Photodegradation

In order to enhance the photocatalytic performance of spinel ferrites nanoparticles, the transmission of surface charges must be accelerated, as well as the holes and electrons recombination rate must be reduced. An alternative approach is to dope these spinel ferrite nanoparticles catalysts with transition metals. Metal doping becomes another effective strategy for modifying the optical band gaps of photocatalysts. As a measure of a longer period of photocatalytic reactions, the main fraction of photogenerated electrons and holes lean toward recombination before they enter the surface of a photocatalyst [55]. Consequently, only a small number of electrons and holes will penetrate the surface of the photocatalyst. Transition-metal ions as dopants will modify host lattice defects, crystallinity, inhibit the recombination of hole and electron pairs, prolong the existence of the charge carrier, and prolong the absorption area in semiconductor photocatalysts [43].

The doping of spinel ferrite nanoparticles with transition metals like Ag, V, Zn, Mo, Ni, Cu (Table 4.5) can overcome the issue of rapid recombination associated with the carrier charges which normally occurs in nanoseconds. This takes place through capturing and continually passing photo-excited electrons to the surface of the photo-catalyst, leading to a reduced recombination rate of holes and electrons [64].

Ricardo *et al.*, [6] effectively synthesized rectangular nanoplate forms of magnetic zinc/calcium ferrite nanoparticles coated with silver. They tested their photocatalytic performance against various dyes, such as rhodamine B and industrial textile dyes. The nanoparticles produced showed impressive results for industrial usage in visible light photo remediation of effluents, with the potential of magnetic recovery.

Table 4.5 Effect of depositing noble and transition metal on metal ferrites for photocatalytic degradation.

Name of nanoparticles	Shape	Size	Pollutants	Percentage degradation (%)	Refs.
$Ag@mTiO_2@ CoFe_2O_4$	-	-	MB	100	[65]
$Ni_xCo_{1-x}Fe_2O_4$	cubic	~(17–36 nm)	MB	79 %	[70]
$CoRu_xFe_{2-x}O_4$	spherical	17.77–18.45 nm	RDR	94 %	[67]
$NiFe_2O_4$: Mg^{2+}	flakes	18–22 nm	IC and Phenol	84.6 % and 79.4 %	[68]
$NiAl_xFe_{2-x}O_4$	spherical	19 and 38	RB	63 % and 99.8 %	[69]
$Zn_{1x}Co_xFe_2O_4$	Spherical	11 ± 3 nm	MB	95.4 %	[71]

*MB-Methylene Blue *RB-Rose Bengal *RDR-Remazol Deep Red

Nanosized $Ag@mTiO_2@ CoFe_2O_4$ nanocomposites were previously studied toward the photocatalytic degradation of Methylene Blue (MB) [65]. According to the study, the photodegradation of MB by these nanocomposites photocatalyst showed excellent photocatalytic performance as a result of the presence of Ag nanoparticle in $Ag@mTiO_2@ ZnFe_2O_4$, which improved the light absorption capacity of the catalyst in visible light range.

A study by Ibrahim *et al.*, [66], reported that highly active photocatalysts can be formed through the coupling of two semiconductors with silver nanoparticles having different band gaps. It was obtained that the introduction of silver nanoparticles onto the titania/ferrite composite had a potential contribution to the photocatalytic reduction of Cr^{+6} species. Based on this mechanism, relatively efficient charge separation was obtained as a result of the photoinduced electrons that were transferred away from the photocatalyst.

Singh *et al.*, [67] synthesized Ru doped cobalt ferrite nanoparticles ($CoRu_xFe_{2-x}O_4$) through the sol-gel approach for the degradation of remazol deep red under visible light illumination. The photocatalytic results indicated that cobalt ferrite when doped with ruthenium exhibits an enhanced photocatalytic performance. Furthermore, $CoRu_xFe_{2-x}O_4$ photocatalyst is strongly ferromagnetic, thus enabling it for better separation and repeated usage of photocatalyst.

Nadumane *et al.*, [68] synthesized nickel ferrite and Mg doped nickel ferrite photocatalysts via a modified green sol-gel approach using *Aloe Vera* gel as a natural template. It was shown that $NiFe_2O_4$: Mg^{2+} exhibited the best photocatalytic performance for the degradation of recalcitrant contaminants. Their excellent activity was due to the morphology, crystallinity, defects, band gap, dopant amount, as well as the combined facets.

Madhukara Naik *et al.*, [69] prepared the nanocrystalline nickel ferrite and aluminium-doped nickel ferrite nanoparticles with an average particle sizes of 19 and 38 nm via the sol–gel auto-combustion technique. The synthesized $NiFe_{2x}Al_xO_4$ nanoparticles were found to be an effective photocatalyst for the degradation of RB, MG, AZ, MB, and TY dyes over the visible light irradiation. The nanoparticles exhibited some important biological activities toward the human pathogens and were, therefore, seen as a promising candidate for biomedical applications.

From these studies, it is clear that the deposition of noble or transition metal greatly enhanced the photocatalytic degradation of the metal ferrites. It is important to note though that other factors, such as particle size, morphology, and percentage loading or deposition of these metals, also greatly influence the photocatalytic activity of these ferrites.

4.5.2 Effect of Carbon Deposited on Metal Ferrites for Photocatalytic Degradation

Carbon-based nanomaterials with diverse structures, like those of nanofibers, carbon nanotubes, graphene, as well as nanospheres, have been extensively studied in recent years owing to their unique properties, as a result of their specific compositional structures [72]. Carbon nanospheres are among the innovative carbon nanostructures known for their possible usage in several applications, including lithium-ion batteries, supercapacitors, adsorbent materials as well as catalytic converters [73]. Carbon nanospheres offer an effective structure for future development, such as a stable phase. This is attributed to their outstanding characteristics, which include conductivity, strong durability, large surface area, and low toxicity [74]. Carbon nanospheres provide an interesting and affordable alternative for effective remediation of different organic pollutants from wastewater, owing to their large surface area, as well as porous nature.

To enhance the photocatalytic efficiency of ferrites, as well as to allow effective use of visible light spectrum, the higher aspect ratio and the spherical nanometric dimensions of carbon nanospheres (CNSs) render themselves as potential candidate toward the photocatalytic activity. Decoration of CNSs surfaces with foreign materials, in particular metal ferrite, expands their field of application and strengthens their mechanical, electrical, chemical, and thermal properties, which can be used in advanced nanotechnology [75].

Hassani *et al.*, [76] successfully prepared the nanocomposite of mesoporous graphitic carbon nitride/cobalt ferrite ($CoFe_2O_4$/mpg-C_3N_4) through the thermal decomposition and sonication route. The magnetic properties of the as-prepared nanocomposites were determined through the M-H and M-T loops and blocking temperature, as well as the magnetic saturation of these nanocomposites were obtained to be 269 K and 6.1 emu/g, respectively. The nanocomposites were further studied for the photocatalytic activity of Malachite Green (MG), which exhibited significantly high photocatalytic performance through the UV-light exposure. In addition to MG, the photocatalytic activity of many other organic dyes such as rhodamine B, methylene blue and acid orange 7 have been studied in order to demonstrate the efficiency of these nanocomposites under optimized conditions.

Singh *et al.*, [77] fabricated the nanocomposites of $Ni_{1-x}Co_xFe_2O_4$/MWCNTs through the microemulsion approach. The nanocomposites exhibited ferromagnetic behavior when characterized using vibrating sample magnetometer (VSM). The saturation magnetization was found to increase with an increase in the cobalt ion concentration, due to higher magnetic moment of cobalt ions relative to the nickel ions. The photocatalytic performance of the incorporated nanocomposites was also investigated toward the photocatalytic decomposition of Rhodamine B (RhB) dye over the visible-light exposure. It was found that the degradation occurred within 15 to 25 min, which was attributed to the CNTs present in the composite.

Chen *et al.*, [78] demonstrated preparation of zinc ferrite/multi-walled carbon nanotubes ($ZnFe_2O_4$/MWCNTs) composites using the hydrothermal methods. The comparative studies of pure $ZnFe_2O_4$, the $ZnFe_2O_4$/MWCNTs composite showed a higher absorption in visible-light spectrum as well as an enhanced photocatalytic performance, with 99%

methylene blue (MB) decomposition under the visible-light exposure for 6 hours in the presence of H_2O_2. The improved photocatalytic efficiency was due to the strong interaction between $ZnFe_2O_4$ and MWCNTs that induces the recombination of photo-generated carrier charges, as well as the production of highly active species •OH on the MWCNTs surface. Hence, more radicals were formed toward the decomposition of the MB dye.

4.5.3 Effect of Coupling Metal Oxide Semiconductors with Metal Ferrites for Photocatalytic Degradation

Over the last two decades, significant emphasis has been put on the degradation of organic pollutants using semiconductor materials, such as TiO_2 and ZnO, for the treatment of harmful pollutants and polluted underground water. Because of their high performance in using solar energy, visible-light–induced photocatalysts have gained tremendous attention. TiO_2 and ZnO, on the other hand, provide little photocatalytic activity when exposed to visible light due to their relatively large bandgap energy of around 3.2 eV. Several changes were made to such UV-active oxides in order to expand their wavelength spectrum into the visible region. Other visible-light sensitive photocatalysts, including CdS, WO_3, as well as Fe_2O_3, are also being investigated; however, their low photocatalytic activity restricts their functional use [71, 83].

TiO_2 and ZnO are widely used because of their outstanding photoactivity, low cost, chemical and physical stability, however, they require a high degree of ultraviolet radiation. A bandgap of 2.0 or less is expected for a material to become active in the presence of sunlight. Metal oxides are usually constrained by rapid electron-hole recombination, even though they have a low bandgap [84].

Doping/coupling or metal incorporation (Table 4.6) has become an alternative approach for overcoming these obstacles and bringing absorption toward the visible range to increase the environmental potential application of materials. The synthesis of these composites increases the surface activity as well as the semiconducting nature of the nanomaterials.

Zou *et al.*, [85] fabricated Cds/$ZnFe_2O_4$ nanowires composites with an average crystallite size in the range of 40 to 50 nm through a low-temperature hydrothermal approach for the photocatalytic degradation of MB, RhB, and MO over the visible light exposure. The Cds/$ZnFe_2O_4$ composites showed an excellent photocatalytic performance than that of Cds and blank $ZnFe_2O_4$. The predominant scavengers involved in the Cds/$ZnFe_2O_4$

Table 4.6 Carbon nanomaterials deposited on metal ferrites for photocatalytic degradation.

Name of nanoparticles	Shape	Size	Pollutants	Percentage degradation (%)	Refs.
$ZnFe_2O_4$/ MWCNT	Cubic	2 and 4 nm	MB	99%	[79]
$CoFe_2O_4$-graphene	Spherical	5.53 nm	Rh B, MB, MO, active red RGB and active black BL-G	94%, 71.66%, 61%	[80]
Activated carbon/ $CoFe_2O_4$	-	14–20 nm	MG	-	[81]
$ZnFe_2O_4$/ graphene	Spherical	5 nm	MB, MO and RhB	~100%, 96%, 22%	[82]
$ZnFe_2O_4$/ MWCNTs	Flaky	15 nm	MB	99%	[78]
Ni1-xCoxFe$_2$O$_4$/ MWCNTs	Tubes	10–15 nm	Rh B	-	[77]
$CoFe_2O_4$/ mpg-C_3N_4	Spherical	10 nm	MG	93.41%	[76]

*MB- Methylene Blue *MG- Malachite Green EB- Evans Blue
*RB- Rose Bengal *DR81- Direct Red 81 DY-Direct Yellow
*Rh-B- Rhodamine B dye * MO- Methyl Orange

process were O^{2-} and ·OH radicals. The enhanced activity may have been due to a rapid charging separation by creating a built-in region.

Boutra *et al.*, [86] synthesized magnetically separable $MnFe_2O_4$/TA/ZnO nanocomposite photocatalyst via the hydrothermal technique for photocatalytic enhancement in organic dyes. The efficiency of the composite was tested against CR dyes under visible light. The degradation capacity of the composite against CR dye was 84.2%, furthermore, the composite could be easily separated from the solution medium using an external magnet, proving that it was efficient and recyclable.

Zouhier *et al.*, [87] prepared a novel $ZnFe_2O_4$/ZnO composite photocatalyst through the combustion route for photocatalytic degradation of MB and RBB under both visible light and UV-light irradiation. The composites showed excellent performance under visible light, and although pristine ZnO revealed no behavior, it was able to transform initial dye levels by up to 80 percent in about 120 min. Increased photocatalytic efficiency of the composite photocatalyst regarding the pristine ZnO was as a result of $ZnFe_2O_4$ formation in conjunction with ZnO, which has a small band gap value which leads to the absorption of visible photons with stronger splitting pathway for the photo-generated charge carriers.

Farhadi *et al.*, [88] synthesized magnetic $CoFe_2O_4$@ZnS core-shell nanocomposite with an average crystallite size of 18 nm using a one-step hydrothermal decomposition. The results indicated that methylene blue (MB) (25 mg/L) was completely degraded in 70 minutes, with 0.5 g/L $CoFe_2O_4$@ZnS and H_2O_2 (4 mM). Furthermore, $CoFe_2O_4$@ZnS' photocatalytic activity was greater than pure ZnS and $CoFe_2O_4$, which demonstrated that a combined ZnS and $CoFe_2O_4$ may have been an appropriate way to increase photocatalytic activity. The trapping tests showed that the most active species for the degradation of the dyes was the ·OH radicals. Additionally, these nanocomposites could also be magnetically isolated and could be recycled for up to 5 cycles without losing stability.

Table 4.7 Effect of coupling metal oxide semiconductors with metal ferrites for photocatalytic degradation.

Name of nanoparticles	**Shape**	**Size**	**Pollutants**	**Percentage degradation (%)**	**Ref**
CdS/ $ZnFe_2O_4$	nanowires	40–50 nm	MB, RhB and MO	95 %	[85]
$MnFe_2O_4$/TA/ZnO	Irregular nanoplates and spherical	-	CR	84.2 %	[86]
$ZnFe_2O_4$/ZnO	Flowers and nanosheets	8.6/36.8	MB and RBB	80 %	[87]
$CoFe_2O_4$@ZnS	microspheres	18 nm	MB, RhB and MO	100 %, 74 % and 44 %	[88]
CuS/$CoFe_2O_4$	Sphere-like	17 nm	MB, RhB and MO	100 %, 83 % and 72 %	[89]

*MB-Methylene Blue *RBB-Remazol Brilliant Blue *Rh-B-Rhodamine B dye* *MO-Methyl Orange *CR-Congo Red

From these studies as demonstrated in Table 4.7, it can be seen that different morphologies of various shapes are formed as it is clear that in most cases it is a physical synthesis strategy whereby individual shapes are maintained. Also, in all these studies, there is vast improvement in the degradation especially against MB dye when these semiconductors have been coupled as compared to being used individually. To further understand where these materials can be used in, biological application (Section 5.6) were also highlighted.

4.5.4 Biological Applications of Plant-Derived Metal Ferrites

Biosynthesized metal ferrite nanoparticles though they have been used vastly in catalysis and as sensors, they have also found use in biological applications especially as antimicrobial agents (Table 4.8). Their antimicrobial activity is mostly attributed to the phytochemicals that are found in plants, such as tannis, polyphenols, glycosides, etc. Furthermore, the antibacterial behaviors can also be attributed to the particles sizes as well, as they are known to be inversely related to the average sizes of nanoparticles. Though this type of synthesis is still mostly at a lab scale level, there is a high need for expanding research facility-based work toward the industrial level, moving more toward bioinformatics methods for structure prediction and understanding the thorough mechanism of formation [90].

Gigasu *et al.*, [91] examined the antimicrobial, cytotoxic and antibiofilm properties using the nanocrystalline $CoFe_2O_4$ (cobalt ferrite) with an average size of 3 to 20 nm. Cobalt ferrite nanoparticles were found to have excellent antimicrobial potential. Gigasu concluded that the usage of seed extract (sesame) was indeed a sustainable and environmentally friendly alternative.

Mahajan *et al.*, [92] reported on the green synthesized cobalt ferrite and silver-doped cobalt ferrite nanoparticles utilizing *Ocimum* extract (*tulsi* seeds) and *Allium Sativum* (garlic cloves). The synthesized nanoparticles showed a good antibacterial activity and were also found to be the most effective toward the gram-positive strains.

Kombaiah *et al.*, [93] synthesized copper ferrite nanoparticles through microwave combustion and the conversational route utilizing the plant extract of *Hibiscus rosa* as a solvent. The spherically shaped nanoparticles with an average particle size in the range of about 50 to 200 nm have were formed. They investigated the photoluminescence analysis together

Table 4.8 Biological applications of plant-derived metal ferrites.

Name of nanoparticles	Plant species	Shape	Size	Application	Refs.
$Ag_xCo_1-_xFe_2O_4$	*Allium Sativum and Ocimum Sanctum*	Cubes	21.6 nm, 18.0 nm and 19.05 nm	Antibacterial activities	[92]
$CoFe_2O_4$	(*Sesamum indicum* L) seeds	Rounded and spherical	3–20.45nm	Anti-biofilm, cytotoxic and antimicrobial activities	[91]
$CoFe_2O_4$	*Okra (A. esculentus)*	Spherical	45–55 nm	Antimicrobial activity	[93]
$ZnFe_2O_4$	Sugar cane	Spherical	22.13 nm	Antibacterial activities	[59]
Fe_3O_4-Ag	*Vitis vinifera* (grape) stem	Spherical	50 nm	Antimicrobial activity	[96]
$NiFe_2O_4$	*Urtica*	Rod-like platelet	51.23 nm	Cytotoxic activity	[97]

(*Continued*)

Table 4.8 Biological applications of plant-derived metal ferrites. (*Continued*)

Name of nanoparticles	Plant species	Shape	Size	Application	Refs.
$ZnFe_2O_4$	*Aegle marmelos* leaves	Spherical	-	Drug delivery and antibacterial activities	[46]
$NiFe_2O_4$	*Salvia Rosmarinus*	Rods	10–28 nm	Biomedical property	[98]
$CuFe_2O_4$	*Hibiscus Rosa Sinensis*	Spherical	25–62 nm	Magnetic Hyperthermia property	[93]
$NiFe_2O_4$	Rosemary	Rods	10–28 nm	Cytotoxicity	[94]

with the magnetic performance of the synthesized nanoparticles. The photoluminescence studies indicated that the as-prepared nanoparticles can absorb light from the ultra-violet to visible region and the magnetic analysis indicated a ferromagnetic property whereby magnetism was influenced by the preparation technique as well as the calcination temperature of 1100°C.

Alijani *et al.*, [94] synthesized nickel ferrite nanorods particles using the Rosemary extract. Nickel ferrite nanoparticles were rod-shaped with an average particle size in the range of 10 to 28 nm. It was observed that the prepared nickel-ferrite nanoparticles exhibited excellent cytotoxicity effect toward the MCF-7 breast cancer cells, which implies that the nanoparticles can be implemented as one of the new anticancer agents.

Kumar *et al.*, [95] prepared nickel-zinc ferrite ($Ni_xZn_{1-x}Fe_2O_4$, x=2.5, 4.5, 6.5, 8.5) nanoparticles through the modified sol-gel technique utilizing metal nitrates with high purity, as well as the plant extract of *aloe vera* as a fuel. The synthesized nanocrystalline particles having the average particle size from the range of 9 nm to 20 nm have been obtained. Particle size has been found to decrease with an increase in Ni content and saturation magnetism values rise as Ni amount increases with a limit of Ni =0.65.

4.6 Challenges of these Materials and Photocatalysis

The application of solar-driven heterogeneous catalysis is considered to be more effective than the traditional techniques for the degradation of fractious organic pollutants. To enhance the potential effectiveness of this technology, substantial steps are required to resolve some of the complexities. There are many other research findings in the previous studies that used dye as an organic pollutant in the photocatalytic reaction model. There is still a need for a greater understanding of the degradation processes as well as the interactions between photocatalytic species and contaminants. Secondly, further advances in the design and fabrication of nanostructures with reproducible morphology and improved exposed crystal facet for tweaking the physicochemical characteristics and consequently enhancing the reactivity and selectivity of photocatalysts. Thus, by efficiently tweaking the ratio of various crystal facets, the photocatalytic reactivity could be harmoniously transformed. The crystal facet with a great ratio of undercoordinated atoms owns a greater reactivity in comparison to that comprising a low ratio of undercoordinated atoms [99, 100]. The crystal facet with low ratio of undercoordinated atoms is unstable during the crystal growth, due to reducing total surface energy of crystals. Thus, realizing a superior ratio of reactive facets through exposed crystal facet engineering is extremely vital for humanizing the photocatalytic reactivity.

In certain instances, photocatalytic reaction efficiencies are determined by the photocatalysts. There have been some functional challenges resulting from the fabrication of photocatalytic frameworks, which include the durability and strength of the polymeric membrane, the leaching of photocatalyst species from supports, as well as the cost of the production. Also, technical issues, such as the recovery and loss of the photocatalytic species during post-treatment, and also the photocatalytic performance of reused photocatalysts need to be tackled. In general, further comprehensive research is required to build and validate mathematical techniques for photocatalytic systems for wastewater treatment in order to predict their optimum conditions, kinetics as well as the quantum yields.

4.7 Conclusion: Future Perspectives

Green synthesis of metal ferrite nanoparticles has been proven to be one of the safest and most environmentally friendly methods compared to the chemical and physical syntheses. The use of plant extracts has been shown to be more efficient compared to fungi and bacteria. The cost-effectiveness and ease in the synthesis process make it a more viable method. Plant extracts contain various phytochemicals, which include tannins, proteins, saponins enzymes, alkaloids, oils, and phenolics that have medicinal benefits, and they can serve as both capping and reducing agents for the formation of metal ferrites nanoparticles.

Studies have shown that various methods can influence the morphology and particle size of the nanoparticles. Various attempts have been made in order to enhance photo-response of metal ferrite nanoparticles. It can be concluded that methods, such as metal doping, metal ferrites coupling with other semiconductors, and carbon nanomaterials, can enhance the photocatalytic efficiency of metal ferrite in various applications. These enhancements are due to the shifts in the energy bandgaps, suppression of the recombination rate of the photogenerated electron-hole pairs, increased charge separation rate, the improved production efficiency of hydroxyl radicals, as well as the narrowed particle size with a high specific surface area. Besides, regardless of the significant work carried out, a systematic study is required to investigate the influence of the integration of possible combinations of various kinds of dopants in a controlled morphology and faceted ternary photocatalysts are desirable in order to realize the possible commercialization of the nanoferrite photocatalyst. As result, further perspectives on the enduring challenges and potential prospect directions in the area of nanoferritte-based photocatalysis are also envisioned.

References

1. Ismail, M., Akhtar, K., Khan, M.I., Kamal, T., Khan, M.A., M Asiri, A., Seo, J., Khan, S.B., Pollution, toxicity and carcinogenicity of organic dyes and their catalytic bioremediation. *Curr. Pharm. Des.*, 25, 34, 3645–3663, 2019.
2. Ameta, R., Benjamin, S., Ameta, A., Ameta, S.C., Photocatalytic degradation of organic pollutants: A review. *Mat. Sci. Forum*, 734, 247–272, 2013.
3. Amiri, M., Salavati-Niasari, M., Akbari, A., Gholami, T., Removal of malachite green (a toxic dye) from water by cobalt ferrite silica magnetic nanocomposite: Herbal and green sol-gel autocombustion synthesis. *Int. J. Hydrogen Energy*, 42, 39, 24846–24860, 2017.
4. Alnuaimi, M.M., Rauf, M.A., Ashraf, S.S., Comparative decolouration study of neutral red by different oxidative processes. *Dyes Pigm.*, 72, 3, 367–371, 2007.
5. Ali, N., Zada, A., Zahid, M., Ismail, A., Rafiq, M., Riaz, A., Khan, A., Enhanced photodegradation of methylene blue with alkaline and transition-metal ferrite nanophotocatalysts under direct sunlight irradiation. *J. Chin. Chem. Soc.*, 66, 4, 402–408, 2019.
6. Fernandes, R.J., Magalhães, C.A., Amorim, C.O., Amaral, V.S., Almeida, B.G., Castanheira, E., Coutinho, P.J., Magnetic Nanoparticles of Zinc/Calcium Ferrite decorated with silver for photodegradation of dyes. *Materials*, 12, 21, 3582, 2019.
7. Hartley, T.W., Public perception and participation in water reuse. *Desalination*, 187, 1–3, 115–126, 2006.
8. Amini, M. and Ashrafi, M., Photocatalytic degradation of some organic dyes under solar light irradiation using TiO_2 and ZnO nanoparticles. *Nanochem. Res.*, 1, 1, 79–86, 2016.
9. Udhaya, P.A. and Meena, M., Albumen assisted green synthesis of $NiFe_2O_4$ nanoparticles and their physico-chemical properties. *Mater. Today: Proc.*, 9, 528–534, 2019.
10. Tatarchuk, T., Bououdina, M., Macyk, W., Shyichuk, O., Paliychuk, N., Structural, optical, and magnetic properties of Zn-Doped $CoFe_2O_4$ nanoparticles. *Nanoscale Res. Lett.*, 12, 1–11, 2017.

11. Reddy, D.H.K. and Yunang, Y.-S., Spinel ferrite magnetic adsorbents: Alternative future materials for water purification? *Coord. Chem. Rev.*, 315, 90–111, 2016.
12. Kombaiah, K., Vijaya, J.J., Kennedy, L.J., Bououdina, M., Al-Lohedan, H.A., Ramalingam, R.J., Studies on *Opuntia dilenii haw* mediated multifunctional $ZnFe_2O_4$ nanoparticles: Optical, magnetic and catalytic applications. *Mater. Chem. Phys.*, 194, 153–164, 2017.
13. Jeseentharani, V., George, M., Jeyaraj, B., Dayalan, A., Nagaraja, K.S., Synthesis of metal ferrite (MFe_2O_4, M= Co, Cu, Mg, Ni, Zn) nanoparticles as humidity sensor materials. *J. Exp. Nanosci.*, 8, 3, 358–370, 2013.
14. Dutta, V., Sharma, S., Raizada, P., Hosseini-Bandegharaei, A., Gupta, V.K., Singh, P., Review of augmentation in photocatalytic activity of $CoFe_2O_4$ via heterojunction formation for photocatalysis of organic pollutants in water. *J. Saudi Chem. Soc.*, 23, 1119–1136, 2019.
15. Naik, M.M., Naik, H.B., Nagaraju, G., Vinuth, M., Naika, H.R., Vinu, K., Green synthesis of zinc ferrite nanoparticles in *Limonia acidissima* juice: Characterization and their application as photocatalytic and antibacterial activities. *Microchem. J.*, 146, 1227–1235, 2019.
16. Vidya, C., Prabha, M.C., Raj, M.A., Green mediated synthesis of zinc oxide nanoparticles for the photocatalytic degradation of Rose Bengal dye. *Environ. Nanotechnol. Monitor. Manage.*, 6, 134–138, 2016.
17. Ishak, N.M., Kamarudin, S.K., Timmiati, S.N., Green synthesis of metal and metal oxide nanoparticles via plant extracts: An overview. *Mater. Res. Express*, 6, 11, 112004, 2019.
18. Din, M., Javed, M., Hussain, Z., Khalid, R., Ameen, S., Slow catalytic pyrolysis of *Saccharum munja* using biochemically synthesized nickel ferrite nanoparticles for the production of high yield biofuel. *Eur. J. Sustain. Dev. Res.*, 4, 3, 1–9, 2020.
19. Goodarz Naseri, M., Saion, E.B., Abbastabar Ahangar, H., Shaari, A.H., Hashim, M., Simple synthesis and characterization of cobalt ferrite nanoparticles by a thermal treatment method. *J. Nanomater.*, 2010, 8, 2010.
20. Manimozhi, V., Partha, N., Sivakumar, E.K.T., Jeeva, N., Jaisankar, V., Preparation and characterization of ferrite nanoparticles for the treatment of industrial wastewater. *Dig. J. Nanomater. Biostruct.*, 11, 3, 1017–1027, 2016.
21. Sheikh, A. and Jain, P., A thorough study of zinc ferrite nanoparticles with reference to green synthesis. *Int. J. Nanomed. Nanosurg.*, 2, 3, 2470–3206, 2016.
22. Matussin, S., Harunsani, M.H., Tan, A.L., Khan, M.M., Plant-extract-mediated SnO_2 nanoparticles: Synthesis and applications. *ACS Sustain. Chem. Eng.*, 8, 8, 3040–3054, 2020.
23. Siddiqi, K.S. and Husen, A., Fabrication of metal nanoparticles from fungi and metal salts: Scope and application. *Nanoscale Res. Lett.*, 11, 1, 98, 2016.

24. Srivastava, N. and Mukhopadhyay, M., Biosynthesis of SnO_2 nanoparticles using bacterium *Erwinia herbicola* and their photocatalytic activity for degradation of dyes. *Ind. Eng. Chem. Res.*, 53, 36, 13971–13979, 2014.
25. Menchaca-Nal, S., Londoño-Calderón, C.L., Cerrutti, P., Foresti, M.L., Pampillo, L., Bilovol, V., Candal, R., Martínez-García, R., Facile synthesis of cobalt ferrite nanotubes using bacterial nanocellulose as a template. *Carbohydr. Polym.*, 137, 726–731, 2016.
26. Ren, Y., Li, S., Dai, B., Huang, X., Microwave absorption properties of cobalt ferrite-modified carbonized bacterial cellulose. *Appl. Surf. Sci.*, 311, 1–4, 2014.
27. Prozorov, T., Palo, P., Wang, L., Nilsen-Hamilton, M., Jones, D., Orr, D., Mallapragada, S.K., Narasimhan, B., Canfield, P.C., Prozorov, R., Cobalt ferrite nanocrystals: Outperforming magnetotactic bacteria. *ACS Nano*, 1, 3, 228–233, 2007.
28. Agarwal, H., Kumar, S.V., Rajeshkumar, S., A review on green synthesis of zinc oxide nanoparticles–An eco-friendly approach. *Resour.-Effic. Technol.*, 3, 4, 406–413, 2017.
29. Yeary, L.W., Moon, J.W., Rawn, C.J., Love, L.J., Rondinone, A.J., Thompson, J.R., Chakoumakos, B.C., Phelps, T.J., Magnetic properties of bio-synthesized zinc ferrite nanoparticles. *J. Magn. Magn. Mater.*, 323, 23, 3043–3048, 2011.
30. Liu, S., Yan, W., Cao, X., Zhou, Z., Yang, R., Bacterial-cellulose-derived carbon nanofiber-supported $CoFe_2O_4$ as an efficient electrocatalyst for oxygen reduction and evolution reactions. *Int. J. Hydrogen Energy*, 41, 11, 5351–5360, 2016.
31. Coker, V.S., Telling, N.D., van der Laan, G., Pattrick, R.A., Pearce, C.I., Arenholz, E., Tuna, F., Winpenny, R.E., Lloyd, J.R., Harnessing the extracellular bacterial production of nanoscale cobalt ferrite with exploitable magnetic properties. *ACS Nano*, 3, 7, 1922–1928, 2009.
32. Sriplai, N., Mongkolthanaruk, W., Pinitsoontorn, S., Synthesis and magnetic properties of bacterial cellulose—ferrite (MFe_2O_4, M=Mn, Co, Ni, Cu) nanocomposites prepared by co-precipitation method. *Adv. Nat. Sci.: Nanosci. Nanotechnol.*, 8, 3, 035005, 2017.
33. Narayanan, K.B., Park, H.H., Han, S.S., Synthesis and characterization of biomatrixed-gold nanoparticles by the mushroom *Flammulina velutipes* and its heterogeneous catalytic potential. *Chemosphere*, 141, 169–175, 2015.
34. Du, L., Xu, Q., Huang, M., Xian, L., Feng, J.X., Synthesis of small silver nanoparticles under light radiation by fungus *Penicillium oxalicum* and its application for the catalytic reduction of methylene blue. *Mater. Chem. Phys.*, 160, 40–47, 2015.
35. Jha, A.K. and Prasad, K., Biological synthesis of cobalt ferrite nanoparticles. *Nanotechnol. Dev.*, 2, 1, e9–e9, 2012.
36. Ahmad, S., Munir, S., Zeb, N., Ullah, A., Khan, B., Ali, J., Bilal, M., Omer, M., Alamzeb, M., Salman, S.M., Ali, S., Green nanotechnology: A review

on green synthesis of silver nanoparticles An ecofriendly approach. *Int. J. Nanomed.*, 14, 5087, 2019.
37. Ravichandran, V., Vasanthi, S., Shalini, S., Shah, S.A.A., Tripathy, M., Paliwal, N., Green synthesis, characterization, the antibacterial, the antioxidant and photocatalytic activity of *Parkia speciosa* leaves extract mediated silver nanoparticles. *Results Phys.*, 15, 102565, 2019.
38. Pati, R., Mehta, R.K., Mohanty, S., Padhi, A., Sengupta, M., Vaseeharan, B., Goswami, C., Sonawane, A., Topical application of zinc oxide nanoparticles reduces bacterial skin infection in mice and exhibits antibacterial activity by inducing oxidative stress response and cell membrane disintegration in macrophages. *Nanomed.: Nanotechnol. Biol. Med.*, 10, 6, 1195–1208, 2014.
39. Sastry, M., Ahmad, A., Khan, M.I., Kumar, R., Biosynthesis of metal nanoparticles using fungi and actinomycete. *Curr. Sci.*, 85, 2, 162–170, 2003.
40. Mohan, S. and Devan, M.V., Photocatalytic activity of Ag/Ni bimetallic nanoparticles on textile dye removal. *Green Process. Synth.*, 8, 1, 895–900, 2019.
41. El-Sayed, E.S.R., Abdelhakim, H.K., Zakaria, Z., Extracellular biosynthesis of cobalt ferrite nanoparticles by *Monascus purpureus* and their antioxidant, anticancer and antimicrobial activities: Yield enhancement by gamma irradiation. *Mater. Sci. Eng.: C*, 107, 110318, 2020.
42. Kahzad, N. and Salehzadeh, A., Green synthesis of $CuFe_2O_4$@Ag nanocomposite using the *Chlorella Vulgaris* and evaluation of its effect on the expression of norA efflux pump gene among *Staphylococcus Aureus* strains. *Biol. Trace Elem. Res.*, 1–12, 2020.
43. Gingasu, D., Mindru, I., Patron, L., Calderon-Moreno, J.M., Mocioiu, O.C., Preda, S., Stanica, N., Nita, S., Dobre, N., Popa, M., Gradisteanu, G., Green synthesis methods of $CoFe_2O_4$ and Ag-$CoFe_2O_4$ nanoparticles using *hibiscus* extracts and their antimicrobial potential. *J. Nanomater.*, 2016, 1–12, 2016.
44. Sorbiun, M., Shayegan Mehr, E., Ramazani, A., Mashhadi Malekzadeh, A., Biosynthesis of metallic nanoparticles using plant extracts and evaluation of their antibacterial properties. *Nanochem. Res.*, 3, 1, 1–16, 2018.
45. Surendra, B.S., Nagaswarupa, H.P., Hemashree, M.U., Khanum, J., *Jatropha* extract mediated synthesis of $ZnFe_2O_4$ nanopowder: Excellent performance as an electrochemical sensor, UV photocatalyst and antibacterial activity. *Chem. Phys. Lett.*, 739, 136980, 2020.
46. Sriramulu, M., Shukla, D., Sumathi, S., *Aegle marmelos* leaves extract mediated synthesis of zinc ferrite: Antibacterial activity and drug delivery. *Mater. Res. Express*, 5, 11, 115404, 2018.
47. Matinise, N., Kaviyarasu, K., Mongwaketsi, N., Khamlich, S., Kotsedi, L., Mayedwa, N., Maaza, M., Green synthesis of novel zinc iron oxide ($ZnFe_2O_4$) nanocomposite via *Moringa Oleifera* natural extract for electrochemical applications. *Appl. Surf. Sci.*, 446, 66–73, 2018.
48. Laokul, P., Amornkitbamrung, V., Seraphin, S., Maensiri, S., Characterization and magnetic properties of nanocrystalline $CuFe_2O_4$, $NiFe_2O_4$, $ZnFe_2O_4$

powders prepared by the *Aloe vera* extract solution. *Curr. Appl. Phys.*, 11, 1, 101–108, 2011.

49. Bashir, A.K.H., Matinise, N., Sackey, J., Kaviyarasu, K., Madiba, I.G., Kodseti, L., Ezema, F.I., Maaza, M., Investigation of electrochemical performance, optical and magnetic properties of NiFe2O4 nanoparticles prepared by a green chemistry methos. *Physica E: Low Dimens. Syst. Nanostruct.*, 119, 114002, 2020.
50. Karunakaran, G., Jagathambal, M., Van Minh, N., Kolesnikov, E., Kuznetsov, D., Green synthesis of $NiFe_2O_4$ spinel-structured nanoparticles using *Hydrangea paniculata* flower extract with the excellent magnetic property. *J. Miner.*, 70, 7, 1337–1343, 2018.
51. Manikandan, A., Durka, M., Antony, S.A., *Hibiscus rosa-Sinensis* leaf extracted green methods, magneto-optical and catalytic properties of spinel $CuFe_2O_4$ nano- and microstructures. *J. Inorg. Organomet. Polym. Mater.*, 25, 5, 1019–1031, 2015.
52. Korotkova, A.M., Polivanova, O.B., Gavrish, I.A., Baranova, E.N., Lebedev, S.V., Green synthesis of zinc-based nanoparticles zinc ferrite by Petroselinum crispum, in: *IOP Conference Series: Earth and Environmental Science*, vol. 341, p. 012175, 2019, October.
53. Hafez Ghoran, S., Fadaei Dashti, M., Maroofi, A., Shafiee, M., Zare-Hoseinabadi, A., Behzad, F., Mehrabi, M., Jangjou, A., Jamali, K., Biosynthesis of zinc ferrite nanoparticles using polyphenol-rich extract of *Citrus aurantium* flowers. *Nanomed. Res. J.*, 5, 1, 20–28, 2020.
54. Sarala, E., Madhukara Naik, M., Vinuth, M., Rami Reddy, Y.V., Sujatha, H.R., Green synthesis of *Lawsonia inermis*-mediated zinc ferrite nanoparticles for magnetic studies and anticancer activity against breast cancer (MCF-7) cell lines. *J. Mater. Sci.: Mater. Electron.*, 31, 11, 8589–8596, 2020.
55. Sharma, S., Dutta, V., Raizada, P., Hosseini-Bandegharaei, A., Thakur, V., Nguyen, V.H., VanLe, Q., Singh, P., An overview of heterojunctioned $ZnFe_2O_4$ photocatalysts for enhanced oxidative water purification. *J. Environ. Chem. Eng.*, 9, 5, 105812, 2021.
56. Ajormal, F., Moradnia, F., Taghavi Fardood, S., Ramazani, A., Zinc ferrite nanoparticles in photo-degradation of dye: Mini-review. *J. Chem. Rev.*, 2, 2, 90–102, 2020.
57. Mahmoodi, N.M., Photocatalytic degradation of textile dyes using ozonation and magnetic nickel ferrite nanoparticle. *Prog. Color Color. Coat.*, 9, 3, 163–171, 2016.
58. Prasad, C., Sreenivasulu, K., Gangadhara, S., Venkateswarlu, P., Bio inspired green synthesis of Ni/Fe_3O_4 magnetic nanoparticles using *Moringa oleifera* leaves extract: A magnetically recoverable catalyst for organic dye degradation in aqueous solution. *J. Alloys Compd.*, 700, 252–258, 2017.
59. Patil, S.B., Naik, H.B., Nagaraju, G., Viswanath, R., Rashmi, S.K., Sugarcane juice mediated eco-friendly synthesis of visible light active zinc ferrite

nanoparticles: Application to degradation of mixed dyes and antibacterial activities. *Mater. Chem. Phys.*, 212, 351–362, 2018.
60. Alfina, A., Stiadi, Y., Lee, H.J., Green synthesis and characterization of ZnO-$CoFe_2O_4$ semiconductor photocatalysts prepared using rambutan (*Nephelium lappaceum* L.) peel extract. *Mat. Res.*, 22, 5, 2019.
61. Fardood, S.T., Moradnia, F., Mostafaei, M., Afshari, Z., Faramarzi, V., Ganjkhanlu, S., Biosynthesis of MgFe2O4 magnetic nanoparticles and their application in photodegradation of malachite green dye and kinetic study. *Nanochem. Res.*, 4, 1, 86–93, 2019.
62. Lagashetty, A., Pattar, A., Ganiger, S.K., Synthesis, characterization and antibacterial study of Ag doped magnesium ferrite nanocomposite. *Heliyon*, 5, 5, e01760, 2019.
63. Balasubramanian, M. and Murali, K.R., Biosynthesis of zinc ferrite ($ZnFe_2O_4$) nanoparticles using flower extract of *nyctanthes arbor-tristis* and their photocatalytic activity. *Ferroelectrics*, 555, 1, 1–14, 2020.
64. Tahir, N., Zahid, M., Bhatti, H.N., Mansha, A., Zia, K.M., Mustafa, G., Soomro, M.T., Qazi, U.Y., Silver-doped ternary compounds for wastewater remediation, in: *Silver Nanomaterials for Agri-Food Applications*, pp. 623–653, 2021.
65. Ghosh, B.K., Moitra, D., Chandel, M., Lulla, H., Ghosh, N.N., Ag nanoparticle immobilized mesoporous TiO_2-cobalt ferrite nanocatalyst: A highly active, versatile, magnetically separable and reusable catalyst. *Mater. Res. Bull.*, 94, 361–370, 2017.
66. Ibrahim, I., Kaltzoglou, A., Athanasekou, C., Katsaros, F., Devlin, E., Kontos, A.G., Ioannidis, N., Perraki, M., Tsakiridis, P., Sygellou, L., Antoniadou, M., Magnetically separable TiO_2/$CoFe_2O_4$/Ag nanocomposites for the photocatalytic reduction of hexavalent chromium pollutant under UV and artificial solar light. *Chem. Eng. J.*, 381, 122730, 2020.
67. Singh, S. and Singhal, S., Transition metal-doped cobalt ferrite nanoparticles: Efficient photocatalyst for photodegradation of textile dye. *Mater. Today: Proc.*, 14, 453–460, 2019.
68. Nadumane, A., Shetty, K., Anantharaju, K.S., Nagaswarupa, H.P., Rangappa, D., Vidya, Y.S., Nagabhushana, H., Prashantha, S.C., Sunlight photocatalytic performance of Mg-doped nickel ferrite synthesized by a green sol-gel route. *J. Sci.: Adv. Mater. Devices*, 4, 1, 89–100, 2019.
69. Naik, M.M., Naik, H.B., Nagaraju, G., Vinuth, M., Vinu, K., Rashmi, S.K., Effect of aluminium doping on structural, optical, photocatalytic and antibacterial activity on nickel ferrite nanoparticles by the sol-gel auto-combustion method. *J. Mater. Sci.: Mater. Electron.*, 29, 23, 20395–20414, 2018.
70. Lassoued, A. and Li, J.F., Magnetic and photocatalytic properties of Ni-Co ferrites. *Solid-State Sci.*, 104, 106199, 2020.
71. Fan, G., Tong, J., Li, F., Visible-light-induced photocatalyst based on cobalt-doped zinc ferrite nanocrystals. *Ind. Eng. Chem. Res.*, 51, 42, 13639–13647, 2012.

72. Xiang, Y., Xu, X., Wei, Y., Zhou, Y., Yang, X., Yang, Y., Yang, J., Luo, L., Zhou, Z., Carbon-based materials as absorbent for antibiotics removal: Mechanisms and influencing factors. *J. Environ. Manage.*, 237, 128–138, 2019.
73. Wang, C., Wang, Y., Graser, J., Zhao, R., Gao, F., O'Connell, M.J., Solution-based carbohydrates synthesis of individual solid, hallow, and porous carbon nanospheres using spray pyrolysis. *ACS Nano.*, 7, 12, 11156–11165, 2013.
74. Singhal, S., Dixit, S., Shukla, A.K., Structural analysis of carbon nanospheres synthesized by CVD: An investigation of surface charges and its effect on the stability of carbon nanostructures. *Appl. Phys. A*, 125, 2–11, 2019.
75. Zampiva, R.Y.S., Junior, C.G.K., Pinto, J.S., Panta, P.C., Alves, A.K., Bergmann, C.P., 3D CNT macrostructure synthesis catalyzed by $MgFe_2O_4$ nanoparticles—A study of surface area and spinel inversion influence. *Appl. Surf. Sci.*, 422, 321–330, 2017.
76. Hassani, A., Eghbali, P., Metin, O., Sonocatalytic removal of methylene blue from water solution by cobalt ferrite/mesoporous graphitic carbon nitride ($CoFe_2O_4$/mpg-C_3N_4) nanocomposites: Response surface methodology approach. *Environ. Sci. Pollut. Res.*, 25, 32, 32140–32155, 2018.
77. Singh, C., Bansal, S., Kumar, V., Singhal, SS., Beading of cobalt substituted nickel ferrite nanoparticles on the surface of carbon nanotubes: A study of theit synthesis mechanism, structure, magnetic, optical and their application as photocatalysts. *Ceramic Int.*, 41, 3, 3595–3604, 2015.
78. Chen, X., Shen, S., Guo, L., Mao, S.S., Semiconductor-based photocatalytic hydrogen generation. *Chem. Rev.*, 110, 11, 6503–6570, 2010.
79. Singhal, S., Sharma, R., Singh, C., Bansal, S., Enhanced photocatalytic degradation of methylene blue using/MWCNT composite synthesized by hydrothermal method. *Ind. J. Mater. Sci.*, 2013, 356025, 2013.
80. Tatarchuk, T., Peter, A., Al-Najar, B., Vijaya, J., Bououdina, M., Photocatalysis: Activity of nanomaterials. *Nano. Environ. Sci.*, 209–292, 2018.
81. Ai, L., Huang, H., Chen, Z., Wei, X., Jiang, J., Activated carbon/$CoFe_2O_4$ composites: Facile synthesis, magnetic performance and their potential application for the removal of malachite green from water. *Chem. Eng. J.*, 156, 2, 243–249, 2010.
82. Lu, D., Zhang, Y., Lin, S., Wang, L., Wang, C., Synthesis of magnetic $ZnFe_2O_4$/graphene composite and its application in photocatalytic degradation of dyes. *J. Alloys Comp.*, 579, 336–342, 2013.
83. Choi, Y.W., Kim, E.J., Hahn, S.H., Photocatalytic activity of Au-Buffered WO_3 thin films prepared by RF magnetron sputtering. *Chem. Eng. J.*, 161, 285, 2010.
84. Rani, M. and Shanker, U., Efficient photocatalytic degradation of Bisphenol A by metal ferrites nanoparticles under sunlight. *Environ. Technol. Innov.*, 19, 100792, 2020.
85. Zou, L., Wang, H., Jiang, X., Yuan, G., Wang, X., Enhanced photocatalytic efficiency in degrading organic dyes by coupling CdS nanowires with $ZnFe_2O_4$ nanoparticles. *Sol. Energy*, 195, 271–277, 2020.

86. Boutra, B., Güy, N., Özacar, M., Trari, M., Magnetically separable $MnFe_2O_4$/TA/ZnO nanocomposites for photocatalytic degradation of Congo Red under visible light. *J. Magn. Magn. Mater.*, 497, 165994, 2020.
87. Zouhier, M., Tanji, K., Navio, J.A., Hidalgo, M.C., Jaramillo-Páez, C., Kherbeche, A., Preparation of $ZnFe_2O_4$/ZnO composite: Effect of operational parameters for photocatalytic degradation of dyes under UV and visible illumination. *J. Photochem. Photobiol. A: Chem.*, 390, 112305, 2020.
88. Farhadi, S., Siadatnasab, F., Khataee, A., Ultrasound-assisted degradation of organic dyes over magnetic $CoFe_2O_4$@ZnS core-shell nanocomposite. *Ultrason. Sonochem.*, 37, 298–309, 2017.
89. Siadatnasab, F., Farhadi, S., Khataee, A., Sonocatalytic performance of magnetically separable CuS/$CoFe_2O_4$ nanohybrid for efficient degradation of organic dyes. *Ultrason. Sonochem.*, 44, 359–367, 2018.
90. Kadyrzhanov, K.K., Egizbek, K., Kozlovskiy, A.L., Zdorovets, M.V., Synthesis and properties of ferrite-based nanoparticles. *Nanomaterials*, 9, 8, 1079, 2019.
91. Gingasu, D., Mindru, I., Mocioiu, O.C., Preda, S., Stanica, N., Patron, L., Ianculescu, A., Oprea, O., Nita, S., Paraschiv, I., Popa, M., Synthesis of nanocrystalline cobalt ferrite through soft chemistry methods: A green chemistry approach using *sesame* seed extract. *Mater. Chem. Phys.*, 182, 219–230, 2016.
92. Mahajan, P., Sharma, A., Kaur, B., Goyal, N., Gautam, S., Green synthesized (*Ocimum sanctum* and *Allium sativum*) Ag-doped cobalt ferrite nanoparticles for antibacterial application. *Vacuum*, 161, 389–397, 2019.
93. Kombaiah, K., Vijaya, J.J., Kennedy, L.J., Bououdina, M., Ramalingam, R.J., Al-Lohedan, H.A., *Okra* extract-assisted green synthesis of $CoFe_2O_4$ nanoparticles and their optical, magnetic, and antimicrobial properties. *Mater. Chem. Phys.*, 204, 410–419, 2018.
94. Alijani, H.Q., Pourseyedi, S., Torkzadeh-Mahani, M., Seifalian, A., Khatami, M., Bimetallic nickel-ferrite nanorod particles: Greener synthesis using rosemary and its biomedical efficiency. *Artif. Cells Nanomed. Biotechnol.*, 48, 1, 242–251, 2020.
95. Kumar, S.A., Sharma, A.S., Singh, M., Dhiman, P.O., Kotnala, R.K., Size controlled synthesis and magnetic properties of Ni–Zn ferrite nanoparticles by using *aloe vera* extract solution. *Nano Vision*, 1, 101–114, 2011.
96. Venkateswarlu, S., Kumar, B.N., Prathima, B., Anitha, K., Jyothi, N.V.V., A novel green synthesis of Fe_3O_4-Ag core-shell recyclable nanoparticles using *Vitis vinifera* stem extract and its enhanced antibacterial performance. *Phys. B: Condens. Matter*, 457, 30–35, 2015.
97. Amiri, M., Pardakhti, A., Ahmadi-Zeidabadi, M., Akbari, A., Salavati-Niasari, M., Magnetic nickel ferrite nanoparticles: Green synthesis by *Urtica* and therapeutic effect of frequency magnetic field on creating a cytotoxic response in neural cell lines. *Colloids Surf. B: Biointerfaces*, 172, 244–253, 2018.

98. Alijani, H.Q., Iravani, S., Pourseyedi, S., Torkzadeh-Mahani, M., Barani, M., Khatami, M., Biosynthesis of spinel nickel ferrite nanowhiskers and their biomedical applications. *Sci. Rep.*, 11, 1, 1–7, 2021.
99. Gong, X.Q. and Selloni, A., Reactivity of anatase TiO_2 nanoparticles: The role of the minority (001) surface. *J. Phys. Chem. B*, 109, 19560–19562, 2005.
100. Vittadini, A., Casarin, M., Selloni, A., Chemistry *of* and *on* TiO_2-anatase surfaces by DFT calculations: A partial review. *Theor. Chem. Acc.*, 17, 663–671, 2006.

Part 2

ADVANCED OXIDATION PROCESSES

5

Selected Advanced Oxidation Processes for Wastewater Remediation

Nhamo Chaukura[1]*, Tatenda C. Madzokere[2,3] and Themba E. Tshabalala[1]

[1]*Department of Physical and Earth Sciences, Sol Plaatje University, Kimberley, South Africa*
[2]*Department of Metallurgy and Materials Engineering, Faculty of Mining and Mineral Processing Engineering, Midlands State University, Gweru, Zimbabwe*
[3]*University of Johannesburg, Department of Metallurgy, Faculty of Engineering and Built Environment P.O. Box Auckland, Johannesburg, South Africa*

Abstract
Advanced oxidation processes (AOPs) can effectively degrade persistent compounds in wastewater by enhancing biodegradation and minimizing toxicity. They achieve this by generating reactive free radical species, such as $H^\bullet$, $OH^\bullet$, $HOO^\bullet$, $SO_4^\bullet$, which attack organic and inorganic compounds and degrade or mineralize them to less harmful products. Combining photocatalysis with techniques, such as hydrodynamic cavitation, ultrasound, and sonoelectrochemical oxidation, creates hybrid AOPs with high oxidation capabilities. This chapter synthesizes literature on selected AOP techniques, including hybrid strategies and a brief discussion on membrane-based AOPs. The principles and properties of selected AOPs are described and future research directions suggested.

Keywords: Free radicals, mineralization, photodegradation, wastewater

5.1 Introduction

Advanced oxidation processes (AOPs) have been widely explored for the sequestration of pollutants from wastewater. These processes include

**Corresponding author*: nhamo.chaukura@spu.ac.za

Elvis Fosso-Kankeu, Sadanand Pandey, and Suprakas Sinha Ray (eds.) Photoreactors in Advanced Oxidation Processes: The Future of Wastewater Treatment, (153–174) © 2023 Scrivener Publishing LLC

ozonation, photocatalysis, photo-Fenton, and sonolysis, and depend on the *in situ* production of reactive oxygen species (ROS) for example H_2O_2, $OH^•$ radicals, O_3, $O_2^•$, ClO_2, which can degrade organic contaminants in wastewater [1–3]. Despite being effective and only requiring mild reaction conditions, the application of these processes is dependent on the physicochemical characteristics of the pollutants, operating conditions, and the chemical composition of the wastewater. Generally, the advantages of AOPs include low chemical demand, rapid reaction kinetics, and ease of installation in existing systems. However, the major limitation is that AOPs are pH-dependent, expensive, and have a high-energy demand [4].

Judicious choice of the AOP is crucial for maximum pollutant removal [2, 3]. Among the various ROS, $OH^•$ radicals exhibit high oxidation potential (2.8 V) and can easily oxidize organic pollutants. This oxidation is not selective and normally generates smaller degradates, or ultimately mineralize the pollutants into CO_2 and H_2O [5]. Photocatalysis is one AOP that has tremendous appeal owing to the potential of using solar radiation to generate $OH^•$ radicals, avoiding the high chemical demand required for other AOPs. Overall, the major aspect of AOPs is their capacity to completely mineralize organic pollutants under ambient temperature and pressure [5].

Combining photocatalysis with techniques, such as hydrodynamic cavitation, ultrasound, and sonoelectrochemical oxidation, creates hybrid AOPs with high oxidation capabilities. This chapter synthesizes literature on selected AOP techniques, including hybrid strategies, and a brief discussion on integrated AOPs and membrane technologies.

5.2 Photocatalysis and Ozonation

5.2.1 Photocatalysis

Despite being extensively researched, photocatalysis has not been commercialized. This is mainly due to high cost associated with the technology, and inadequate efficiency. Nevertheless, a significant body of literature has shown that photocatalysis is effective in degrading endocrine-disrupting chemicals (EDCs), natural organic matter (NOM), perfluorinated compounds, pesticides, and pharmaceutical and personal care products (PPCPs) at concentrations in the range of ng/L to μg/L [5]. Problems, such as slow kinetics and challenges associated with scaling-up, deserve further research. For practical applications, TiO_2 is the photocatalyst with the most promise because it is abundant, economical, non-toxic, is chemically stable, and exhibits excellent photocatalytic activity [6]. TiO_2 has three crystal structures, namely anatase, brookite, and rutile. Of these, the anatase phase is the most photocatalytically active due to the indirect bandgap of the anatase phase, which inhibits electron transitions from the conduction band to the valence band [6, 7]. This prolongs the lifetime and consequently lengthens the diffusion path of the excited electrons and holes, thus improving the likelihood of successful photocatalytic reactions [7, 8]. Moreover, the anatase phase is the first phase formed during synthesis, making the preparation process easier. At temperatures exceeding 500°C, the probability of transition from the anatase to rutile phase increases. However, the major challenge in using TiO_2 is that its performance under solar irradiation is limited by the wide bandgap of 3.2 eV, which is only capable of capturing UV light [8].

The mechanism involved in photocatalysis can be illustrated using TiO_2 (Scheme 5.1). When TiO_2 particles are exposed to UV radiation, pairs of excited electrons (e^-) and holes (h^+) are generated [68]. The holes react with H_2O to generate $OH^\bullet$ radicals, which in turn oxidize pollutant molecules to produce organic radicals. With E^o=2.7 V, $OH^\bullet$ radicals attack electron-rich sites in the chemical structure of the pollutant via electron transfer, electrophilic addition, hydrogen atom abstraction, or radical-radical addition [9]. Initiation reactions will produce peroxyl radicals ($ROO^\bullet$), which will start chain reactions that produce

Generation of free radicals

Reaction 1: $TiO_2 + hv \longrightarrow TiO_2^+ + e^-$

Reaction 2: $TiO_2^+ \longrightarrow TiO_2 + h^+$

Reaction 3: $HO_2 + h^+ \longrightarrow OH^\cdot + H^+$

Degradation of organic pollutants

Reaction 4: $OH^\cdot + O_2^{\bullet-}$ + organic pollutant $\longrightarrow$ degradation products

Degradation of Microbials

Reaction 5: $OH^\cdot + O_2$ + microbe $\longrightarrow$ cell wall break down

Reaction 6: $OH^\cdot$ + cell constituents $\longrightarrow$ dead biomass

Scheme 5.1 The reaction mechanism involved in photocatalysis.

CO_2 and H_2O along with other stable byproducts [10]. However, the byproducts are sometimes recalcitrant to degradation and can be carcinogenic. Thus, the efficacy of the photodegradation process has to be assessed taking the stability of byproducts into consideration [10].

The mechanism can be summarized as: (1) visible light irradiation on the photocatalyst initiates the reaction by pumping electrons from valence to conduction band; (2) oxidation of OH^- in water by photo-induced holes will generate free $OH^•$ radicals; (3) the reduction reaction between transitioning electrons and O_2 in solution generate superoxide radical anions ($^•O_2^-$); and (5) $OH^•$ radicals are generated from $^•O_2^-$. Subsequently, $OH^•$ radicals initiate chain reactions that degrade targeted compound to form non-toxic compounds [6]. The major problem associated with photocatalysts is the recombination of the separated electrons. However, this can be prevented by the use of a porous support, such as zeolites or activated carbon.

5.2.2 Ozonation

Strong oxidants, such as O_3, Cl_2, ClO_2, H_2O_2 with O_3 (peroxone) and combinations of these, can be used to improve the sequestration of organic pollutants from wastewater [11, 12]. Among these, O_3 is the most powerful oxidant, and has been used for disinfection, and removal of color and turbidity, and oxidation of toxic metals [13]. In addition, a previous study reported that O_3 can effectively remove *Cryptosporidium oocysts* and *Giardia cysts*, since it deactivates the cysts and oocysts, which are unaffected by Cl_2 attack [13]. Generally, ozonation can be used as a pretreatment step before the coagulation step, the purpose of which is to degrade compounds, such as phenolics, and other organic pollutants [14]. Preozonation can degrade aromatic compounds into smaller molecules that are more biodegradable and reduces chemical and operational costs [3]. However, this results in the production of oxidative byproducts, such as aldehydes, bromate, and carboxylic acids, dissolved organic carbon (DOC), and disinfection by-product (DBP) precursors [15, 16].

Mechanistically, the reaction of O_3 occurs either: (1) directly via oxidation, which is a very selective molecular process, or (2) indirectly with free radicals generated upon the decomposition of O_3 (Scheme 5.2) [17]. The formation of $OH^•$ radicals increases with an increase in pH, whereas direct oxidation is dominant at lower pH.

$$O_3 + H_2O \longrightarrow HO_3^+ + OH^-$$

$$HO_3^+ + OH^- \longrightarrow 2HO_2$$

$$O_3 + HO_2 \longrightarrow HO + 2O_2$$

$$HO + HO_2 \longrightarrow H_2O + O_2$$

Scheme 5.2 The decomposition of ozone in water.

In addition to photocatalysis and ozonation technology, recent advances have demonstrated the effectiveness of hydrodynamic cavitation, sonoelectrochemical oxidation, and sonophotocatalytic degradation processes in degrading pollutants in wastewater. The next section will explore the use of these techniques in wastewater treatment.

5.3 Hybrid AOP Technologies

Photocatalysis and ozonation are limited by the potential generation of toxic stable products. Besides, the kinetics can be sluggish and thus difficult to use in practical applications. In this regard, hybrid advanced oxidation processes (HAOP) are becoming more popular in wastewater treatment and other environmental remediation applications [18]. The HAOP technologies come in different configurations, such as sonoelectrochemical oxidation and sonophotocatalytic degradation [18, 19]. The next sections explore each of these configurations.

5.3.1 Hydrodynamic Cavitation

Cavitation is a physical phenomenon involving rapid pressure changes in a liquid resulting in the formation of miniscule vapor-filled voids followed by the collapse of the voids within a short space of time (μs) [20]. This transient event releases a lot of energy over a very small area, generating confined temperature and pressure hotspots in the range of 1000 to 15000 K, and 500 to 5000 bars, respectively. This process generates highly reactive species, such as free radicals, coupled with severe turbulence that results in biological, chemical, and physical processes, such as degradation of organic and inorganic pollutants [20].

In hydrodynamic cavitation (HC), voids are produced due to liquid pressure variations induced via geometric constrictions, such as throttle valves or orifice plates in a liquid flow system. The Bernoulli principle governs cavity formation in HC reactors. The extent of cavitation in an HC reactor can be expressed with respect to a cavitation number, C_v (Equation 5.1) [20]:

$$C_v = \frac{p_2 - p_v}{(1/2)\rho_{vth}^2} \tag{5.1}$$

where ρ is the density of the liquid, p_2 is recovered pressure after the constriction, p_v is the liquid vapor pressure, v_{th} is the velocity at the constriction throat.

An understanding of bubble dynamics and modeling techniques HC processes is imperative for effective application of the technology in HAOP operations. For instance, the solution of the fundamental bubble dynamics equation (Rayleigh-Plesset equation) (Equation 5.2) gives valuable information about cavity radius and collapse pressure [20]:

$$R\frac{d^2R}{dt^2} + \frac{3}{2}\left(\frac{dR}{dt}\right)^2 = \frac{1}{\rho}\left[P_i - \frac{2\sigma}{R} - \frac{4\mu}{R}\frac{dR}{dt} - P_\infty\right] \tag{5.2}$$

where dR/dt is the bubble wall velocity, d^2R/dt^2 is the bubble acceleration/cavity, μ is the medium viscosity, σ is the liquid surface tension, P_i and P_∞ are the pressure inside the bubble and pressure after the constriction, respectively.

The pollutant degradation mechanisms in HC involve breaking down chemical bonds, pollutant decomposition at high temperature, liquid-phase combustion, and pyrolysis of organic compounds inside the cavities [21]. HC has potential advantages in wastewater treatment due to the simplicity in reactor design, ease of operation, easy scaling-up capabilities, and a lower energy demand [18, 21]. Operationally, HC depends on system pH, pollutant concentration, bubble radius, hydrodynamic properties of the water, inlet pressure, and the catalyst used [18, 21]. For instance, a study on the photodegradation of acid red dye reported that degradation increased as the initial dye concentration increased, and became more pronounced after addition of H_2O_2 and Fe/

TiO_2 catalyst [18]. In another study, low pH favored more production of OH· radicals in addition to a higher oxidation potential environment, which reduced the recombination rate of OH· radicals [22]. The optimal pH will, therefore, depend on the dissociation constant (pK_a) of the pollutant. using a hybrid system HC reactors and Fenton process, acidic environments have also been reported to improve the photodegradation of azo dyes, such as Orange G where high degradation (99%) was achieved at pH 2 in contrast to low degradation (12.3%) under neutral pH conditions [22]. However, in a related investigation, neutral and alkaline pH environments were favorable for the degradation of Brilliant Cresyl Blue dye at 20°C [23]. A previous study demonstrated that the type of acid used for pH adjustment influenced the radical oxidation processes as well as potential production of secondary toxic pollutants [24]. In addition, HC is more energy efficient than acoustic cavitation and exhibited higher degradation rates for the same power dissipation [20].

Recently, HC reactors have been investigated in large-scale wastewater treatment processes involving pollutants, such as dyes, pharmaceuticals, pesticides, and phenolics [25]. An increase in pressure increased pollutant degradation, i.e., 32.32% degradation with a rate constant of 3.41×10^{-3} min^{-1} at 5 bar over 2 hours. Operating beyond the optimal pressure resulted in 'chocked cavitation' accompanied by a reduction in OH• radical generation, and consequently a reduction in the rate of degradation.

5.3.2 Hybrid AOP Systems Based on Hydrodynamic Cavitation

Integrating HC with other external agents, such as O_3, $Na_2S_2O_8$, H_2O_2, surfactants, plasma, and catalysts increased the degradation efficiency. These integrated systems are sometimes referred to as hybrid AOPs [23, 25]. Previous research reported hybrid AOPs to be effective in decolorizing organic pollutants [20]. Common hybrid systems include HC/H_2O_2, HC/O_3, HC/UV, and HC/Fenton, which exhibit remarkable synergy compared to stand-alone HC systems. Recently, a study demonstrated the potential of HC coupled with plasma in wastewater treatment in a continuous flow system concurrently achieving microbial suppression and organic pollutant degradation [25]. Moreover, HC has been combined with acoustic systems to form hybrid hydrodynamic acoustic cavitation for enhanced performance [26]. In such systems, the bubbles are more violently expanded and collapsed than bubbles generated in the

HC and acoustic cavitation separately. This results in higher pollutant degradation efficiency. However, the energy demand and overall cost of these systems has not been adequately investigated, and, as such, deserve further research.

5.3.3 Hybrid AOP Systems Based on Ultrasound Radiation

Sonochemistry refers to the influence of sound waves with frequencies above 20 kHz or 20,000 cycles per second on the transformation of molecules through a chemical reaction. When ultrasound is applied on liquid micro-sized bubbles under suitable conditions the bubbles grow bigger and subsequently collapse violently, releasing a large amounts of energy within the margins of the bubble (Figure 5.1). This process is called cavitation [27]. During the cavitation process microbubbles formed act as reactors with the core of the bubble being a gaseous phase possessing volatile and hydrophobic molecules. These molecules will undergo pyrolytic degradation due to the high temperature and pressure in the core of the bubble resulting in the formation of $OH^\bullet$ radicals [28]. The $OH^\bullet$ radicals formed in the core of the bubble will migrate into the gas-liquid region of the bubble, and may subsequently react with each other to form H_2O_2. The presence of free radicals is predominant in the bulk liquid phase due to the migration of radicals from the gas-liquid region, and these free radicals will be significant in the sonochemistry.

The application of sonochemistry in water treatment is based on the phenomenon that heat energy generated from the cavitation process decomposes water into reactive hydrogen atoms and $OH^\bullet$ radicals [29].

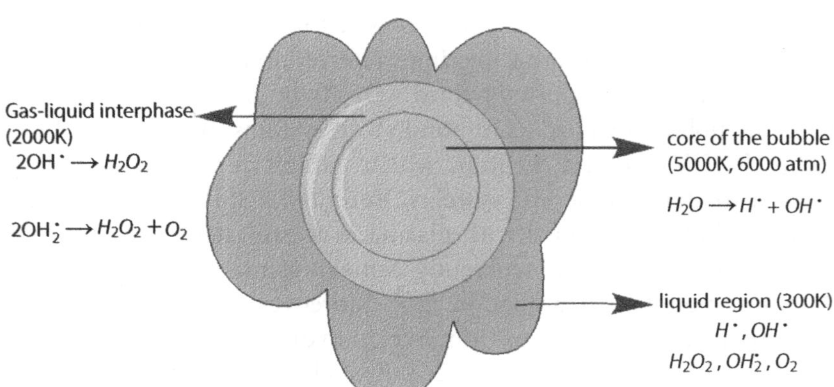

Figure 5.1 Schematic representation of radical generation in a bubble cavity.

Under the conditions of high temperature and pressure, highly reactive species, such as H•, OH•, HOO• radicals, and H_2O_2, are formed [30]. With the secondary molecules present in the water, there are possible chemical reactions that might occur between the secondary molecules and the radicals generated from the decomposition of a water molecule [31]. Owing to the need to reduce organic pollutants in industrial effluents, sonochemistry has provided an alternative approach to degrade various organic pollutants in aqueous systems.

5.3.3.1 Sonoelectrochemical Oxidation

Sonoelectrochemical oxidation (SECO) techniques have been hugely employed in wastewater treatment and environmental remediation. Electrochemical based technologies have numerous advantages, including cost-effectiveness, adaptability, and environmental compatibility [32]. Earlier research has shown the great potential of SECO in decontaminating ammonia and pathogenic contaminants in municipal wastewater, organic dye removal, and EDCs, such as parabens and bisphenol A from industrial effluents [32].

Owing to their high aspect ratio and desirable pollutant removal properties, nanostructured materials have been used to enhance SECO processes. For instance, nanocoated electrodes have been reported to perform better in the degradation of methylene blue in wastewater than pristine electrodes by enhancing the generation of OH• radicals [33]. Supersonic waves in the system enhanced mass transfer on electrode surface coated with nanomaterials leading to efficient degradation of pollutants. Another study investigating the degradation of ibuprofen in municipal wastewater found that current intensity and residence time in the reactor were the major parameters governing the degradation rate [34]. Optimal degradation (90%) was achieved at residence time and current density of 110 min and 4.09 A, respectively. Other factors, which affect the performance of SECO, include the chemical composition of wastewater, flow rate, electrode cell configuration, and type of electrode [69]. In other studies, dimensionally stable anodes have been used to disinfect *E. coli* [35]. Recently, biobased electrochemical systems have demonstrated a potential for the cost-effective removal of total ammonium nitrogen in wastewater [36].

In a study involving the treatment of wastewater containing methylene blue dye using $Ti/Ta_2O_5{-}SnO_2$ electrodes, COD reduction was 80% while decolorization was at 96% efficiency at neutral pH [37]. Another study reported a total organic carbon (TOC) reduction of 95.5% in organic wastewater containing amaranth dye accompanied by a nearly complete degradation of the dye (99%) [38].

Lately, it has been demonstrated that SECO is a potential method for degrading emerging pollutants, such as pharmaceuticals, recreational drugs, steroids, personal care products, flame retardants, hormones, plasticizers, and surfactants [39]. Some of these pollutants are associated with antimicrobial resistance with a high capability of bioaccumulating in vital organs, such as the kidneys, liver, and gills of aquatic animals.

5.3.3.2 Sonophotocatalytic Degradation

Sonophotocatalytic (SPC) degradation is a hybrid advanced oxidation technique that simultaneously uses photocatalysis and ultrasound. This technique is based on sonochemistry, which involves physicochemical processes occurring in a solution using ultrasound energy and enhanced by incorporating radiation in catalytic reactions [40]. The resultant effects of the presence of ultrasound are the cavitation phenomena described previously [20, 40]. SPC transforms molecules by enhancing chemical reactions through the formation of radicals using a combination of ultrasonic sound waves, ultraviolet radiation, and a catalyst. The commonly used catalysts are semiconductors, such as TiO_2, ZnO, ZnS, Fe_2O_3, and CdS. These semiconductors are regarded as photocatalysts due to the electronic structure of the metal atom, which is characterized by a filled valence band and an empty conduction band [70]. They have been extensively used in photocatalysis due to their advantages, such as low cost, commercial availability, stability, catalyst regeneration, and high efficiency. Compared with other AOPs, SPC has numerous advantages including: (1) enhances the absorption of light by catalyst/semiconductor particles, (2) increases the dispersion of particles in water and further accelerate the splitting of H_2O_2 and/or H_2O to form $HO^\bullet$ and $HOO^\bullet$ radicals, (3) enhances the catalyst surface through the reduction of particle size, (4) under ultrasound, there is an increase in the continuous desorption of products from the surface of the catalyst, which minimizes the deactivation rate of the catalyst, and (5) enhances the mass transport of organic pollutants.

The incorporation of a semiconductor to facilitate the photocatalytic aspect of SPC introduces an additional component of $HO^\bullet$ radical generation through the bandgap excitation of the respective semiconductor (Figure 5.2). Upon irradiation of the semiconductor with UV light with energy higher or equal to the bandgap of the semiconductor, an electron

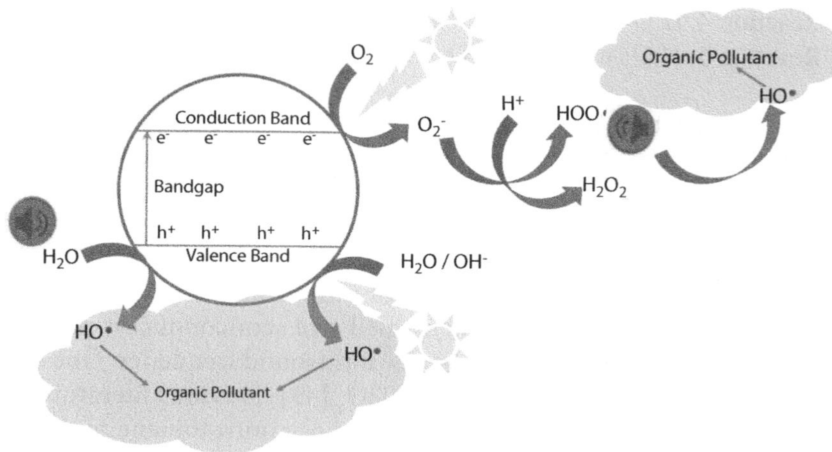

Figure 5.2 The degradation of an organic pollutant under sonophotocatalysis.

from the valence band is promoted to the conduction band with simultaneous generation of a hole in the valence band (Reaction 1). The generated valence hole can react with OH- or H_2O to form $HO^\bullet$ radicals (Reaction 2), while the excited electrons in the conduction band can react with the dissolved oxygen present in the reaction medium (i.e. H_2O) to produce $O_2^{\bullet-}$ radical ions, which subsequently react with protons to form additional $HO^\bullet$ radicals (Reaction 3).

Reaction 1: $TiO_2 + hv \rightarrow TiO_2(e^-) + TiO_2(h^+)$
Reaction 2: $h^+ + OH^- \rightarrow HO^\cdot$
Reaction 3: $e^- + O_2 \rightarrow O_2^{\bullet-} \rightarrow HO^\bullet$

The $HO^\bullet$ radicals generated through photocatalysis may react with the organic pollutant in the bulk medium. Most of the organic pollutants are hydrophilic, therefore they would reside in the bulk medium (i.e., H_2O) and the degree of degradation of the pollutant is dependent on the proximity of $HO^\bullet$ radicals produced and pollutant molecule. Furthermore, the $HO^\bullet$ radicals are highly reactive and have a short lifespan; they may react with water or recombine to form H_2O_2 without reacting with the targeted pollutant molecule. Hence the ultrasound radiation will increase the concentration of $HO^\bullet$ radicals in the bulk medium by facilitating the H_2O and H_2O_2 split reaction through sonolysis (Reactions 4 and 5). This will further minimize the distance between $HO^\bullet$ radicals and the pollutant molecule leading to high degradation efficiency.

Reaction 4: $H_2O + US))) \rightarrow H^{\bullet} + HO^{\bullet}$
Reaction 5: $H_2O_2 + US))) \rightarrow HO^{\bullet} + HO^{\bullet}$

SPC degradation is an important technique for pollutant degradation as it enhances oxidation intensity and degradation kinetics with major drawbacks associated with operational costs and deficiency of cost-effective reactor design [41]. However, with further research, this technique has proved to be effective in degrading emerging pollutants [42]. Recently, the synergistic effects of ultrasound and photocatalysis have been more pronounced when small size semiconductor particles are incorporated into the system at low ultrasound frequency. One such material, which has proven useful, is TiO_2 [43]. Previous literature has reported the oxidative degradation of 2,4,6-trinitrotoluene and dinitrotoluene in wastewater using TiO_2-based SPC [71]. TiO_2 powder enhanced the degradation process by facilitating cavitation of bubbles. Another investigation showed that a hybrid SPC system was more effective in waste degradation than stand-alone sonolysis and photocatalysis systems [44]. First-order kinetic rate constants for the hybrid were 10 times and two times higher compared to sonolysis and photocalataysis, respectively. Higher rate constants have also been observed in the degradation of Reactive Brilliant dye in synthetic wastewater coupled with COD reduction [45]. The increase in rate constants was ascribed to the occupation of active sites on the TiO_2 catalyst. Several other studies demonstrated the effectiveness of TiO_2 in enhancing pollutant degradation using SPC [43]. Further, the use TiO_2 catalyst in two different phases (rutile and anatase) was effective in degrading 2,4,6 trichlorophenol [46]. The difference in the catalytic activities of the two phases is due to the position of the conduction band on the two phases. It was observed that the anatase phase exhibited a higher photogeneration capability.

Other metal oxide catalysts like ZnO have also been incorporated into SPC. For example, SPC and ZnO catalyst have been investigated for the degradation of phenol in wastewater with a resultant generation of H_2O_2 which acted as a hole scavenger and electron acceptor consequently driving the degradation process [10]. Owing to their electronic and optical properties, nanostructured materials in combination with SPC systems have gained increased research interest [47–49]. A composite of ZnO nanoparticles and multi-walled carbon nanotubes (MWCNTs) was used in SPC system for the removal of Rhodamine B in wastewater [50]. Degradation kinetics were fast for the hybrid system mainly due to the larger surface area offered by the nanocomposite catalyst, which resulted in more reactive radicals being produced. In addition, the COD of the textile wastewater was markedly reduced. Recently, a Fe_3O_4/SnO_2 nanocomposite catalyst deposited on grapheme platelets in a SPC system was effective in degrading methylene blue dye [51]. The catalyst enhanced dye degradation considerably, and was largely stable and efficient as it was used for four cycles without the need for regeneration. In another study, ZnO-based SPC effectively reduced the TOC for textile wastewater (92.55%), while UV+ZnO achieved a 71.08% reduction at neutral pH [48]. Furthermore, the nanocatalyst had high recyclability and efficiency, in agreement with other researchers who have employed nanocatalysis in SPC systems [51]. Enhanced SPC degradation (92%) of an emerging pollutant, bisphenol A was achieved in 60 min using bimetal sulfide-intercalated MXenes, 2D/2D nanocomposite [49]. Another study used copper oxide nanoparticles with SPC for the treatment of landfill leachate, and observed a significant reduction of COD (85.82%) in 10 min at pH 6.9 and a nanoparticle dosage of 0.05 g L^{-1} [47]. Panchangam *et al.* [52] reported a perfluorooctanoic acid decomposition of above 65% under the combination of ultrasound and photocatalysis. The removal of pesticides from aquatic systems has also received extensive interest, and SPC has a significant role in treating wastewater from the agro-processing plants. Different catalysts used under SPC conditions have shown to effectively degrade various pesticides, such as dichlorvos using TiO_2 catalyst [53], simazine over Au/TiO_2 catalyst [54], and diazinon using Fe_3O_4/MOF nanocomposite catalyst [55].

Overall, combining SPC with other techniques, such as photocatalysis or nanomaterials, significantly enhances pollutant removal. Although this is a viable wastewater treatment approach, it is limited by the difficulties of separating the photocatalyst particles from treated water. To avoid this challenge, the nanoparticles could be immobilized on a reactor or some such support as a membrane [8].

5.4 Membrane-Based AOPs

A number of challenges, such as the difficulty in separating the suspended photocatalyst from the treated wastewater and low recyclability, have to be solved for the uptake of photocatalytic degradation techniques in large-scale wastewater treatment. Postreaction, the suspended photocatalyst requires separation to enable its efficient recovery [56]. In view of this, it is advantageous to integrate photocatalytic degradation with membrane separation. This can be achieved through: (1) mixed matrix membranes [57], (2) photocatalysts immobilized on the membrane surface, or (3) membranes for the subsequent separation of spent photocatalysts from the treated wastewater [58]. Immobilized nanoparticles allow continuous treatment of the wastewater and avoid the requirement for removing the nanoparticles after treatment. Owing to the increased efficiency, self-cleaning and anti-fouling properties, photocatalytic membranes represent an energy sustainable and ecofriendly wastewater treatment approach [59, 60]. However, the preparation of effective photocatalytic membranes is challenging because loading the photocatalyst on the membrane surface can result in layer delamination and pore blocking because of repetitive processes [59].

To address these challenges, a number of strategies have been used to fabricate photocatalytic membranes. For instance, a natural zeolite was used to prepare a mesoporous cylindrical membrane using the extrusion technique [61]. The membrane was used to degrade RB5-dye, and a total organic carbon removal $TOC_{90}\%$ = 60.0 and $TOC_{180}\%$ = 66.7 in the retentate was achieved, while $TOC_{180}\%$ = 89.5 was attained in the permeate. Using the phase inversion technique, a polyethersulfone membrane was prepared onto which poly-acrylic acid (PAA) was then grafted [57]. The PAA coating introduced carboxylic moieties suitable for surface modification using a nitrogen-doped carbon quantum dots-titanium dioxide photocatalyst through self-assembly. The membrane degraded 94.41% of methylene blue dye within 1 hour of visible-light irradiation. A high flux recovery (83.92%) demonstrated the antifouling capabilities of the modified membrane. Another study chemically grafted conjugated polyelectrolytes (CPEs) on a polyvinylidene fluoride (PVDF) membrane surface to produce visible light-active highly stable hydrophilic photocatalytic membranes [59]. Using chemical tethering of CPE on the membrane and subsequently anion exchange, the hydrophilic character of the membrane was significantly improved without a reduction in porosity and water permeability. The photocatalytic membrane showed high visible light removal of azo dyes, photoreduction of Cr(VI), and photocatalytic inactivation of a biofilm. The anti-fouling property achieved more than 97 % flux recovery in repeat cycles. To simultaneously address the challenges associated with the separation of suspended photocatalyst and membrane fouling, a PVDF membrane coated with a catalytic layer of $ZnIn_2S_4$ was prepared [62].

The photodegradation performance for fluvastatin and TOC removal improved significantly relative to the pristine PVDF membrane. At the same time, the membrane showed better antifouling capabilities compared to the pristine PVDF membrane.

Numerous photocatalytic membrane fabrication strategies have exploited the photocatalytic properties of metal oxides, especially TiO_2. For example, a photoactive layer made up of TiO_2 was deposited onto the permeate side of an alumina membrane by dipping the membrane into a TiO_2 and smectite suspension [63]. Following thermal treatment, the coated hybrid was attached to the macroporous matrix and used for the photodegradation of methylene blue and phenol in aqueous solution. The hybrid membrane degraded 0.007 and 0.023 mmol/L of methylene blue and phenol, respectively. Another study prepared a Au-TiO_2 nanocomposite photocatalyst via sodium citrate reduction, sol-gel synthesis, and high-temperature hydrothermal reaction [64]. Thereafter, Au-TiO_2/PVDF visible light-active photocatalytic membrane was prepared via a phase conversion method, and the Au-TiO_2 nanocomposite and PVDF were physically blended into a casting solution. The photodegradation performance of the membrane attained 75% under 2 hours. A related study fabricated Ti^{3+}-doped TiO_2 photocatalytic membrane via a combination of hydrothermal synthesis and cryogenic vacuum-activation [65]. In addition to anti-fouling capabilities, the membrane showed excellent tetracycline degradation under visible light. Owing to its branched structure, the photocatalytic membrane also exhibited high emulsion separation. A PVDF membrane modified with TiO_2 was used to treat dye effluent [66]. A 70% Rhodamine B dye degradation, which remained constant for a long time, was achieved after 5 hours 20 minutes in a continuous contactor. A similar study fabricated a composite membrane with antifouling, photocatalytic, and self-cleaning characteristics via surface coating with a nitrogen-doped carbon quantum dots modified TiO_2 photocatalyst using PAA as a linker [67]. The membrane had high resistance to methylene blue fouling and recovered 93.7% of original water flux upon 15 min of visible-light irradiation. Moreover, the membrane had a high (90.9%) methylene blue retention.

For efficient photodegradation performance, the immobilized photocatalytic materials on a membrane substrate must have a large surface area, which should be completely irradiated and in good contact with the polluted wastewater, otherwise, the mass transport of the pollutant to the surface of the photocatalyst will limit the reaction kinetics [8]. In this regard, an alternative could be a flow-through photocatalytic membrane contactor, which is made from a thin coating of nanoporous photocatalyst on the inside wall of a porous membrane. Such a structure enjoys an intrinsic large aspect ratio, which increases the reaction kinetics relative to the mass-transport-limited bulk contactor [65]. Moreover, it provides a uniform contact time that sustains the output at the desired degradation rate. In addition, the passing wastewater refreshes the active surface and transports the degradates further down the system and enhancing contactor stability [8, 56]. Factors that affect the performance of the system include reactor configuration, physico-chemical properties of the photocatalyst, operating conditions, and the characteristics of the membrane matrix [8].

Blending or coating nanoparticles directly onto the membrane substrate however, compromises the stability of the system owing to agglomeration of the nanoparticles, detachment, membrane damage, and pore blockage [8, 65]. Moreover, since the photocatalytic process is not specific, the photocatalyst may degrade polymeric membranes. Thus, despite their high cost, ceramic membranes might be an attractive alternative [8].

5.5 Conclusion and Future Perspectives

This chapter has demonstrated the effectiveness of AOPs in reclaiming wastewater. However, the technology has still not been taken up by the wider wastewater treatment industry. This could be ascribed to lack of convincing experimental data and information on the cost-effectiveness of the technology. Moreover, recovering photocatalysts from treated wastewater for multiple uses is a challenge. Therefore, design of systems that permit easy photocatalyst recovery, are cost-effective, and stable could be an attractive strategy. For instance, the use of metal oxide activators, carbon-rich materials, and integrating AOPs with other wastewater treatment technologies should be considered. Further, combining empirical studies with theoretical modelling could be useful in the elucidation of the degradation mechanisms. The production of potentially toxic byproducts and the design of a suitable photocatalytic system is complex and thus a limitation.

The key challenge with photocatalysts is the recombination of the separated electrons. However, this can be avoided by using a porous substrate, such as zeolites or activated carbon. Operational conditions influence the rate of mass transport and thus govern the degradation of pollutants in wastewater. Besides, agglomeration of the nanoparticles at high concentrations decreases the number of reactive sites and their transparency, thus limiting the penetration of radiation and mass transport within the agglomerates.

Several treatment technologies cannot effectively remove emerging pollutants from wastewater. Lately, it has been demonstrated that SECO can degrade emerging pollutants, although the technology is still developing. Further research thus needs to explore the up-scaling of such technologies and decrease their cost to facilitate uptake by industry. A particularly interesting approach is to combine AOPs with membranes. In this regard, ceramic membranes are a good candidate due to their mechanical and chemical stability. The major disadvantage, however, is the cost associated with such systems.

References

1. Dewil, R., Mantzavinos, D., Poulios, I., Rodrigo, M.A., New perspectives for advanced oxidation processes. *J. Environ. Manage.*, 195, 93–99, 2017.
2. Miklos, D.B., Remy, C., Jekel, M., Linden, K.G., Drewes, J.E., Hübner, U., Evaluation of advanced oxidation processes for water and wastewater treatment—A critical review. *Water Res.*, 139, 118–131, 2018.
3. Oller, I., Malato, S., Sánchez-pérez, J.A., Combination of advanced oxidation processes and biological treatments for wastewater decontamination — A review. *Sci. Total Environ.*, 409, 20, 4141–4166, 2011.
4. Bhatnagar, A. and Sillanpää, M., Removal of natural organic matter (NOM) and its constituents from water by adsorption – A review. *Chemosphere*, 166, 497–510, 2017.
5. Sharma, A., Ahmad, J., Flora, S.J.S., Application of advanced oxidation processes and toxicity assessment of transformation products. *Environ. Res.*, 167, 223–233, 2018.
6. Li, C., Sun, Z., Ma, R., Xue, Y., Zheng, S., Fluorine doped anatase TiO_2 with exposed reactive (001) facets supported on porous diatomite for enhanced visible-light photocatalytic activity. *Microporous Mesoporous Mater.*, 243, 281–290, 2017.
7. Todorova, N., Giannakopoulou, T., Vaimakis, T., Trapalis, C., Structure tailoring of fluorine-doped TiO2 nanostructured powders. *Mater. Sci. Eng. B*, 152, 5054, 2008.
8. Berger, T.E., Regmi, C., Schäfer, A.I., Richards, B.S., Photocatalytic degradation of organic dye via atomic layer deposited TiO_2 on ceramic membranes in single-pass flow-through operation. *J. Membr. Sci.*, 604, 118015, 2020.

9. Horne, G.P., Zalupski, P.R., Daubaras, D.L., Rae, C., Mezyk, S.P., Mincher, B.J., Radiolytic degradation of formic acid and formate in aqueous solution: Modeling the final stages of organic mineralization under advanced oxidation process conditions. *Water Res.*, 186, 116314, 2020.
10. Duta, A. and Visa, M., Simultaneous removal of two industrial dyes by adsorption and photocatalysis on a fly-ash–TiO_2 composite. *J. Photochem. Photobiol. A: Chem.*, 306, 21–30, 2015.
11. Deeudomwongsa, P., Phattarapattamawong, S., Andrew Lin, K.Y., Control of disinfection byproducts (DBPs) by ozonation and peroxone process: Role of chloride on removal of DBP precursors. *Chemosphere*, 184, 1215–1222, 2017.
12. Von Gunten, U., Ozonation of drinking water: Part I. Oxidation kinetics and product formation. *Water Res.*, 37, 1443–1467, 2003.
13. Zhong, X., Cui, C., Yu, S., Seasonal evaluation of disinfection by-products throughout two full-scale drinking water treatment plants'. *Chemosphere*, 179, 290–297, 2017.
14. Huang, W., Wang, L., Zhou, W., Lv, W., Hu, M., Chu, H., Dong, B., Effects of combined ozone and PAC pretreatment on ultrafiltration membrane fouling control and mechanisms. *J. Membr. Sci.*, 533, 378–389, 2017.
15. Gassie, L.W. and Englehardt, J.D., Advanced oxidation and disinfection processes for onsite net-zero grey water reuse: A review. *Water Res.*, 125, 384–399, 2017.
16. Qi, W., Zhang, H., Hu, C., Liu, H., Qu, J., Effect of ozonation on the characteristics of effluent organic matter fractions and subsequent associations with disinfection by-products formation. *Sci. Total Environ.*, 610–611, 1057–1064, 2018.
17. Fallis, A., Ozonation of water and wastewater. *J. Chem. Inf. Model.*, 53, 9, 1689–1699, 2013.
18. Saharan, V.K., Pandit, A.B., Kumar, P.S.S., Anandan, S., Hydrodynamic cavitation as an advanced oxidation technique for the degradation of acid red 88 dye. *Ind. Eng. Chem. Res.*, 51, 4, 1981–1989, 2012.
19. Oturan, M.A., Electrochemical advanced oxidation technologies for removal of organic pollutants from water. *Environ. Sci. Pollut. Res.*, 21, 333–8335, 2014.
20. Thanekar, P. and Gogate, Application of hydrodynamic cavitation reactors for treatment of wastewater containing organic pollutants: Intensification using hybrid approaches. *Fluids*, 3, 98–122, 2018.
21. Tao, Y., Cai, J., Huai, X., Guo, Z., Liu, B., Application of hydrodynamic cavitation to wastewater treatment. *Chem. Eng. Technol.*, 39, 8, 1363–1376, 2016.
22. Cai, M., Su, J., Zhu, Y., Wei, X., Jin, M., Zhang, H., Dong, C., Wei, Z., Decolorization of azo dyes Orange G using hydrodynamic cavitation coupled with heterogeneous Fenton process. *Ultrason. Sonochem.*, 28, 302–310, 2016.

23. Cako, E., Gunasekaran, K.D., Soltani, R.D.C., Boczkaj, G., Ultrafast degradation of brilliant cresyl blue under hydrodynamic cavitation based advanced oxidation processes (AOPs). *Water Resour. Ind.*, 24, 100134, 2020.
24. Gągol, M., Cako, E., Fedorov, K., Soltani, R.D.C., Przyjazny, A., Boczkaj, G., Hydrodynamic cavitation based advanced oxidation processes: Studies on specific effects of inorganic acids on the degradation effectiveness of organic pollutants. *J. Mol. Liq.*, 307, 113002, 2020, https://doi.org/10.1016/j.molliq.2020.113002.
25. Abramov, V., Abramova, A., Cravotto, G., Nikonov, R., Fedulov, I., Ivanov, V., Flow-mode water treatment under simultaneous hydrodynamic cavitation and plasma. *Ultrason. Sonochem.*, 70, 105323, 2020. 10.1016/j.ultsonch.2020.105323
26. Wu, H., Sun, J., Qi, D., Zhou, C., Yang, H., Photocatalytic removal of elemental mercury from flue gas using Multiwalled carbon nanotubes impregnated with titanium dioxide. *Fuel*, 230, 218225, 2018.
27. Pollet, B.G. and Ashokkumar, M., *Introduction to ultrasound, sonochemistry and sonoelectrochemistry*, 1 ed., Springer Cham Publishing, Springer Nature Switzerland, 2019.
28. Mondal, J., Li, W., Rezk, A.R., Yeo, L.Y., Lakkaraju, R., Ghosh, P., Ashokkumar, M., Acoustic cavitation at low gas pressures in PZT-based ultrasonic systems. *Ultrason. Sonochem.*, 73, 105493, 2021.
29. Perelshtein, I., Perkas, N., Gedanken, A., Chapter 3 - Making the hospital a safer place by the sonochemical coating of textiles by antibacterial nanoparticles, in: *Surface Chemistry of Nanobiomaterials*, A.M. Grumezescu, (Ed.), William Andrew Publishing, 2016.
30. Joseph, C.G., Puma, G.L., Bono, A., Krishnaiah, D., Sonophotocatalysis in advanced oxidation process: A short review. *Ultrason. Sonochem.*, 16, 583–589, 2009.
31. Spahr, S., Cirpka, O.A., Von Gunten, U., Hofstetter, T.B., Formation of N-Nitrosodimethylamine during chloramination of secondary and tertiary amines: Role of molecular oxygen and radical intermediates. *Environ. Sci. Technol.*, 51, 280–290, 2017.
32. Madhavan, J., Theerthagiri, J., Balaji, D., Sunitha, S., Choi, M.Y., Ashokkumar, M., Hybrid advanced oxidation processes involving ultrasound: An overview. *Molecules*, 24, 18, 3341–3359, 2019.
33. Yang, B., Zuo, J., Tang, X., Effective ultrasound electrochemical degradation of methylene blue wastewater using a nanocoated electrode. *Ultrason. Sonochem.*, 21, 1310–1317, 2014.
34. Cotillas, S., Llanos, J., Cañizares, P., Mateo, S., Rodrigo, M.A., Optimization of an integrated electrodisinfection/electrocoagulation process with Al bipolar electrodes for urban wastewater reclamation. *Water Res.*, 47, 5, 1741–50, 2013.

35. Cotillas, S., Llanos, J., Castro-Ríos, K., Taborda-Ocampo, G., Rodrigo, M.A., Cañizares, P., Synergistic integration of sonochemical and electrochemical disinfection with DSA anodes. *Chemosphere*, 163, 562–568, 2016.
36. Dietrich, M., Marcus, F., Stelter, M., Braeutigam, P., Degradation of endocrine disruptor bisphenol a by ultrasound-assisted electrochemical oxidation in water. *Ultrason. Sonochem.*, 39, 741–749, 2017.
37. Shestakova, M., Graves., J., Sitarz, M., Mika Sillanpa, M., Optimization of Ti/ Ta_2O_5–SnO_2 electrodes and reaction parameters for electrocatalytic oxidation of methylene blue. *J. Appl. Electrochem.*, 46, 3, 349–58, 2016.
38. Steter, J.R., Barros, W.R.P., Lanza, M.R.V., A.J., Electrochemical and sonoelectrochemical processes applied to amaranth dye degradation. *Chemosphere*, 117, 200–207, 2014.
39. Frontistis, Z., Sonoelectrochemical degradation of propyl paraben: An examination of the synergy in different water matrices. *Int. J. Environ. Res. Public Health*, 17, 2621, 2020.
40. Chatel, G., Ultrasound in combination with ionic liquids: Studied applications and perspectives, in: *Sonochemistry: From Basic Principles to Innovative Applications*, J.C. Colmenares, G. Chatel (Eds.), pp. 203–23, Springer International Publishing, Cham, 2017.
41. Gogate, P.R. and Pandit, A.B., Sonophotocatalytic reactors for wastewater treatment: A critical review. *AIChE*, 50, 5, 1051–1079, 2004.
42. Yap, H.C., Pang, Y.L., Lim, A comprehensive review on state-of-the-art photo-, sono-, and sonophotocatalytic treatments to degrade emerging contaminants. *Int. J. Environ. Sci. Technol.*, 16, 601–628, 2019, https://doi.org/10.1007/s13762-018-1961-y.
43. Silva, A.M.T., Nouli, E., Carmo-Apolinário, Â.C., Xekoukoulotakis, N.P., Mantzavinos, D., Sonophotocatalytic/H2O2 degradation of phenolic compounds in agro-industrial effluents. *Catal. Today*, 124, 3, 232–239, 2007.
44. Verma, A., Chhikara, I., Dixit, D., Photocatalytic treatment of pharmaceutical industry wastewater over TiO_2 using immersion well reactor: Synergistic effect coupling with ultrasound. *Desalin. Water Treat.*, 52, 34–36, 6591–6597, 2014.
45. An, T., Gu, H., Xiong, Y., Chen, W., Xihai Zhu, X., Sheng, G., Fu, J., Decolourization and COD removal from reactive dye-containing wastewater using sonophotocatalytic technology. *J. Chem. Technol. Biotechnol.*, 78, 1142–1148, 2003.
46. Joseph, C.G., Liew, Y.L.S., Krishnaiah, D., Bono, A., Application of a semiconductor oxide-based catalyst in heterogenous wastewater treatment: A Green technology approach. *J. Appl. Sci.*, 12, 18, 1996–1971, 2012.
47. Amirian, P., Bazrafshan, E., Payandeh, A., Optimisation of chemical oxygen demand removal from landfill leachate by sonocatalytic degradation in the presence of cupric oxide nanoparticles. *Waste Manag. Res.: J. Sust. Circ. Econ.*, 35, 6, 636–646, 2017.

48. Mahajan, V.K. and Sonwane, G.H., Effective degradation and mineralization of real textile effluent by sonolysis, photocatalysis, and sonophotocatalysis using ZnO Nano catalyst. *Nanochem. Res.*, 1, 2, 58–263, 2016.
49. Vigneshwaran, S., Jun, B., Prabhu, S.B., Elanchezhiyan, S.S.D., Ok, Y.S., Meenakshi, S., Park, C.M., Enhanced sonophotocatalytic degradation of bisphenol a using bimetal sulfide-intercalated MXenes, *2D/2D Nanocomposite, Sep. Purif. Technol.*, 250, 117178, 2020.
50. Ahmad, M., Ahmed, E., Hong, Z.L., Waqar, A., Adbelbary, E., Khalid, N.R., Photocatalytic, sonocatalytic and sonophotocatalytic degradation of Rhodamine B using ZnO/CNTs composites photocatalysts. *Ultrason. Sonochem.*, 21, 2, 761–773, 2014.
51. Paramarta, V. and Saleh, R., Wastewater treatment by sonocatalysis and sonophotocatalysis using Fe3O4/SnO2 composite supported on nanographene platelets. *AIP Conf. Proc.*, 2023, 020013, 2018, https://doi.org/10.1063/1.5064010.
52. Panchangam, S.C., Lin, A.Y., Shaik, K.L., Lin, C.F., Decomposition of perfluorocarboxylic acids (PFCAs) by heterogeneous photocatalysis in acidic aqueous medium. *Chemosphere*, 77, 2, 242–24, 2009.
53. Dogruel, S., Olmex-Hanci, T., Kartal, Z., Arslan-Alaton, I., Orhon, D., Effect of Fenton's oxidation on the particle size distribution of organic carbon in olive mill wastewater. *Water Res.*, 43, 3974–3983, 2009.
54. Sathishkumar, P., Mangaralaja, R.V., Mansilla, H.D., Gracia-Pinilla, M.A., Anandan, S., Sonophotocatalytic (42kHz) degradation of Simazine in the presence of Au–TiO2 nanocatalysts. *Appl. Catal. B: Environ.*, 160–161, 692–700, 2014.
55. Sajjadi, S., Khataee, A., Bagheri, N., Kobya, M., Senocak, A., Demirbas, E., Karaoglu, A.G., Degradation of diazinon pesticide using catalyzed persulfate with Fe_3O_4@MOF-2 nanocomposite under ultrasound irradiation. *J. Ind. Eng. Chem.*, 77, 280–290, 2019.
56. Nasir, A.M., Awang, N., Jaafar, J., Ismail, A.F., Othman, M.H.D., Rahman, M.A., Aziz, F., Yajid, M.A.M., Recent progress on fabrication and application of electrospun nanofibrous photocatalytic membranes for wastewater treatment: A review. *J. Water Process Eng.*, 40, 101878, 2021.
57. Heng, Z.W., Chong, W.C., Pang, Y.L., Koo, C.H., Self-assembling of NCQDs-TiO_2 nanocomposite on poly(acrylic acid)-grafted polyethersulfone membrane for photocatalytic removal and membrane filtration. *Mater. Today: Proc.*, 46, 1901–1907, 2021a, https://doi.org/10.1016/j.matpr.2021.02.021.
58. Nasrollahi, N., Ghalamchi, L., Vatanpour, V., Khataee, A., Photocatalytic-membrane technology: A critical review for membrane fouling mitigation. *J. Ind. Eng. Chem.*, 93, 101–116, 2021.
59. Jeong, E., Byun, J., Bayarkhuu, B., Hong, S.W., Hydrophilic photocatalytic membrane via grafting conjugated polyelectrolyte for visible-light-driven biofouling control. *Appl. Catal. B: Environ.*, 282, 119587, 2021.

60. Zhang, Y., Li, J., Shen, Y., Xu, O., Wan, S., Zhu, X., Construction of stable Ti^{3+}-TiO_2 photocatalytic membrane for enhanced photoactivity and emulsion separation. *J. Membr. Sci.*, 618, 118748, 2021a.
61. Domenzain-Gonzalez, J., Castro-Arellano, J.J., Galicia-Luna, L.A., Rodriguez-Cruz, M., Hernandez-Lopez, R.T., Lartundo-Rojas, L., Photocatalytic membrane reactor based on Mexican Natural Zeolite: RB5 dye removal by photo-Fenton process. *J. Environ. Chem. Eng.*, 9, 105281, 2021.
62. Liu, T., Wang, L., Liu, X., Sun, C., Lv, Y., Miao, R., Wang, X., Dynamic photocatalytic membrane coated with $ZnIn_2S_4$ for enhanced photocatalytic performance and antifouling property. *Chem. Eng. J.*, 379, 122379, 2020.
63. Deepracha, S., Atfane, L., Ayral, A., Ogawa, M., Simple and efficient method for functionalizing photocatalytic ceramic membranes and assessment of its applicability for wastewater treatment in up-scalable membrane reactors. *Sep. Purif. Technol.*, 262, 118307, 2021.
64. Yan, M., Wu, Y., Liu, X., Photocatalytic nanocomposite membranes for high-efficiency degradation of tetracycline under visible light: An imitated core-shell. Au-TiO_2-based design. *J. Alloys Compd.*, 855, 157548, 2021.
65. Zhang, M., Yang, Y., An, X., Hou, L., A critical review of g-C_3N_4-based photocatalytic membrane for water purification. *Chem. Eng. J.*, 412, 128663, 2021b.
66. Vatanpour, V., Darrudi, N., Sheydaei, M., A comprehensive investigation of effective parameters in continuous submerged photocatalytic membrane reactors by RSM. *Chem. Eng. Process.: Process Intensif.*, 157, 108144, 2020.
67. Heng, Z.W., Chong, W.C., Pang, Y.L., Sim, L.C., Koo, C.H., Novel visible-light responsive NCQDs-TiO2/PAA/PES photocatalytic membrane with enhanced antifouling properties and self-cleaning performance. *J. Environ. Chem. Eng.*, 9, 105388, 2021b.
68. Zhang, Y. and Liu, L., Fly ash-based geopolymer as a novel photocatalyst for degradation of dye from wastewater. *Particuology*, 11, 353–358, 2013.
69. Tran, N., Drogui, P., Brar, S.K., De Coninck, A., Synergistic effects of ultrasounds in the sonoelectrochemical oxidation of pharmaceutical carbamazepine pollutant. *Ultrason Sonochem.*, 34, 380–388, 2017.
70. Silva, J., Neto, N.A., Oliveira, M., Ribeiro, R., De Lazaro, S., Gomes, Y., Paskocimas, C., Bomio, M., Motta, F.J.N.J.O.C., Recent progress and approaches on the synthesis of Mn-doped zinc oxide nanoparticles: a theoretical and experimental investigation on the photocatalytic performance. *New J. Chem.*, 44, 8805–8812, 2020.
71. Chen, W.S. and Huang, Y.L., Removal of dinitrotoluenes and trinitrotoluene from industrial wastewater by ultrasound enhanced with titanium dioxide. *Ultrason Sonochem.*, 18, 5, 1232–1240, 2011.

6

Advanced Oxidation Processes-Mediated Removal of Aqueous Ammonia Nitrogen in Wastewater

Mohammad Aslam[1], Ahmad Zuhairi Abdullah[1], Mukhtar Ahmed[1] and Mohd. Rafatullah[2,3]*

[1]*School of Chemical Engineering, Universiti Sains Malaysia, Nebong Tebal, Penang, Malaysia*
[2]*Environmental Technology Division, School of Industrial Technology, Universiti Sains Malaysia, Penang, Malaysia*
[3]*Renewable Biomass Transformation Cluster, School of Industrial Technology, Universiti Sains Malaysia, Penang, Malaysia*

Abstract

Ammonia nitrogen (NH_3/NH_4^+) polluted wastewater raises a serious threat to the safety of aquatic system. In addition to eutrophication, the presence of NH_3/NH_4^+ in the aquatic body reduces chlorine disinfectant efficiency during water treatment. Concerns regarding other conventional NH_3-N removal processes, to tackle these contaminants, advanced oxidation processes (AOPs) have been applied to remove ammonia nitrogen. The AOPs utilizes hydroxyl radical (•OH) for oxidation and have received considerable attention during the last few years in wastewater treatment, research technologies, and development. These processes are used to gradually reduce ammonia nitrogen to innocuous products with the help of high reactivity of hydroxyl radicals. In this chapter, a systematic study is carried out with the focus on the effects of ammonia nitrogen on the growth, physiology, biochemistry, and immune response of aquatic species. Moreover, experimentations and working procedures that can be used to remove ammonia nitrogen by AOP application are been discussed. This chapter also reviews recent findings and observations for the removal of ammonia nitrogen by photocatalysis and ozonation techniques and provide some recommendations for future research works.

**Corresponding author*: mohd_rafatullah@yahoo.co.in; mrafatullah@usm.my

Elvis Fosso-Kankeu, Sadanand Pandey, and Suprakas Sinha Ray (eds.) Photoreactors in Advanced Oxidation Processes: The Future of Wastewater Treatment, (175–214) © 2023 Scrivener Publishing LLC

Keywords: Ammonia nitrogen (NH_3/NH_4^+), advanced oxidation processes, hydroxyl radicals, wastewater treatment, aquatic species, photocatalysis, ozonation

Abbreviations

ACH-50	Alternative complement pathway activities
ACTH	Adrenocorticotropic hormone
AO	Aldehyde oxidase
BAF	B-Cell Activating Factor
C3	Complement 3
C4	Complement 4
CAD	Catalase activities declined
CAT	Catalase
CaM	Calmodulin
cAMP	c adenosine monophosphate receptor
c-GT	c-glutamyl transpeptidase
CEA	Cellular energy allocation
CHH	Crustacean hyperglycaemic hormone
CRH	Corticotropin releasing hormone
CREB	CAMP response element-binding protein
D4	DA receptor 4
GH	Growth hormone
GOT	Glutamic oxalate transaminase
GPT	Glutamic pyruvate transaminase
GPX	Glutathione peroxidase
GRA	Glutathione reductase activities
GSase	Glutamine synthetase
GSH	Glutathione
GST	Glutathione s-transferase
HCO_3	Blood bicarbonate
HSP-70	Heat shock protein 70
HSP-90	Heat shock protein 90
IDH	Isocitrate dehydrogenase
IGF-1	Insulin-like growth factor-1
IgM	Immunoglobulin M
LZM	Lysozyme
MDA	Malondialdehyde
MTP	Mitochondrial transmembrane potential

NF-ҜB	Nuclear factor kappa B
NKCC1	Na+: K+:2Cl$^-$ cotransporter 1
NO	Nitrogen monoxide
P53	Tumor suppressor gene
PCO	Protein carbonyl group
PCO_2	Partial pressure of CO_2
PKA	Protein kinase A
PKC	Protein kinase C
PKG	Protein kinase G
PO	Phenoloxidase
RBC	Red Blood Cell
ROS	Reactive oxygen species
SGR	Specific growth rates
SOD	Superoxide dismutase
TBARS	Thiobarbituric acid reactive substances
THC	Total hemocyte count
TLR4	Toll likes receptor 4
TNF-a	Pleiotropic pro-inflammatory cytokine
WBC	White blood cell
XO	Xanthine oxidase
γ-GCS	γ-glutamyl cysteinyl synthetase
5-HT	5-hydroxytryptamine
5-HT7	5-Hydroxytryptamine 7

6.1 Introduction

Inorganic contaminants, such as aqueous ammonia present, are a serious problem to the environment, since they are harmful to numerous aquatic species, can lead to eutrophication, and can also contribute to numerous environmental impacts. Ammonia is found in varying concentrations in aquatic ecosystems. The residence time of ammonia in the atmosphere is estimated to be about 30 days. Ammonia is readily oxidized to nitrite and nitrate by bacteria and thus consumes dissolved oxygen in water [1]. The overall concentration of ammonia in water is the sum of NH_4^+ (at low pH) and NH_3 (at high pH). Furthermore, ammonia is poisonous to most species of aquatic life, as it blocks the flow of oxygen to the fish gills [2]. The presence of ammonia decreases the effectiveness of existing purification methods, such as chlorination because ammonia utilizes most of the chlorine. Moreover, ammonia is an important source of unwanted odor in sewage and wastewater. Therefore, the removal of ammonia in wastewater is essential to protect the environment before it is discharged into water systems.

Conventional methods used to degrade ammonia usually involve biological nitrification, air stripping, ion exchange, break-point chlorination and chemical precipitation [3, 4]. However, these methods have their own disadvantages and limitations. For instance, break point chlorination can leave chlorine in the treated water causing disinfection by-products, which requires further separation of chlorine while biological nitrification is pH- and temperature-sensitive [2]. In case of air-stripping method, when ammonia is transferred from liquid to gas phase, it generates air pollution [5]. Therefore, new techniques for the degradation of aqueous ammonia have been explored for enhanced the degradation efficiency. In this respect, advanced oxidation processes (AOPs) among the methods that can overcome the disadvantages of the conventional processes.

AOPs have attracted significant consideration in the recent decades in the research and development of wastewater treatment technologies. These processes have been successfully used to remove or degrade emerging contaminants or to convert pollutants into short-chain compounds, which can be further treated using conventional or biological processes [6]. AOPs are widely defined as aqueous phase oxidation methods that are based on the mechanism for the destruction of organic contaminants from intermediate levels of highly reactive species such as hydroxyl radicals (primarily, but not exclusively). In the past, three decades, R&D on the AOPs has been immense for two main reasons (i) technological diversity involved and (ii) potential application in various areas. AOPs include heterogeneous and homogeneous photocatalysis based on ultraviolet (UV) or solar visible irradiation, photo-Fenton ($Fe_2^+/H_2O_2/UV$), ozonation, catalytic

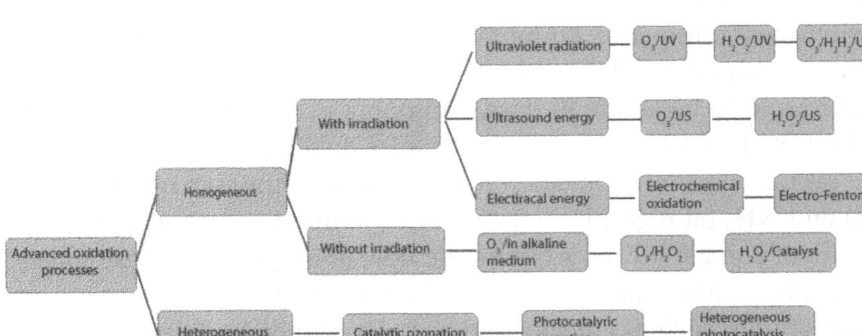

Figure 6.1 Advanced oxidation processes for wastewater treatment.

ozonation, UV/O_3, H_2O_2/UV, $O_3/H_2O_2/UV$. A list of the multiple possibilities delivered by AOP is shown in Figure 6.1.

6.2 Basic Chemistry and Occurrence of Ammonia Nitrogen

6.2.1 Basic Chemistry of Ammonia Nitrogen

Total nitrogen is made up of organic nitrogen (amino sugars, amino acids, or protein), nitrate, nitrite, and aqueous ammonia. Aqueous ammonia (NH_3/NH_4^+) has two basic forms, i.e., un-ionized from (ammonia: NH_3) and protonated form (ammonia: NH_4^+) [7]. The concentrations of both unionized and protonated are collectively known as ammonia nitrogen and commonly expressed as "total ammonia nitrogen." The occurrence of both the form changes, depending on pH and temperature of the solution as illustrated by equation (6.1).

$$NH_3 + H_2O \leftrightarrow NH_4^+ + OH^- \tag{6.1}$$

The equilibrium constants K_a and K_b of above equation are in accordance with equations (6.2) and (6.3) [8].

$$K_a = \frac{[NH_3][H^+]}{[NH_4^+]} = 5.6 \times 10^{-10} \tag{6.2}$$

$$K_b = \frac{[NH_4^+][OH^-]}{[NH_3]} = 1.8 \times 10^{-5} \tag{6.3}$$

At a pH value of 9.25, ammonia and ammonium are in equilibrium. Thus, NH_3 shows dominance toward total ammonia when the pH is higher than 9.25, while NH_4 is the dominant form of total ammonia when the pH is below 9.25 [9].

6.2.2 Sources of Ammonia Nitrogen

The main source of nitrogen nutrients in water is synthetic chemical fertilizers. Most of the nitrogen compounds that are not used in crops are mainly drained by land drainage and surface rivers into water bodies [10].

Moreover, point and nonpoint sources of pollution can also lead to the ammonia concentration in the environment. Ammonia is released into the atmosphere from a variety of wastewater effluents (aquaculture, industrial, municipal and agricultural) as a point source. Runoffs from industrial and agricultural practices include nonpoint sources of ammonia. The severity of pollution impact from these nonpoint sources is affected by the rainfall, land use, vegetation and various human activities [11].

Ammonia, nitrite, and nitrate are the basic forms of the dissolved inorganic nitrogen ions that are important in aquatic life, even though ammonia could also be the major pollutant [12]. Ammonia, as a nutrient salt, is commonly present in seawater and is a significant indicator for measuring environmental pollution [13]. Ammonia in seawater is an integral part of the oceanic nitrogen cycle and a major source for marine phytoplankton. Thus, in natural seawater, the ingestion of ammonium ions through phytoplankton leads to a level of nanomolar ammonia in seawater [14].

In addition, ammonia could also be produced from aquaculture activities by various sources, such as unconsumed feed, excreta, dead aquatic species, culture density, algae, nitrogen-containing exogenous substances, and interchange of ammonia among both aquatic species and the adjacent water. This leads to an accumulation of large amount of ammonia in aquaculture fauna [15]. Mostly, ammonia is absorbed from the body or hemolymph due to the concentration gradients into surrounding environment in shrimp [16]. Therefore, ammonia is the key metabolite in an aquatic environment that usually contains nitrate-containing compounds and a major inorganic pollutant. Moreover, it is a significant environmental factor, which represents survival status of aquaculture species by monitoring or evaluating water quality [17].

6.2.3 Effects of Ammonia Nitrogen on Aquaculture Species

It is reported that when ammonia range surpasses the exposure limit of the aquatic animal, tissues including gills and hepatopancreas can be directly damaged to have an impact on breathing, metabolism, immunity, osmotic regulation, discharge, moulting, and growth [18, 19]. Moreover, ammonia is one of the major water contaminants in shrimp aquaculture and the main limiting factor which ultimately causes rapid growth in shrimp death and consequently leads to considerable economic losses for farmers [20]. In addition to its immediate impacts on several tissues and organs, the toxicity of ammonia also gives rise to inflamation and continues to increase the level of free reactive oxygen species (ROS) in shrimps [21]. Nonionic ammonia can freely enter the shrimp body through cell membranes when

Table 6.1 Effect of aqueous ammonia nitrogen on aquaculture species.

Species	Ammonia concentration range (mg L^{-1})	Size/life stage	Increased factor	Decreased factor	Tissues	Ref.
Ctenopharyngodon idellus	0.5–18.0	Juveline	Antioxidative enzymes and antioxidants	CAT	Liver, gills and muscle	[31]
Hippoglossus hippoglossus	0.06–0.17	Juveline	pH, HCO_3^-	PCO_2	Blood	[32]
Hypophthalmythys nobilis	0.06–0.264	Larvae	SGR, weight, GSH	SOD	Whole-body homogenates	[33]
Megalobrama amblycephala	10	13.80.04 g	ACH50, NO, SOD, CAT, HSP70, HSP90	Cortisol, glucose, MDA	Plasma, liver	[34]
Monopterus albus	50 mmol L^{-1}	150–250 g	-	NKCC1	Brain	[35]
Monopterus cuchia	50 Mm	150–170 g	MDA, H_2O_2, HSP70, HSP90	-	Plasma	[36]
Oreochromis niloticus	5–10	Juveline	PCO, TBARS, XO, AO, CAT, γ-GT and γ-GCS	-	Liver and white muscle	[37]

(Continued)

Table 6.1 Effect of aqueous ammonia nitrogen on aquaculture species. (*Continued*)

Species	Ammonia concentration range (mg L^{-1})	Size/life stage	Increased factor	Decreased factor	Tissues	Ref.
Pelteobagrus fulvidraco	5.70	Juveline	Glutamine and TBARS,	SOD, GPX, GRA,	Brain	[38]
Scophthalmus maximus	20–40	Juveline	CRH, ACTH, SOD, CAT, HSP70, HSP90, MDA	GH, LZM, C3, C4, IgM, GSH, IGF-1	Plasma, liver	[39]
Sebastes schlegelii	0.1–1.0	38.36 ± 3.45 g	Glucose, GOT, GPT	RBC, WBC and Total protein	Serum	[40]
Sebastes schlegelii	0.1–1.0	38.36 ± 3.45 g	SOD, CAT, GST, Cortisol, HSP70	GSH, phagocytosis, lysozyme activity	Liver, plasma	[41]
Takifugu obscurus	1.43–7.13 Mm	25.5 ± 1.8 g	ROS, BAF, TNF-a, HSP90, HSP70, CAT, P53	-	Liver, blood	[42]

(*Continued*)

Table 6.1 Effect of aqueous ammonia nitrogen on aquaculture species. (*Continued*)

Species	Ammonia concentration range (mg L^{-1})	Size/life stage	Increased factor	Decreased factor	Tissues	Ref.
Portunus trituberculatus	1.0–20.0	104.8 9.6 g	-	Phagocytic and antibacterial activity, THC, α_2-M	Hemolymph	[43]
Scylla serrata	10–140	Juveline	pH	Na^+, Ca^{2+}	Hemolymph	[44]
Chlamys farreri	20.0	14.90 ± 1.36 g	IDH, HSP70, HSP90, GSase	CEA	Hemolymph, serum	[45]
Corbicula fluminea	10 and 25	1.5 ± 0.2 g	NF-$_\kappa$B, MAPK	TLR4	Digestive gland	[46]
Ruditapes philippinarum	0.1–0.5	3.5 ± 0.3 g	Apoptosis ratio	MTP, Ca^{2+}-ATPase, H^+-ATPase, K^+-ATPase	Gill, hemocytes	[18]

(*Continued*)

Table 6.1 Effect of aqueous ammonia nitrogen on aquaculture species. (*Continued*)

Species	Ammonia concentration range (mg L^{-1})	Size/life stage	Increased factor	Decreased factor	Tissues	Ref.
Penaeus monodon	0, 40, 80, 100,120, 140	21 ± 1 g	-	Chitinase, Chitinase-1, Chitinase-5	Hepatopancreas, gill	[47]
Penaeus monodon	78.15	8.2 ± 1.0 g	-	Chitinase	Hepatopancreas	[48]
Litopenaeus vannamei	32	Juveline	Calreticulin	Chitinase-5	Hepatopancreas	[49]
Litopenaeus vannamei	0, 5, 10, 15, 20, 25, 30, 35, 40	Postlarvae	Na/K+-ATPase, Na+/ NH_4 + transport	-	Gill	[50]
Portunus pelagicus	0, 20, 40, 60, 80, 100	Juveline	Na/K+-ATPase, Na+/ NH_4 + transport, K+.	-	Gill	[23]
Litopenaeus vannamei	0, 2.5, 5.0, 7.5, 10	63.54 mm	PO, antibacterial, bacteriolytic, glucose, lactate levels	Oxy-hemocyanin	Plasma, hemolymph	[51]

(*Continued*)

Table 6.1 Effect of aqueous ammonia nitrogen on aquaculture species. (*Continued*)

Species	Ammonia concentration range (mg L^{-1})	Size/life stage	Increased factor	Decreased factor	Tissues	Ref.
Litopenaeus vannamei	0.07, 2, 10, 20	5.5 ± 1.0 g	DA, 5-HT, guanylyl, 5-HT_7, cAMP, cGMP, CaM, cyclase, PKA, PKG, CREB, NF-$_{\kappa}$B and CHH	D_4, α_2 adrenergic receptor, PKC, THC, phagocytic, antibacterial, PO	Hemolymph	[52]
Palaemon serratus	10	Juveline	-	Oxygen consumption rate	-	[53]
Penaeus monodon	0, 10, 20, 30	Juveline	C-lysozyme, antibacterial peptide (crustin), antilipopolysaccharide factor	-	Hepatopancreas, muscle, gill	[54]

ammonia exceeds a certain mass concentration in aquaculture water, causing physiological imbalance and growth inhibition [22]. In shrimp and other crustaceans, harmful effects of ammonia have been observed, particularly on osmotic regulation [23] and metabolism [24]. For example, in *L. vannamei*, NH_4^+ influences ammonia-metabolic enzymatic activity and ammonia excretion [25].

Shrimp gills have direct exposure to external environments. Thus, environmental factors, such as ammonia stress, can directly influence them. Although many researchers reported that ammonia can damage crustacean tissue directly and causes anoxia, ammonia also reduces immunity, metabolic disturbance and increases their sensitivity to pathogens [26, 27]. Moreover, exposure of *L. vannamei* to high ammonia stress can cause major effect to the mucosal digestive tract, remove most of the epithelial cells, together with necrosis in the basement membrane [28]. It was observed that apoptosis in hepatopancreatic cells can be caused by high ammonia stress [29], whereas another recent study have shown that the expression of apoptotic genes in shrimp hemocytes is effected by the ammonia oxidative stress [30]. Effect of aqueous ammonia nitrogen on aquaculture species are summarized in Table 6.1. The oxidative effect of ammonia nitrogen on fruit cell is shown in Figure 6.2. In addition to the increased consumption

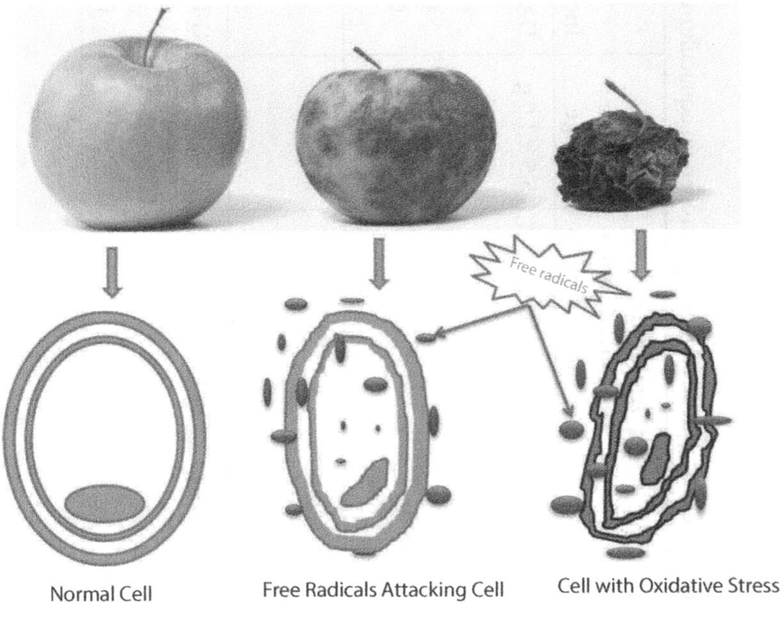

Figure 6.2 Effect of oxidative stress on fruit cell.

of oxygen by crustaceans, the massive amounts of nonionic ammonia in aquatic environment can also reduce oxy-hemocyanin concentrations in these organisms. It can also disrupt the excretory systems and osmotic balance [19].

6.3 Photocatalytic Technique for Removal of Aqueous Ammonia Nitrogen From Wastewater

The involvement of advanced oxidation processes (AOPs) for wastewater treatment has been enormously increasing in recent years [55, 56]. In addition to other advanced oxidation processes, photocatalytic water treatment is cheaper and environmentally friendly. This process is stable semiconducers (such as Titanium dioxide (TiO_2)) for natural light irradiation compared to other techniques e.g. Fenton and photo-Fenton etc. [57, 58]. These processes produce enough hydroxyl radicals (OH·) to oxidize contaminants and purify the water [59]. Reactive oxygen species (ROS) or free radicals are atoms or molecule that has single or more unpaired electrons that are capable of independent existence [60]. These free radicals includes superoxide radical (O_2·), hydroperoxyl radical (HO_2·), hydroxyl radical (OH·), and alkoxyl radical (RO·) and nonradicals (hydrogen peroxide (H_2O_2)), singlet oxygen ($^1O_2$), and hypochlorous acid (HOCl) [61]. The significant concentration of NH_4^+ in the aqueous system requires efficient removal methods caused by harmful effects on human beings and environment. Since existing methods (biological treatment and detreated towers) have limited ability to remove NH_4^+, AOPs are now seen as possible solutions to achieve such objectives [62, 63]. The photocatalytical technique can be seen as a viable and economical alternative as natural light sources can be used easily and it also avoid harmful reagents [64]. Table 6.2 provides a list of the photocatalytic systems applied to remove NH_4^+. TiO_2-based catalyst have commonly used for the removal of NH_4^+. However, Pristine TiO_2 catalyst has limited performance. Therefore, different types of TiO_2-based catalyst were explored for the degradation of NH_4^+. Various non-TiO_2 materials were also assessed for extending photocatalytic technology to renewable energy and addressing the shortcomings in TiO_2-based catalysts (e.g., lower quantum energy efficiency) as summarized in Table 6.2.

6.3.1 TiO_2/TiO_2-Based Photocatalyst

In the late 1990s, the viability of suspended TiO_2 NP's under UV and sun light had been investigated as the first detailed study on the photocatalytic

Table 6.2 Various photocatalytic systems for the remediation of ammonia nitrogen.

S. no.	TiO_2-based photocatalyst	Amount of catalyst (g L^{-1})	Photocatalytic reactor used	Light source	NH_3(aq) concentration (ppm), pH, irradiation time (min)	Removal efficiency (%)	Ref.
1.	TiO_2–P25	0.1	Batch photoreactor for cylindrical Pyrex (1.1 L) with an external cooling jacket	UVA lamp (25 W)	100, 10.5, 360	35	[67]
2.	TiO_2–P25	2	Batch photoreactor cylindrical Pyrex (containing 1.3 L of aqueous slurry)	Hg UV lamp (450 W)	1.7, 10.5, 350	100	[68]
3.	TiO_2–P25	0.0002	Parabolic trough reactor (8 L)	Natural solar light (730 W m^{-2})	8.5, 9.9, 360	40.5	[65]

(Continued)

Table 6.2 Various photocatalytic systems for the remediation of ammonia nitrogen. (*Continued*)

S. no.	TiO_2-based photocatalyst	Amount of catalyst (g L^{-1})	Photocatalytic reactor used	Light source	NH_3(aq) concentration (ppm), pH, irradiation time (min)	Removal efficiency (%)	Ref.
4.	TiO_2 thin film	NA	Photoreactor of borosilicate glass (0.5 L)	UV lamp (360 mW)	700, 7, 120	51	[69]
5.	0.467 wt% Pt–TiO_2	0.1	Batch photoreactor Cylindrical Pyrex (1.1 L)	UVA lamp (25 W)	100, 10–10.5, 360	51.4	[70]
6.	20 wt% Pt/ titanate nanotubes (TNTs)	0.5	Cylindrical quartz batch photoreactor (3 L)	Hg UV lamp (400 W)	20, 10, 180	100	[71]
7.	0.5 wt% Pt/TiO_2	0.5	Pyrex glass batch photoreactor	Hg UV lamp (450 W)	85, 12, 600	100	[72]

(*Continued*)

Table 6.2 Various photocatalytic systems for the remediation of ammonia nitrogen. (*Continued*)

S. no.	TiO_2-based photocatalyst	Amount of catalyst ($g\ L^{-1}$)	Photocatalytic reactor used	Light source	NH_3(aq) concentration (ppm), pH, irradiation time (min)	Removal efficiency (%)	Ref.
8.	Pt/TiO_2	0.5	A quartz window fitted with Pyrex Batch Photoreactor (0.2 L)	Xe-arc lamp (300 W)	1.7, 10, 40	80	[73]
9.	0.5 wt% Ni/TiO_2	4	Photoreactor quartz glass batch (0.008 L)	Xe lamp (500 W)	10030, NA, 180	NA	[74]
10.	1.2 wt % Ce/TiO_2	1	Stainless steel-based mixed batch annular photoreactor	Hg pen-ray lamp (8 W)	827, NA, 600	NA	[75]
11.	La/Fe/TiO_2	1	Photoreactor XPA-7 batch with external jacket cooling	Lamp of mercury (500 W; $\lambda = 365$ nm)	101, 10, 300	64.6	[76]

(*Continued*)

Table 6.2 Various photocatalytic systems for the remediation of ammonia nitrogen. (*Continued*)

S. no.	TiO_2-based photocatalyst	Amount of catalyst (g L^{-1})	Photocatalytic reactor used	Light source	NH_3(aq) concentration (ppm), pH, irradiation time (min)	Removal efficiency (%)	Ref.
12.	0.5 wt % Pt/TiO_2	4	Cylindrical batch photoreactor	Xe lamp (469–515 mW cm^{-2})	10030, 10.7, 360	NA	[77]
13.	3.2 wt.% of Pt/TiO_2 exposed on an acrylic-based material	2.45	Photoreactor for Continuous Flow (0.98 L)	UV lamp of Two Hg vapor (400 W each)	45, 10, 4320	100	[63]
14.	TiO_2 or light-expanded clay aggregate (LECA)	25	Photoreactor of cylindrical plastic batch (10 L)	UV lamp (125 W)	975, 11, 180	58.1	[78]

(*Continued*)

Table 6.2 Various photocatalytic systems for the remediation of ammonia nitrogen. (*Continued*)

S. no.	TiO_2-based photocatalyst	Amount of catalyst ($g\ L^{-1}$)	Photocatalytic reactor used	Light source	NH_3(aq) concentration (ppm), pH, irradiation time (min)	Removal efficiency (%)	Ref.
15.	TiO_2-ZnO/LECA	25	Photoreactor of Pyrex glass (1.5 L)	High pressure UV-C Hg lamp (125 W)	400, 11, 180	95.2	[79]
16.	TiO_2/perlite	11.7	Pyrex glass photoreactor batch (1.5 L) in a metal box	Hg UV lamp (125 W)	170, 11, 180	68	[80]
17.	TiO_2/LECA	40	Photoreactor of Pyrex glass batch (2.6 L)	Hg UV Lamp Medium Pressure (80 W)	170, 11, 300	85	[81]
18.	TiO_2–CuO/hemp stem biochar carbon (HSC)	0.6	Batch photoreactor	UV lamp (25 W)	100, 7, 120	99.7	[82]

(*Continued*)

Table 6.2 Various photocatalytic systems for the remediation of ammonia nitrogen. (*Continued*)

S. no.	TiO_2-based photocatalyst	Amount of catalyst (g L^{-1})	Photocatalytic reactor used	Light source	NH_3(aq) concentration (ppm), pH, irradiation time (min)	Removal efficiency (%)	Ref.
19.	CdS/TNT composite	1	Photoreactor (1 L) with external jacket for cooling	UV lamp (400 W)	10, 10, 360	52.3	[83]
20.	ZnO/oak charcoal	1.6	Pyrex glass photoreactor batch (1.5 L) in a metal box	Hg UV lamp (125 W)	154, 10.5, 240	80	[84]
21.	Bi_2WO_6 nanoplate	1	Batch photoreactor of quartz with an external cooling jacket	Lamp (18 W; λ = 390–780 nm)	10, 11.2, 240	100	[85]
22.	$ZnFe_2O_4$/rGO	2	Beaker with its aluminium foil coated wall (0.1 L)	UV-visible lamp (300 W)	100, 9.5, 480	92.3	[86]

(*Continued*)

Table 6.2 Various photocatalytic systems for the remediation of ammonia nitrogen. (*Continued*)

S. no.	TiO_2-based photocatalyst	Amount of catalyst ($g\ L^{-1}$)	Photocatalytic reactor used	Light source	NH_3(aq) concentration (ppm), pH, irradiation time (min)	Removal efficiency (%)	Ref.
23.	$Bi_2Fe_4O_9$	1	Batch photoreactor	Xe lamp (500 W)	10, 10.8, 360	68	[87]
24.	Graphitic–C_3N_4	0.2	Photoreactor with cylindrical quartz batch (0.1 L)	Xe Short Arc High Pressure Lamp (500 W)	1.5, 11, 360	80	[88]
25.	CuO–ZnO/ pottery plate	NA	Glass frame positioned at 45° from the ground surface (immobilized with the photocatalysts)	Natural solar light	85, 9.3, 240	77.2	[89]

(*Continued*)

Table 6.2 Various photocatalytic systems for the remediation of ammonia nitrogen. (*Continued*)

S. no.	TiO_2-based photocatalyst	Amount of catalyst (g L^{-1})	Photocatalytic reactor used	Light source	NH_3(aq) concentration (ppm), pH, irradiation time (min)	Removal efficiency (%)	Ref.
26.	Photoreactor of cylindrical batch (0.01 L) ($Ru(bpy)_3^{2+}$) as the sensitizer, methyl viologen as an electron (e-) donor, and oxygen (O_2) as the electron gainer	NA	Photoreactor of cylindrical batch (0.01 L)	Visible light from a halogen lamp (100 W)	170000, NA, 540	NA	[90]
27.	TiO_2–P25	0.3	Annular quartz reactor (2L)	UV mercury lamp (75 mW/cm^2, 365 nm)	38, 17, 180	90.12	[91]

(*Continued*)

Table 6.2 Various photocatalytic systems for the remediation of ammonia nitrogen. (*Continued*)

S. no.	TiO_2-based photocatalyst	Amount of catalyst (g L^{-1})	Photocatalytic reactor used	Light source	NH_3(aq) concentration (ppm), pH, irradiation time (min)	Removal efficiency (%)	Ref.
28.	Pt–TiO_2	0.5	Pyrex glass photoreactor	High pressure mercury lamp (450 W)	NA, 12, 240	94	[63]
29.	TiO_2–P25	NA	Pyrex glass batch reactor (1 L)	UV lamp (400 W)	10, 12, 360	97	[92]
30.	Immobilized ZnO nanoparticles on glass plate	NA	Batch reactor covered with circular water jacket	Low pressure UV lamp (6 W)	NA, 6, 116	76	[93]

remediation of ammonia (NH_3) in aqueous phase [65]. Consequently, after a natural light irradiation for 6 hours at a pH value of 9.9 with photocatalyst (0.2 mg L^{-1}), a 40.5% removal rate was obtained for 8.5 ppm concentrated aqueous ammonia [65]. The pH of the solution is one of the key process variables, which regulates the degradation of NH_3 (into NO_2^-, NO_3^-, and nitrogen), as it is shown to have photocatalytical effects (e.g., equilibrium between NH_3 and NH_4^+) [65]. Photocatalytic degradation of NH_4^+/NH_3 was insignificant below pH 7.2. The competition for the catalyst surface exchange site increased in acidic condition with increasing levels of H+ Ions and finally led to decrease the photocatalytic activity [66]. On the other hand, at a 9.9 pH, the threshold criteria were achieved to cause a highest possible degradation of NH_3 (aq) by TiO_2, as adsorption of free/neutral NH_3 onto the surface of TiO_2 controlled the overall oxidation process rate [67].

Photocatalytic performance decreased when the pH was above 10, probably due to competitive OH- ion to NH_3 adsorption on the surface of photocatalyst, combined with partial catalytic dissolution [9, 65]. The photocatalytical efficiency of TiO_2-P25 toward NH_3 (aq) was relatively low at a pH of 10.2 which can be seen in Figure 6.3 [94]. On the other hand, the eradication of aqueous ammonia (NH_3) in the presence of UV rays, best photocatalytical efficiency at pH 3.4 was achieved with thin films dried in an oven at 550°C under UV [95]. Reaction equations (6.4) and (6.5) have been used to support general information on the reaction and formation pathway(s) [70, 76]. The authors reported that the enhanced light absorbance could demonstrate the enhancement in photocatalytical efficiency of TiO_2 toward aqueous ammonia under strong acidic environment. The presence of negative radicals/ions in solution and positive charged ions on the catalysts surface can demonstrate such an impact [70]. However, additional clarifications as well as theoretical and experimental scientific evidence must be supported.

$$TiOH + H^+ \rightarrow TiOH^{2+} \tag{6.4}$$

$$TiOH + OH^- \rightarrow TiO^- + H_2O \tag{6.5}$$

6.3.2 Modified TiO_2 Photocatalyst

Incorporation of metallic species, like platinum (Pt), cerium (Ce), lanthanum (La), nickel (Ni), and iron (Fe) has been revealed as propitious alternatives for boosting the effectiveness of pristine TiO_2 particles (Table 6.2). In view of the possible weakness of low quantum effectiveness and low selectivity toward oxidation of harmless N_2 (in spite of dangerous nitrate and nitrite radicals), it is necessary to increase photocatalytical efficiency of pristine TiO_2. In a practical perspective, single stage photocatalytic transformation of aqueous ammonia to nitrogen is appealing in comparison to biological nitrification/denitrification process as it is repetitive, receptive and cost effective two stage procedure. In this regard, a platinum-assisted TiO_2-brookite catalyst was used for the photocatalytic degradation of NH_3 (aq) at pH 10 and was synthesized by deposition-precipitation method (Pt(DP)/B) [71]. The system showed improved photocatalytical activity toward NH_3 (aq) removal. Brookite and anatase form of TiO_2 catalysts recorded the significant improvements in photocatalytic efficiency on platinum nanoparticle deposition compared to P25, probably because of dissimilarities in the charge-separation properties of these intrinsic crystal structures [71]. Similarly, in oxidative degradation of NH_3 (aq), 0.5 wt% of Pt-TiO_2 particles showed strong photocatalytic capability at a pH of 12 (Table 6.2). Compared with pristine counterparts, platinum modified TiO_2 photocatalysts displayed more removal efficiency of NH_3 (aq) and NO_3^- formation. The concentration of NO_3^- in the water increases with increase in time. Meanwhile, the concentration of NO_2^- attained a peak and then decreases with time [72].

Additionally, a significant improvement of photocatalytic efficiency against NH_3 was achieved at a pH of 10 when titanium oxide (TiO_2) catalysts were doped with metals such as La or Fe or TiO_2 [76]. However, the photocatalytic performance of La/Fe/TiO_2 was substantially reduced with the existence of inorganic anions along with ammonia [76]. NO_3^- radicals functioned as internal filters to reduce the severity of the ultraviolet rays that reaches inside the batch process, resulting in a decrease in the photocatalytic performance [96]. Bi-carbonate or carbonate (HCO_3^-/CO_3^{2-}) radicals have strong affinity toward OH^- so that they could react to produce carbonate radicals with OH· radicals, which in turn have poor oxidation capacities and do not react with the majority of pollutants. Hence, the presence of HCO_3^-/CO_3^{2-} ions was considerably reduced the photocatalytic performance of La/Fe/TiO_2 [76, 96] (Figure 6.3). In addition, Cl^- and SO_4^{2-} ions are similar to the HCO_3^-/CO_3^{2-} ions with OH· radicals [97]. However, because of their relatively low reaction capacity, Cl^- and SO_4^{2-} ions have a lesser impact on the La/Fe/TiO_2 photocatalytic performance as compared to NO_3^- and HCO_3^-/CO_3^{2-} ions.

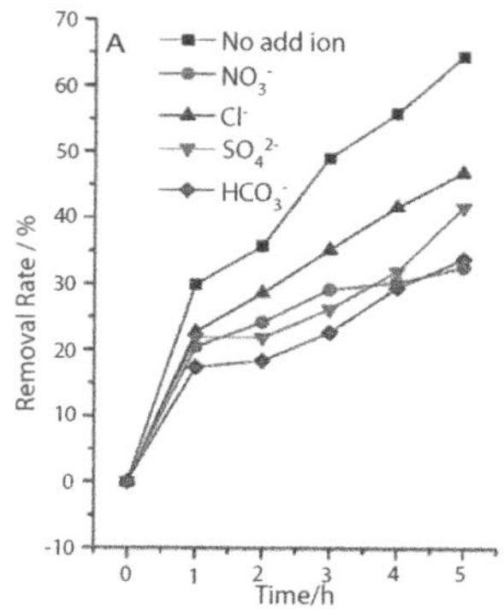

Figure 6.3 Effect of ions on NH_3 degradation by La/Fe/TiO_2 in the aquatic system [76].

6.4 Ozonation Technique for Removal of Aqueous Ammonia Nitrogen From Wastewater

Ozone is often used widely as an oxidizing agent to degrade contaminants from wastewater because of its high oxidation and reactivity potential [98]. Ozone and/or hydroxyl radicals (·OH) produced by ozone decomposition can oxidize organic contaminants during ozonation [99]. In comparison to ozone, ·OH must be responsible for the majority of various pollutants degradation in ozonation treatment of wastewater. Ozone is an oxidizing agent that reacts preferably by organics including olephins, active aromatic systems or amines, while ·OH is a less selective oxidant and can react immediately with the majority of organic moieties [100]. The low effectiveness of the ozone decomposition in a single ozone system usually leads to insufficient ·OH generation. Hence, several ozone-based advanced oxidation processes, such as ozone/H_2O_2 [99], UV/ozone [101, 102] and catalytic ozone [103], have been modified to improve free radical oxidation for the treatment of wastewater. It is assumed that for ozone/H_2O_2 systems, H_2O_2 increases its ·OH production in the process of ozone transformation, while being conveniently used without causing secondary emissions [104]. Adsorption-catalysis impact on the catalyst's surface can improve oxidation for O_3/catalyst systems [105].

6.4.1 Noncatalytic Ozonation of Ammonia Nitrogen

Singer and Zilli reported the first document regarding the effect of noncatalytic ozonation of ammonia nitrogen [106]. The study claimed that in

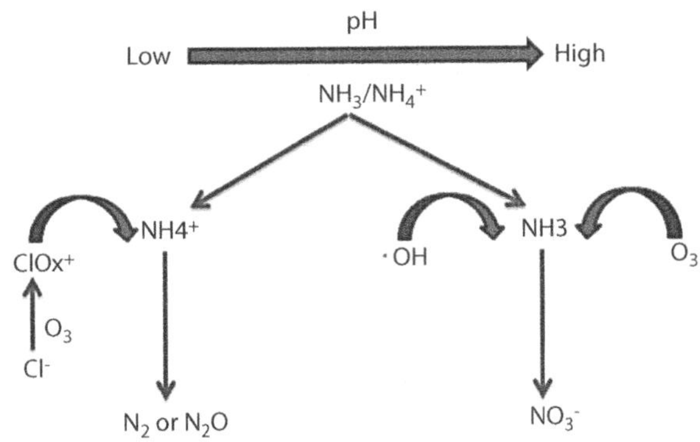

Figure 6.4 The basic reaction mechanism of oxidation of ammonia nitrogen by O_3.

accordance with the following equation, ammonia nitrogen was oxidized into nitrate under mild-alkaline conditions.

$$NH_3 + 4O_3 \rightarrow NO_3^- + 4O_2 + H_2O + H^+ \tag{6.6}$$

The reaction was influenced by the pH, where the rate of reaction increased with an increase in the solution's pH. Evidently, the first-order rate constant of ammonia nitrogen at pH 9 was approximately 1.5 times greater than that at pH 7 [106]. This became clear that the alkalinity had an effect on the rate when the ozone and OH- reaction may occur in such conditions as to pre-oxidized ammonia nitrogen [107].

Another study discusses the reason behind the dependence of noncatalytic ozonation of ammonia nitrogen on pH [108]. The basic reaction mechanism of oxidation of ammonia nitrogen by O_3 is shown in Figure 6.4. The authors soon determined that this was simply a change in the reaction mechanism based on pH, since direct oxidation with O_3 was predominant at pH < 9, while free radicals including OH· at pH > 9 were involved. In addition, as a consequence of ozone degradation, OH· readily forms in alkaline conditions with OH- [108]. This shows that the oxidation process was only regulated by the reaction (Eq. 6.6) which directly oxidizes the ammonium nitrogen if the pH of the solution was between 7 and 9. While, at pH above 9, the reaction between ammonia nitrogen, primarily NH_3, predominates with free radicals. It is, therefore, reasonable because when the reaction is performed in alkaline medium the reaction mechanism follows either direct oxidation with ozone or free-radical pathway.

6.4.2 Catalytic Ozonation of Ammonia Nitrogen

The first paper of homogeneous catalytic ozonation for degradation of ammonia nitrogen was reported by [109]. The reaction was performed at pH values between 7.1 and 8.4. The involvement of Br– in the solution of reaction was revealed to substantially increase the oxidation rate of ammonia nitrogen with ozone. The pH of the sample solutions had little impact on the bromide-catalyzed reaction. For example, the first-order rate constant of ammonia nitrogen at pH 8.4 was the same as that at pH 8.1 (k = 0.30 mg $L^{-1}{\cdot}min^{-1}$) and the initial concentrations of ammonia nitrogen and bromide were 200 and 20 μM, respectively. Though the rate was slightly reduced at 7.1 pH (k = 0.21 mg $L^{-1}{\cdot}min^{-1}$). In addition, because the decomposition rate of ammonia nitrogen was linearly associated with bromine concentration and NH_2Br production was proposed in the solution, it was suggested that $Br^-$was first oxidized with ozone in the reaction solution to form OBr^- at a very rapid rate (k = 160 $M^{-1}{\cdot}s^{-1}$), that further oxidized ammonia nitrogen into NH_2Br. NH_2Br was ultimately reacted with molecular ozone to produce NO_3^- and Br^- as the final product. Equations 6.7–6.10 [110] shows the whole reaction mechanism involved throughout the process.

$$O_3 + Br^- \rightarrow OBr^- + O_2 \tag{6.7}$$

$$H^+ + OBr^- \rightarrow HOBr \tag{6.8}$$

$$HOBr + NH_3 \rightarrow NH_2Br + H_2O \tag{6.9}$$

$$3O_3 + NH_2Br \rightarrow 2H^+ + NO^{3-} + Br^- + 3O_2 \tag{6.10}$$

Recently, another study proposed a degradation mechanism of ozonation for ammonia nitrogen in the presence of Cl^- at 283 K at different pH values [111]. Ammonia nitrogen was oxidized under alkaline medium by direct oxidative effects and free radicals, such as OH·. Moreover, at low pH, the involvement of Cl^- in the reaction solution gives rise to another reaction pathway. They observed that Cl^- was first oxidized by ozone into ClOx in the reaction solution, which then further oxidized NH_4^+ to N_2 and N_2O. Equations 6.11–6.16 describe the reaction mechanism involved [111].

$$Cl^- + O_3 \rightarrow ClO^- + O_2 \tag{6.11}$$

$$ClO_3^- \rightarrow ClO^- + O_2 \tag{6.12}$$

$$3ClO^- + 2NH_{4^+} \rightarrow N_2 + 3Cl^- + 2H^+ + 3H_2O \tag{6.13}$$

$$4ClO^- + 2NH_{4^+} \rightarrow N_2O + 4Cl^- + 2H^+ + 3H_2O \tag{6.14}$$

$$3ClO_{3^-} + 2NH_{4^+} \rightarrow N_2 + Cl^- + 2H^+ + 3H_2O \tag{6.15}$$

$$4ClO_{3^-} + 6NH_{4^+} \rightarrow 3N_2O + 4Cl^- + 6H^+ + 9H_2O \tag{6.16}$$

The NH_4^+ reacts with $ClOx^-$ to regenerate Cl^- which means that Cl^- supported the reaction catalytically as illustrated by equations 6.13–6.16. The end product was primarily N_2 and N_2O by the catalytic reaction. The study stated that the oxidation rate of ammonia nitrogen was insignificant for Cl^- while the NO_3^- was produced as the main product [106]. This variation proposed that when the pH of the solution was acidic, the inclusion of Cl^- had a significant effect on the reaction mechanism and the products, indicating that NH_4^+ was present predominantly. While the participation of Cl^- was far less substantial under acidic condition, and Cl^- did not affect the reaction of NH_3 with O_3 as well as free radicals [111]. Consequently, in the vicinity of Cl^-, the pH of the solution not only affected the reaction mechanism of oxidation of ammonia nitrogen by ozone but also the products.

The catalytic ozonation research on heterogeneous catalysis has been investigated intensively over the last two decades, and the study on this topic has been well explained by [112]. Metal oxides, such as MgO, NiO, Co_3O_4, CuO, ZnO, Mn_3O_4, Fe_2O_3, SnO_2, and Al_2O_3, demonstrated ozone ammonia nitrogen decomposition activity in the presence of Cl^- at 333 K under mild acidic condition [113]. MgO showed the best catalytic activity but the lowest gaseous nitrogen selectivity, on the other hand Co_3O_4 has shown highest selectivity with moderate activity.

Owing to the high selectivity of gaseous nitrogen compounds, the deterministic study was based on the reaction to Co_3O_4, since Co_3O_4 has been the most potentially applicable for realistic usages. It was revealed that Cl^- was necessary to continue with the oxidation reaction with ozone under the reaction conditions. It was therefore suggested that Cl^- was first oxidized to produce ClO^- by O_3 over Co_3O_4, which then oxidized to

NH_4^+ [113]. However, another finding revealed that the catalytic performance of Co_3O_4 for the ozonation of ammonia nitrogen in the vicinity of Cl^- was considerably increased with each re-use under the reaction conditions [113]. The -NHx group formation on Co_3O_4 was deemed the cause behind the enhancement. Meanwhile the impact of calcination temperature of Co_3O_4 on enhancement showed that the enhancement occurred only after the Co_3O_4 was calcinated to lower temperature, i.e. the hydroxyl surface group had played a significant role in the enhancement. However, the formation of -NHx on the surface was not directly demonstrated. Thus, the accurate mechanism for such an enhancement is still not clear.

Another group of authors later confirmed that the Co_3O_4 and MgO (Mg:Co = 4:1) composite was a much more effective catalyst for ammonia nitrogen decomposition in mild-alkaline conditions than pristine Co_3O_4 and could be operated at room temperature [114]. It was suggested that ozone first produced OH· under alkaline conditions by reacting with OH^-, which then oxidized NH_3 to give N_2 and NO_3 as the final product. The drawback of this research is that the leaching behavior of MgO was not carefully examined, and there was no clarification about how the pH of the solution was maintained under mild alkaline conditions during protons generations.

6.5 Conclusion and Future Prospects

The continued increase in marine aquaculture involves issues including insufficient culture technology, outbreak of diseases, water contamination and environmental damaging associated problems. One of these is the ammonia level, which over the last few years has increased significantly, primarily as a result of agriculture and other anthropogenic activities, which has consequently affected aquaculture life, especially shrimp. When the levels of ammonia nitrogen exceed the tolerance limits in aquaculture pond, it directly prevent the chitinase expression, moulting, growth, phenoloxidase and hemolymph antimicrobial activity, and thus reduces innate shrimp immunity. To overcome this problem, advanced oxidation processes is one of the methods through which the degradation of aqeuos ammonia nitrogen is successfully accomplished. Among the AOPs reviewed, photocatalytic degradation and ozone-based oxidation are the two most frequently and widely used methods for the removal of ammonia nitrogen.

The implementation of photocatalytic systems to wastewater treatment and remediation has gained extensive researchers' attention in past years due to the performance and economically and environmentally nonthreatening factors (in terms of end products such as nitrogen in the case of NH_3). The comparative study of all existing data from aqueous matrices shows the best possible quantitative efficiency in NH_3 removal for modified TiO_2 materials which combine the efficiency of adsorption and photocatalysis. However, thin films of pristine TiO_2 can also attain significant photocatalytical efficiency. Moreover, it was observed that the amount of the NH_3 degradation was controlled by pH, with the highest possible performances under alkali conditions (pH ranging from 7 to 12). NO_2^-, NO_3^-, and N_2O were reported to be the significant value added products along with nitrogen in ammonia degradation process.

Ozonation is another promising technique for the removal of ammonia nitrogen from wastewater because the reaction can occur at low temperature and atmospheric pressure. Moreover, ammonia nitrogen can be oxidized effectively in mild-alkaline conditions with ozone and oxidized at high pH with free radicals (including OH). While homogenous Br^- catalytic ozonation can work at neutral to slightly alkaline pH, only acidic conditions can allow the Cl^--catalyzed reaction to proceed. Although several articles have been published, but there is still a poor understanding of the reaction mechanism, particularly for the catalyst reaction. Heterogeneous catalysis with Co_3O_4 followed $ClOx^-$ pathway. Finally, only few articles are concluded because not so many routes to oxidize ammonia nitrogen can be taken. The door of heterogeneous catalysis remains wide open, particular for oxidation reaction, which can be carried out in the absence of Cl^- or Br^-. Further research should be conducted to identify the maximum potential of these processes.

Acknowledgments

The authors would like to express their appreciation to Ministry of Higher Education Malaysia for Fundamental Research Grant Scheme with Project Code: FRGS/1/2019/STG07/USM/02/12 and Long Term Grant Scheme with Project Code: 67215001.

References

1. Huang, L., Li, L., Dong, W., Liu, Y., Hou, H., Removal of ammonia by OH radical in aqueous phase. *Environ. Sci. Technol.*, 42, 8070–8075, 2008.

2. Khuntia, S., Majumder, S.K., Ghosh, P., Removal of ammonia from water by ozone microbubbles. *Ind. Eng. Chem. Res.*, 52, 1, 318–326, 2013.
3. Delwiche, C.C., *Denitrification, nitrification, and atmospheric nitrous oxide.* New York, NY, Wiley, 1981.
4. Adams, C.E., Removing nitrogen from wastewater. *Environ. Sci. Technol.*, 8, 696–701, 1994.
5. Obaid-ur-Rehman, S. and Beg, S.A., Ammonia removal by air stripping – from origin to present state of technology. *J. Environ. Sci. Health A*, 25, 343–364, 1990.
6. Anjaneyulu, Y., Chary, N.S., Raj, D.S.S., Decolourization of industrial effluents—available methods and emerging technologies—A review. *Rev. Environ. Sci. Biotechnol.*, 4, 245–273, 2005.
7. Zhang, W., Fu, R., Wang, L., Zhu, J.W., Feng, J., Yan, W., Rapid removal of ammonia nitrogen in low-concentration from wastewater by amorphous sodium titanate nano-particles. *Sci. Total Environ.*, 668, 815–824, 2019.
8. Rezakazemi, M., Shirazian, S., Ashrafizadeh, S.N., Simulation of ammonia removal from industrial wastewater streams by means of a hollow-fiber membrane contactor. *Desalination*, 285, 31, 383–392, 2012.
9. Vohra, M.S., Selimuzzaman, S.M., Al-Suwaiyan, M.S., NH_4^+-NH_3 removal from simulated wastewater using UV-TiO_2 photocatalysis: Effect of co-pollutants and pH. *Environ. Technol.*, 31, 6, 641–654, 2010.
10. Zhou, Q., Yin, H., Wang, A., Si, Y., Preparation of hollow B–SiO_2@TiO_2 composites and their photocatalytic performances for degradation of ammonia-nitrogen and green algae in aqueous solution. *Chin. J. Chem. Eng.*, 27, 2535–2543, 2019.
11. Lu., X., Zhou., Q., Yin., H., Wang, A., Meng., F., Synthesis of hollow B-SiO_2@ $CaTiO_3$ nanocomposites and their photocatalytic performance in ammonia nitrogen degradation. *Water Air Soil Pollut.*, 231, 102, 2020.
12. Romano, N. and Zeng, C.S., Toxic effects of ammonia, nitrite, and nitrate to decapod crustaceans: A review on factors influencing their toxicity, physiological consequences, and coping mechanisms. *Rev. Fish. Sci. Aquac.*, 21, 1–21, 2013.
13. O'Connor, S.L., Almeida, M., Bassett, C., McKelvie, I.D., Kolev, S.D., Gas-diffusion-based passive sampler for ammonia monitoring in marine waters. *Talanta*, 181, 52–56, 2018.
14. Olowe, K.O. and Kumarasamy, M., Development of the hybrid cells in series model to simulate ammonia nutrient pollutant transport along the Umgeni River. *Environ. Sci. Pollut. Res. Int.*, 24, 22967–22979, 2017.
15. Hong, M., Chen, L., Sun, X., Gu, S., Zhang, L., Chen, Y., Metabolic and immune responses in Chinese mitten-handed crab (*Eriocheir sinensis*) juveniles exposed to elevated ambient ammonia. *Comp. Biochem. Physiol. C. Toxicol. Pharmacol.*, 145, 363–369, 2007.

16. Habaki, H., Tajiri, S., Egashira, R., Sato, K., Uptake rate of ammonia-nitrogen with *Sterile ulva* sp for water quality control of intensive shrimp culture ponds in developing countries. *Chem. Biochem. Eng. Q.*, 25, 341–349, 2011.
17. Bouwman, A.F., Pawlowski, M., Liu, C., Beusen, A.H.W., Shumway, S.E., Glibert, P.M., Overbeek, C.C., Global hindcasts and future projections of coastal nitrogen and phosphorus loads due to shellfish and seaweed aquaculture. *Rev. Fish. Sci. Aquac.*, 19, 331–357, 2011.
18. Zhou, K., Zhou, F., Huang, J., Yang, Q., Jiang, S., Qiu, L., Yang, L., Jiang, S., Characterization and expression analysis of a chitinase gene (PmChi-4) from black tiger shrimp (Penaeus monodon) under pathogen infection and ambient ammonia nitrogen stress. *Fish Shellfish Immunol.*, 62, 31–40, 2017.
19. Qiu, L., Shi, X., Yu, S., Han, Q., Diao, X., Zhou, H., Changes of ammonia-metabolizing enzyme activity and gene expression of two strains in shrimp Litopenaeus vannamei under ammonia stress. *Front. Physiol.*, 9, 211, 2018.
20. Zhao, M., Yao, D., Li, S., Zhang, Y., Aweya, J.J., Effects of ammonia on shrimp physiology and immunity: A review. *Rev. Aquacult.*, 2194-2211, 1–18, 2020.
21. Zhang, Y., Ye, C., Wang, A., Zhu, X., Chen, C., Xian, J., Sun, Z., Isolated and combined exposure to ammonia and nitrite in giant freshwater pawn (*Macrobrachium rosenbergii*): Effects on the oxidative stress, antioxidant enzymatic activities and apoptosis in haemocytes. *Ecotoxicology*, 24, 1601–1610, 2015b.
22. Xiao, J., Li, Q.Y., Tu, J.P., Chen, X.L., Chen, X.H., Liu, Q.Y., Liu, H., Zhou, X.Y., Zhao, Y.Z., Wang, H.L., Stress response and tolerance mechanisms of ammonia exposure based on transcriptomics and metabolomics in Litopenaeus vannamei. *Ecotoxicol. Environ. Saf.*, 180, 491–500, 2019.
23. Cui, Y., Ren, X., Li, J., Zhai, Q., Feng, Y., Xu, Y., Ma, L., Effects of ammonia-N stress on metabolic and immune function via the neuroendocrine system in Litopenaeus vannamei. *Fish Shellfish Immunol.*, 64, 270–275, 2017.
24. Liang, Z., Liu, R., Zhao, D., Wang, L., Sun, M., Wang, M., Song, L., Ammonia exposure induces oxidative stress, endoplasmic reticulum stress and apoptosis in hepatopancreas of pacific white shrimp (Litopenaeus vannamei). *Fish Shellfish Immunol.*, 54, 523–528, 2016.
25. Pinto, M.R., Lucena, M.N., Faleiros, R.O., Almeida, E.A., McNamara, J.C., Leone, F.A., Effects of ammonia stress in the Amazon river shrimp Macrobrachium amazonicum (Decapoda, Palaemonidae). *Aquat. Toxicol.*, 170, 13–23, 2016.
26. Sung, Y.Y., MacRae, T.H., Sorgeloos, P., Bossier, P., Stress response for disease control in aquaculture. *Rev. Aquac.*, 3, 120–137, 2011.
27. Zhou, K., Zhou, F., Huang, J., Yang, Q., Jiang, S., Qiu, L., Yang, L., Zhu, C., Jiang, S., Characterization and expression analysis of a chitinase gene (PmChi-4) from black tiger shrimp (Penaeus monodon) under pathogen infection and ambient ammonia nitrogen stress. *Fish Shellfish Immunol.*, 62, 31–40, 2017.

28. Duan, Y., Liu, Q., Wang, Y., Zhang, J., Xiong, D., Impairment of the intestine barrier function in Litopenaeus vannamei exposed to ammonia and nitrite stress. *Fish Shellfish Immunol.*, 78, 279–288, 2018a.
29. Liang, Z., Liu, R., Zhao, D., Wang, L., Sun, M., Wang, M., Song, L., Ammonia exposure induces oxidative stress, endoplasmic reticulum stress and apoptosis in hepatopancreas of pacific white shrimp (Litopenaeus vannamei). *Fish Shellfish Immunol.*, 54, 523–528, 2016.
30. Liu, F., Li, S., Yu, Y., Sun, M., Xiang, J., Li, F., Effects of ammonia stress on the hemocytes of the pacific white shrimp Litopenaeus vannamei. *Chemosphere*, 239, 124759, 2020.
31. Yao, D., Wang, Z., Wei, M., Zhao, X., Aweya, J.J., Zhong, M., Li, S., Zhang, Y., Analysis of Litopenaeus vannamei hemocyanin interacting proteins reveals its role in hemolymph clotting. *J. Proteomics*, 201, 57–64, 2019.
32. Paust, L.O., Foss, A., Imsland, A.K., Effects of chronic and periodic exposure to ammonia on growth, food conversion efficiency and blood physiology in juvenile Atlantic halibut (*Hippoglossus hippoglossus* L.). *Aquaculture*, 315, 400–406, 2011.
33. Sun, H., Lu, K., Minter, E.J., Chen, Y., Yang, Z., Montagnes, D.J., Combined effects of ammonia and microcystin on survival, growth, antioxidant responses, and lipid peroxidation of bighead carp Hypophthalmythys nobilis larvae. *J. Hazard. Mater.*, 221–222, 213–219, 2012.
34. Zhang, C.N., Li, X.F., Tian, H.Y., Zhang, D.D., Jiang, G.Z., Lu, K.L., Liu, G.X., Liu, W.B., Effects of fructooligosaccharide on immune response, antioxidant capability and HSP70 and HSP90 expressions of blunt snout bream (Megalobrama amblycephala) under high ammonia stress. *Fish. Physiol. Biochem.*, 41, 203–217, 2015a.
35. Ip, Y.K., Hou, Z., Chen, X.L., Ong, J.L., Chng, Y.R., Ching, B., Hiong, K.C., Chew, S.F., High brain ammonia tolerance and down-regulation of Na+: K+:2Cl- cotransporter 1b mRNA and protein expression in the brain of the Swamp Eel, Monopterus albus, exposed to environmental ammonia or terrestrial conditions. *PLoS One*, 8, 9, 69512, 2013.
36. Hangzo, H., Banerjee, B., Saha, S., Saha, N., Ammonia stress under high environmental ammonia induces Hsp70 and Hsp90 in the mud eel, Monopterus cuchia. *Fish. Physiol. Biochem.*, 43, 77–88, 2017.
37. Hegazi, M.M., Attia, Z.I., Ashour, O.A., Oxidative stress and antioxidant enzymes in liver and white muscle of Nile tilapia juveniles in chronic ammonia exposure. *Aquat. Toxicol.*, 99, 118–125, 2010.
38. Li, M., Gong, S., Li, Q., Yuan, L., Meng, F., Wang, R., Ammonia toxicity induces glutamine accumulation, oxidative stress and immunosuppression in juvenile yellow catfish Pelteobagrus fulvidraco. *Comp. Biochem. Physiol. C Toxicol. Pharmacol.*, 183–184, 1–6, 2016.
39. Rui, J., Liu, B.L., Cen, H., Huang, B., Lei, J.L., Effects of ammonia exposure on stress and immune response in juvenile turbot (Scophthalmus maximus). *Aquac. Res.*, 48, 3149–3162, 2016.

40. Shin, K.W., Kim, S.H., Kim, J.H., Hwang, S.D., Kang, J.C., Toxic effects of ammonia exposure on growth performance, hematological parameters, and plasma components in rockfish, Sebastes schlegelii, during thermal stress. *Can. J. Fish. Aquat. Sci.*, 19, 44, 2016.
41. Kim, S.H., Kim, J.H., Park, M.A., Hwang, S.D., Kang, J.C., The toxic effects of ammonia exposure on antioxidant and immune responses in rockfish, Sebastes schlegelii during thermal stress. *Environ. Toxicol. Pharm.*, 40, 954–959, 2015.
42. Chang, Z.W., Chiang, P.C., Cheng, W., Chang, C.C., Impact of ammonia exposure on coagulation in white shrimp, Litopenaeus vannamei. *Ecotoxicol. Environ. Saf.*, 118, 98–102, 2015.
43. Yue, F., Pan, L., Xie, P., Jian, L.I., Effects of ammonia exposure on prophenoloxidase system and immune parameters of swimming crab Portunus trituberculatus. *J. Fish Aquat. Sci.*, 17, 761–770, 2010a.
44. Romano, N. and Zeng, C., Acute toxicity of ammonia and its effects on the haemolymph osmolality, ammonia-N, pH and ionic composition of early juvenile mud crabs, Scylla serrate (Forskal). *Comp. Biochem. Physiol., Part A Mol. Integr. Physiol.*, 148, 278–285, 2007a.
45. Wang, X., Wang, L., Yao, C., Qiu, L., Zhang, H., Zhi, Z., Song, L., Alternation of immune parameters and cellular energy allocation of *Chlamys farreri* under ammonia-N exposure and *Vibrio anguillarum* challenge. *Fish Shellfish Immunol.*, 32, 741–749, 2012.
46. Zhang, T., Yan, Z., Zheng, X., Fan, J., Wang, S., Wei, Y., Yang, L., Wang, P., Guo, S., Transcriptome analysis of response mechanism to ammoniastress in Asian clam (*Corbicula fluminea*). *Aquat. Toxicol.*, 214, 105235, 2019.
47. Cong, M., Wu, H., Cao, T., Ji, C., Lv, J., Effects of ammonia nitrogen on gill mitochondria in clam Ruditapes philippinarum. *Environ. Toxicol. Pharmacol.*, 65, 46–52, 2019.
48. Li, C., Wang, F., Aweya, J.J., Yao, D., Zheng, Z., Huang, H., Li, S., Zhang, Y., Trypsin of *Litopenaeus vannamei* is required for the generation of hemocyanin-derived peptides. *Dev. Comp. Immunol.*, 79, 95–104, 2018a.
49. Lu, X., Luan, S., Dai, P., Meng, X., Cao, B., Luo, K., Kong, J., iTRAQ-based comparative proteome analysis for molecular mechanism of defense against acute ammonia toxicity in pacific white shrimp *Litopenaeus vannamei. Fish Shellfish Immunol.*, 74, 52–61, 2018.
50. Valencia-Castaneda, G., Frias-Espericueta, M.G., Vanegas-Perez, R.C., Perez-Ramirez, J.A., Chavez-Sanchez, M.C., Paez-Osuna, F., Acute toxicity of ammonia, nitrite and nitrate to shrimp *Litopenaeus vannamei* postlarvae in low-salinity water. *Bull. Environ. Contam. Toxicol.*, 101, 229–234, 2018.
51. Romano, N. and Zeng, C.S., Importance of balanced Na^+/K^+ ratios for blue swimmer crabs, Portunus pelagicus, to cope with elevated ammonia-N and differences between *in vitro* and *in vivo* gill Na^+/K^+-ATPase responses. *Aquaculture*, 318, 154–161, 2011.

52. Zhang, L., Pan, L., Xu, L., Si, L., Effects of ammonia-N exposure on the concentrations of neurotransmitters, hemocyte intracellular signaling pathways and immune responses in white shrimp Litopenaeus vannamei. *Fish Shellfish Immunol.*, 75, 48–57, 2018a.
53. Kir, M. and Oz, O., Effects of salinity on acute toxicity of ammonia and oxygen consumption rates in common prawn, *Palaemon serratus* (Pennat, 1777). *J. World Aquac. Soc.*, 46, 76–82, 2015.
54. Yang, L.S., Yang, Q.B., Jiang, S.G., Li, Y., Zhou, F.L., Li, T., Huang, J., Metabolic, immune responses in prawn (Penaeus monodon) exposed to ambient ammonia. *Aquat. Int.*, 23, 1049–1062, 2015.
55. Alvarez, M.A., Ruidiaz-Martinez, M., Cruz-Quesada, G., Lopez-Ramon, M.V., Rivera-Utrilla, J., Sanchez-Polo, M., Mota, A.J., Removal of parabens from water by UV-driven advanced oxidation processes. *Chem. Eng. J.*, 379, 122334, 2020.
56. Ma, S., Kim, K., Chun, S., Moon, S.Y., Hong, Y., Plasma- assisted advanced oxidation process by a multi-hole dielectric barrier discharge in water and its application to wastewater treatment. *Chemosphere*, 243, 125377, 2020.
57. Horikoshi, S. and Serpone, N., Can the photocatalyst TiO_2 be incorporated into a wastewater treatment method? Background and prospects. *Catal. Today*, 340, 334–346, 2020.
58. Chauhan, A., Sharma, M., Kumar, S., Thirumalai, S., Kumar, R.V., Vaish, R., TiO_2@C core@shell nanocomposites: A single precursor synthesis of photocatalyst for efficient solar water treatment. *J. Hazard. Mater.*, 381, 120883, 2020.
59. Cooper, W.J., Cramer, C.J., Martin, N.H., Mezyk, S.P., O'Shea, K.E., Sonntag, C.V., Free radical mechanisms for the treatment of methyl tert-butyl ether (MTBE) via advanced oxidation/reductive processes in aqueous solutions. *Chem. Rev.*, 109, 3, 1302–1345, 2009.
60. Gomes, A. and Fernandes, E., Fluorescence probes used for detection of reactive oxygen species. *J. Biochem. Biophys. Methods*, 65, 2-3, 45–80, 2005.
61. Tai, C., Gu, X., Zou, H., Guo, Q., A new simple and sensitive fluorometric method for the determination of hydroxyl radical and its application. *Talanta*, 58, 4, 661–667, 2002.
62. Poblete, R. and Pérez, N., Use of sawdust as pretreatment of photo-fenton process in the depuration of landfill leachate. *J. Environ. Manage*, 253, 109697, 2020.
63. Sun, D., Hong, X., Wu, K., Hui, K.S., Du, Y., Hui, K.N., Simultaneous removal of ammonia and phosphate by electro- oxidation and electrocoagulation using RuO_2–IrO_2/Ti and microscale zero-valent iron composite electrode. *Water Res.*, 169, 115239, 2020.
64. Utsunomiya, A., Okemoto, A., Nishino, Y., Kitagawa, K., Kobayashi, H., Taniya, K., Ichihashi, Y., Nishiyama, S., Mechanistic study of reaction mechanism on ammonia photodecomposition over Ni/TiO_2 photocatalysts. *Appl. Catal. B*, 206, 378–383, 2017.

65. Bonsen, E.-M., Schroeter, S., Jacobs, H., Broekaert, J.C., Photocatalytic degradation of ammonia with t102 as photocatalyst in the laboratory and under the use of solar radiation. *Chemosphere*, 35, 7, 1431–1445, 1997.
66. Shavisi, Y., Sharifnia, S., Hosseini, S.N., Khadivi, M.A., Application of TiO_2/perlite photocatalysis for degradation of ammonia in wastewater. *J. Ind. Eng. Chem.*, 20, 1, 278–283, 2014.
67. Altomare, M., Chiarello, G.L., Costa, A., Guarino, M., Selli, E., Photocatalytic abatement of ammonia in nitrogen-containing effluents. *Chem. Eng. J.*, 191, 394–401, 2012.
68. Zhu, X., Castleberry, S.R., Nanny, M.A., Butler, E.C., Effects of pH and catalyst concentration on photocatalytic oxidation of aqueous ammonia and nitrite in titanium dioxide suspensions. *Environ. Sci. Technol.*, 39, 10, 3784–3791, 2005.
69. Gong, X., Wang, H., Yang, C., Li, Q., Chen, X., Hu, J., Photocatalytic degradation of high ammonia concentration wastewater by TiO_2. *Future Cities Environ.*, 1, 1, 12, 2017.
70. Altomare, M., Dozzi, M.V., Chiarello, G.L., Di Paola, A., Palmisano, L., Selli, E., High activity of brookite TiO_2 nanoparticles in the photocatalytic abatement of ammonia in water. *Catal. Today*, 252, 184–189, 2015.
71. Ou, H.-H., Hoffmann, M.R., Liao, C.-H., Hong, J.-H., Lo, S.-L., Photocatalytic oxidation of aqueous ammonia over platinized microwave-assisted titanate nanotubes. *Appl. Catal., B*, 99, 1, 74–80, 2010.
72. Shibuya, S., Aoki, S., Sekine, Y., Mikami, I., Influence of oxygen addition on photocatalytic oxidation of aqueous ammonia over platinum-loaded TiO_2. *Appl. Catal., B*, 138–139, 294–298, 2013.
73. Lee, J., Park, H., Choi, W., Selective photocatalytic oxidation of NH_3 to N_2 on platinized TiO_2 in water. *Environ. Sci. Technol.*, 36, 24, 5462–5468, 2002.
74. Utsunomiya, A., Okemoto, A., Nishino, Y., Kitagawa, K., Kobayashi, H., Taniya, K., Ichihashi, Y., Nishiyama, S., Mechanistic study of reaction mechanism on ammonia photodecomposition over Ni/TiO_2 photocatalysts. *Appl. Catal., B*, 206, 378–383, 2017.
75. Reli, M., Ambrozova, N., Sihor, M., Matejova, L., Capek, L., Obalova, L., Matej, Z., Kotarba, A., Koci, K., Novel cerium doped titania catalysts for photocatalytic decomposition of ammonia. *Appl. Catal., B*, 178, 108–116, 2015.
76. Luo, X., Chen, C., Yang, J., Wang, J., Yan, Q., Shi, H., Wang, C., Characterization of La/Fe/TiO_2 and its photocatalytic performance in ammonia nitrogen wastewater. *Int. J. Environ. Res. Public Health*, 12, 11, 14626–14639, 2015.
77. Nemoto, J., Gokan, N., Ueno, H., Kaneko, M., Photodecomposition of ammonia to dinitrogen and dihydrogen on platinized TiO_2 nanoparticules in an aqueous solution. *J. Photochem. Photobiol., A*, 185, 2, 295–300, 2007.
78. Pretzer, L.A., Carlson, P.J., Boyd, J.E., The effect of Pt oxidation state and concentration on the photocatalytic removal of aqueous ammonia with Pt-Modified titania. *J. Photochem. Photobiol., A*, 200, 2, 246–253, 2008.

79. Shavisi, Y., Sharifnia, S., Zendehzaban, M., Mirghavami, M.L., Kakehazar, S., Application of solar light for degradation of ammonia in petrochemical wastewater by a floating TiO_2/LECA photocatalyst. *J. Ind. Eng. Chem.*, 20, 5, 2806–2813, 2014.
80. Mohammadi, Z., Sharifnia, S., Shavisi, Y., Photocatalytic degradation of aqueous ammonia by using TiO_2ZnO/LECA hybrid photocatalyst. *Mater. Chem. Phys.*, 184, 110–117, 2016.
81. Zendehzaban, M., Sharifnia, S., Hosseini, S.N., Photocatalytic degradation of ammonia by light expanded clay aggregate (LECA)-Coating of TiO_2 nanoparticles. *Korean J. Chem. Eng.*, 30, 3, 574–579, 2013.
82. Peng, X., Wang, M., Hu, F., Qiu, F., Dai, H., Cao, Z., Facile fabrication of hollow biochar carbon-doped TiO_2/CuO composites for the photocatalytic degradation of ammonia nitrogen from aqueous solution. *J. Alloys Compd.*, 770, 1055–1063, 2019.
83. Chen, Y.-C., Lo, S.-L., Ou, H.-H., Chen, C.-H., Photocatalytic oxidation of ammonia by cadmium sulfide/titanate nanotubes synthesised by microwave hydrothermal method. *Water Sci. Technol.*, 63, 3, 550–557, 2011.
84. Tafreshi, N., Sharifnia, S., Moradi Dehaghi, S., Box–Behnken experimental design for optimization of ammonia photocatalytic degradation by ZnO/Oak charcoal composite. *Process Saf. Environ. Prot.*, 106, 203–210, 2017.
85. Zhang, L., Wang, W., Sun, S., Photocatalytic oxidation of ammonia by Bi_2WO_6 nanoplates using fluorescent light. *Chin. Sci. Bull.*, 59, 18, 2181–2185, 2014.
86. Liu, S.-Q., Zhu, X.-L., Zhou, Y., Meng, Z.-D., Chen, Z.-G., Liu, C.-B., Chen, F., Wu, Z.-Y., Qian, J.-C., Smart photocatalytic removal of ammonia through molecular recognition of zinc Ferrite/Reduced graphene oxide hybrid catalyst under visible- light irradiation. *Catal. Sci. Technol.*, 7, 15, 3210–3219, 2017.
87. Sun, S., Wang, W., Zhang, L., Shang, M., Visible light- induced photocatalytic oxidation of phenol and aqueous ammonia in flowerlike Bi2Fe4O9 suspensions. *J. Phys. Chem. C*, 113, 29, 12826–12831, 2009.
88. Wang, H., Su, Y., Zhao, H., Yu, H., Chen, S., Zhang, Y., Quan, X., Photocatalytic oxidation of aqueous ammonia using atomic single layer graphitic-C_3N_4. *Environ. Sci. Technol.*, 48, 20, 11984–11990, 2014.
89. Shavisi, Y., Sharifnia, S., Mohamadi, Z., Solar-light-harvesting degradation of aqueous ammonia by CuO/ZnO immobilized on pottery plate: Linear kinetic modeling for adsorption and photocatalysis process. *J. Environ. Chem. Eng.*, 4, 3, 2736–2744, 2016.
90. Kaneko, M., Katakura, N., Harada, C., Takei, Y., Hoshino, M., Visible light decomposition of ammonia to dinitrogen by a new visible light photocatalytic system composed of sensitizer (Ru- (bpy)32+), electron mediator (methylviologen) and electron acceptor (dioxygen). *Chem. Commun.*, 27, 3436–3438, 2005.

91. Zhang, F., Feng, C., Jin, Y., Li, W., Hao, G., Cui, J., Photocatalytic degradation of ammonia nitrogen with suspsended TiO2, in: *Bioinformatics and Biomedical Engineering, 3rd International Conference*, pp. 1–4, 2009.
92. Mikami, I., Aoki, S., Miura, Y., Photocatalytic oxidation of aqueous ammonia in the presence of oxygen over platinum-loaded TiO_2. *Chem. Lett.*, 39, 7, 704–705, 2010.
93. Rezaee, A., Soltani, R.D.C., Khataee, A.R., Godini, H., Optimization of combined photocatalytic involving immobilized ZnO nanoparticles and electrochemical processes for ammoniacal nitrogen removal from aqueous solutions. *J. Mater. Environ. Sci.*, 3, 5, 955–966, 2012.
94. Mohammadi, Z., Sharifnia, S., Shavisi, Y., Photocatalytic degradation of aqueous ammonia by using TiO_2Zno/LECA hybrid photocatalyst. *Mater. Chem. Phys.*, 184, 110–117, 2016.
95. Peng, X., Wang, M., Hu, F., Qiu, F., Dai, H., Cao, Z., Facile fabrication of hollow biochar carbon-doped TiO_2/CuO composites for the photocatalytic degradation of ammonia nitrogen from aqueous solution. *J. Alloys Compd.*, 770, 1055–1063, 2019.
96. Kim, D.H. and Anderson, M.A., Solution factors affecting the photocatalytic and photoelectrocatalytic degradation of formic acid using supported TiO_2 thin films. *J. Photochem. Photobiol., A*, 94, 2, 221–229, 1996.
97. Von Sonntag, C. and Von Gunten, U., Chemistry of ozone in water and wastewater treatment, in: *From Basic Principles to Applications*, IWA Publishing, UK, 2012.
98. KasprzykeHordern, B., Andrzejewski, P., Dabrowska, A., Czaczyk, K., Nawrocki, J., MTBE, DIPE, ETBE and TAME degradation in water using perfluorinated phases as catalysts for ozonation process. *Appl. Catal. B Environ.*, 51, 51–66, 2004.
99. Wang, H.W., Li, X.Y., Hao, Z.P., Sun, Y.J., Wang, Y.N., Li, W.H., Tsang, Y., Transformation of dissolved organic matter in concentrated leachate from nanofiltration during ozone-based oxidation processes (O_3, O_3/H_2O_2 and O_3/UV). *J. Environ. Manage.*, 191, 244–251, 2017a.
100. Wang, H., Zhan, J., Yao, W., Wang, B., Deng, S., Huang, J., Yu, G., Wang, Y., Comparison of pharmaceutical abatement in various water matrices by conventional ozonation, peroxone (O_3/H_2O_2), and an electro-peroxone process. *Water Res.*, 30, 127–138, 2017b.
101. Ruppert, G., Bauer, R., Heisler, G., UV-O_3, UV-H_2O_2, UV-TiO_2 and the photofenton reaction-comparison of advanced oxidation processes for wastewater treatment. *Chemosphere*, 28, 1447–1454, 1994.
102. Roshani, B., McMaster, M., Rezaei, E., Soltan, J., Catalytic ozonation of benzotriazole over alumina supported transition metal oxide catalysts in water. *Sep. Purif. Technol.*, 135, 158–164, 2014.
103. Von Gunten, U., Ozonation of drinking water: Part I. Oxidation kinetics and product formation. *Water Res.*, 37, 1443–1467, 2003.

104. Ernst, M., Lurot, F., Schrotter, J.C., Catalytic ozonation of refractory organic model compounds in aqueous solution by aluminum oxide. *Appl. Catal. B Environ.*, 47, 15–25, 2004.
105. Singer, P.C. and Zilli, W.B., Ozonation of ammonia in wastewater. *Water Res.*, 9, 127–134, 1975.
106. Hewes, C.G. and Davison, R.R., Kinetics of ozone decomposition and reaction with organics in water. *AIChE J.*, 17, 141–147, 1971.
107. Hoigne, J. and Bader, H., Ozonation of water: Kinetics of oxidation of ammonia by ozone and hydroxyl radicals. *Environ. Sci. Technol.*, 12, 79–84, 1978.
108. Haag, W.R., Hoigne, J., Bader, H., Improved ammonia oxidation by ozone in the presence of bromide ion during water treatment. *Water Res.*, 18, 1125–1128, 1984.
109. Hofmann, R. and Andrews, R.C., Ammoniacal bromamines: A review of their influence on bromate formation during ozonation. *Water Res.*, 35, 599–604, 2001.
110. Liu, H., Chen, L., Ji, L., Ozonation of ammonia at low temperature in the absence and presence of MgO. *J. Hazard. Mater.*, 376, 125–132, 2019.
111. Nawrocki, J. and Kasprzyk-Hordern, B., The efficiency and mechanisms of catalytic ozonation. *Appl. Catal. B Environ.*, 99, 27–42, 2010.
112. Ichikawa, S., Mahardiani, L., Kamiya, Y., Catalytic oxidation of ammonium ion in water with ozone over metal oxide catalysts. *Catal. Today*, 232, 192–197, 2014.
113. Mahardiani, L. and Kamiya, Y., Enhancement of catalytic activity of cobalt oxide for catalytic ozonation of ammonium ion in water with repeated use. *J. Jpn. Pet. Inst.*, 59, 31–34, 2016.
114. Chen, Y., Wu, Y., Liu, C., Guo, L., Nie, J., Chen, Y., Qiu, T., Low-temperature conversion of ammonia to nitrogen in water with ozone over composite metal oxide catalyst. *J. Environ. Sci. (China)*, 66, 265–273, 2018.

Part 3

DESIGN AND MODELLING OF PHOTOREACTORS

7

Recent Advances in Photoreactors for Water Treatment

Jean Bedel Batchamen Mougnol[1], Shelter Maswanganyi[1], Rashi Gusain[2,3], Neeraj Kumar[2,3], Elvis Fosso-Kankeu[4*], Suprakas Sinha Ray[2,3†] and Frans Waanders[1]

[1]*Water Pollution Monitoring and Remediation Initiatives Research Group, School of Chemical and Minerals Engineering, North West University, Potchefstroom, South Africa*
[2]*Centre for Nanostructures and Advanced Materials, DSI-CSIR Nanotechnology Innovation Centre, Council for Scientific and Industrial Research, Pretoria, South Africa*
[3]*Department of Chemical Sciences, University of Johannesburg, Doornfontein, Johannesburg, South Africa*
[4]*Department of Mining Engineering, College of Science Engineering and Technology, University of South Africa, Florida Science Campus, Johannesburg, South Africa*

Abstract
Water pollution and wastewater management is a global issue and gaining attention worldwide. Among several other wastewater treatment technologies, photocatalytic treatment is a promising and sustainable approach to recycle wastewater into clean water for various applications. Several semiconductor photocatalysts have been proven successful candidates to convert the toxic organic and inorganic water contaminant molecules into non-toxic molecules under the irradiation of light. Besides several key parameters of photocatalytic reactions (viz. light intensity, pH, photocatalyst concentration, temperature), the selection or designing of a photoreactor also significantly influence the efficiency of photocatalysis. In this regard, several types of photoreactors, such as slurry/membrane/rotating-drum/annular/closed-loop step photoreactor and microphotoreactor have been designed and employed for wastewater treatment. Heterogeneous, homogenous, and combined photocatalyst systems have been exploited in the reactors to achieve effective photocatalysis. Therefore, knowledge of photoreactor designing and different photocatalyst systems help to scale up the photocatalysis process at an

**Corresponding author*: elvisfosso.ef@gmail.com
†*Corresponding author*: rsuprakas@csir.co.za

Elvis Fosso-Kankeu, Sadanand Pandey, and Suprakas Sinha Ray (eds.) Photoreactors in Advanced Oxidation Processes: The Future of Wastewater Treatment, (217–246) © 2023 Scrivener Publishing LLC

industrial scale. This chapter is mainly focused on the fundamentals of photocatalysis and configurations of different photoreactor for the mineralization of toxic water contaminants.

***Keywords*:** Photocatalysis, photoreactors, monitoring and treatment, water purification, green environment

7.1 Introduction

Photocatalysis has been identified as a promising method for wastewater treatment to break down the organic pollutants and disinfection of toxic microorganisms [1–6]. In the photocatalytic process, reactive species, such as hydroxyl radicals (˙OH), generates and reacts to degrade the toxic components under light illumination with a suitable catalyst. In spite the discovery of photocatalysis, the application of this technology is still facing some serious challenges, such as degradation rate, reproducibility, and cost [7]. The efficiency of the photocatalysis process is controlled by several factors, such as: (a) by the band gap of suitable photocatalyst [8–11], (b) appropriate light source [12–14], (c) by the recovery of photocatalyst after the reaction for the subsequent reactions [15–17] and (d) by optimizing the photocatalytic reactor and reaction conditions [18].

Photoreactor is a device that has been developed with precise parameters and measurements to be used in different applications of photocatalysis, including wastewater treatment. It is an integral part of the photocatalysis mechanism that is vital to engineering applications. The function of the photoreactor is to enable the photons that will interact with a photocatalyst with an objective to initiate the photocatalytic degradation of pollutants from wastewater. There are two crucial considerations during the dealing with photoreactors: (a) how the photocatalyst can be illuminated effectively (a wide region must be illuminated for high activity) and (b) how photoreactors can be optimized for the use of irradiation from various sources [19]. The first ever used solar photocatalytic reactor were designed and built at the end of the 1980s, which was in the form of a parabolic collector [20]. Thereafter a significant number of photoreactors have been examined and evaluated on the efficacy of their functionality in the elimination of toxins via photodegradation. For industrial-scale purposes, the aimed photoreactors should be a low investment and operating cost with high performance. A suitable designed photocatalytic reactor is one of the crucial challenges in the heterogeneous photocatalysis, which involve development of several mathematical models [21].

To achieve the high photocatalytic efficiency of a photocatalyst, the designing and fabrication of a photoreactor is one of the extremely important factors. For example, for a wide band gap photocatalyst, the photocatalytic reactor must be configured of the UV-light source, whereas, if the photocatalyst is visible-light-driven, it is configured with the visible light source. Additionally, the position of the light source, material, temperature and dimensions of the photoreactor, along with other parameters, also significantly affects the photocatalysis process [18].

This chapter will be focused on the study of configuration and different types of photoreactors used in wastewater treatment on a laboratory or pilot scale. The main parameters, which are essential during the designing of the photoreactor, is also discussed. Moreover, the special attention is focused on the detailed review of photoreactors used in water treatment.

7.2 Photocatalysis Fundamentals and Mechanism

The term photocatalysis is used to illustrate a photo-induced reaction in the presence of a semiconductor material, which acts as a photocatalyst. It is a popular technique in widespread applications, such as water treatment, cancer treatment, self-sterilization, anti-fogging, CO_2 reduction, water-splitting, and so on [22, 23]. This process was first discovered by Fujishima and Honda to evaluate the photocatalytic characteristics of titanium dioxide (TiO_2) [24]. Recently, photocatalysis has been rapidly expanding and winning more attention from researchers due to the cost efficacy, no secondary waste production, use of renewable source (sunlight) and ease of operational efficiency in rectifying environmental challenges (e.g. wastewater treatment) [25].

Photocatalytic processes can be classified into two categories, namely homogenous and heterogeneous photocatalysis [26, 27]. Homogenous photocatalysis involves the photocatalyst material and the reactants in similar phases. The most commonly used, homogenous photocatalysts are metal complexes, such as complexes of Fe, Cu, and Cr. In homogenous photocatalysis, toxic compounds and water pollutants are degraded by the hydroxyl radicals that are generated by the metal ion complex with a higher oxidation state. In contrast, in heterogeneous photocatalysis, reactants and photocatalysts materials exist in different phases. The chemical reaction is driven by the transfer of electrons from the filled valence band to the empty conduction band, unlike in homogenous photocatalysis where an oxidizing agent is required for chemical reactions to take place and also highly dependent on the continuum electronic states of the metals [28].

Moreover, heterogeneous photocatalysis can make use of the light from the sun as the source of irradiation, unlike homogenous processes, which uses UV radiation.

On comparing the homogeneous and heterogeneous photocatalysis, heterogeneous photocatalysis is considered a promising and effective technique for the degradation of pollutants from wastewater due to total mineralization, absence of inhibition by water-found ions, easy separation of photocatalyst, low cost, simple operations and easy maintenance of the equipment [29]. TiO_2, CuO, MoS_2, WO_3, ZnO, ZnS, $BiVO_4$, Bi_2MoO_6, BiOCl, ZrO_2, Fe_2O_3 etc., are the most commonly used semiconductor materials as heterogeneous photocatalysts [30–36]. During the heterogeneous photocatalytic process, semiconducting materials use sensitizers for the light irradiation due to their electronic structure. Figure 7.1 depicts the main steps that are typically involved in heterogeneous photocatalysis. The important steps in heterogeneous photocatalysis are as follows: (a) absorption of reactant species on the photocatalyst surface; (b) adsorption of the photon by photocatalyst; (c) electron-hole pair generation; (d) separation of electron hole-pairs and migration to the surface of the photocatalyst; (e) recombination of electron-hole pairs and/or (f) trapping of electron-hole pairs on the surface via redox reactions; and (g) diffusion or desorption of product from photocatalyst [37, 38].

Nowadays, heterogeneous photocatalysis is the most commonly used technique to degrade the pollutants in wastewater. The fundamental

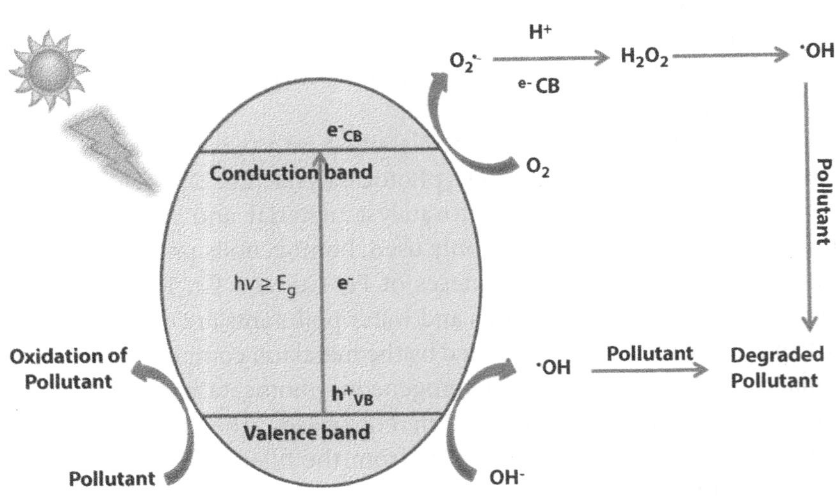

Figure 7.1 General mechanism of photocatalytic wastewater treatment. Reproduced with permission from ref. [38].

mechanism of heterogeneous photocatalysis in wastewater treatment occurs in the following steps:

(a) First, the light radiation with energy equivalent or greater than the band gap energy of the photocatalyst, strike on the surface of the photocatalyst. (b) Second, the electrons in photocatalyst semiconductor material agitated from the valence band and moved to the conduction band, leaving holes in the valence band. (c) These photo-induced electrons and holes either initiate the redox reaction by charge transfer or recombine back and release the energy as heat. The recombination of photo-induced electrons and holes is undesirable as it lowers the efficiency of the photocatalyst and prevents the initiation of the redox reaction. (d) The holes in the valence band can react directly with the pollutant adsorbed on the surface and degrade into smaller nontoxic components. However, the hole is more likely to react with water molecules adsorbed on the photocatalyst surface to degrade the pollutant molecules. The reaction between the hole and water molecules or hydroxide ion (OH^-) is called oxidation reaction and generate the reactive $^{\cdot}OH$, which in turn degrade the toxic components [39]. Conversely, the electrons in the conduction band reduced the oxygen (O_2) to produce reactive oxygen species (ROS) [40]. These reactions prevent the recombination of electron hole-pairs and allow for the accumulation of ROS that can participate in the degradation of the pollutant. The reduction reaction in the conduction band further produces the intermediate hydroperoxyl radical ($HO_2\bullet$), which can be protonated to react with hydrogen ions (H^+), and thus produces hydrogen peroxide (H_2O_2). The formed H_2O_2 behaves as an oxidizing agent and produces more hydroxyl radicals, which help in degrading the organic pollutants. The production of more radical enhances the photocatalytic degradation of the pollutants, as these radicals are the ones responsible for the destruction of the pollutant.

7.3 Configuration of Photoreactor

To perform the photocatalytic reaction efficiently, the effectual configuration of photocatalytic reactor is one of the most important steps. Most of the past research was aimed to focus on the configuration of certain basic photoreactors setups [41]. According to Daisey *et al.* (1982), an effective photoreactor for the photodegradation of an organic pollutant is designed with few elements that are inexpensive, hence easy to build and being operational [42]. During the designing of the photoreactor, the following featured should be considered:

a) The device should be three-dimensional, which can trap the particles trapped in the air or other gasses during irradiation.
b) The device should not limit the time of irradiation.
c) The spectrum of light in the reaction should be as similar as possible to that of sunlight.
d) It should be possible to assess the light intensity of the reactor.
e) It should determine the reproductive efficiency of pollutant degradation.

The designing and configuration of the photoreactor determine the efficiency and reproducibility of the photocatalytic reaction. Therefore, it is essential to have an idea of a photoreactor configuration before constructing it. Several parameters, such as light intensity, light reflection, photocatalyst loading, photoreactor temperature, and depth of photolyte, significantly affect the photocatalytic reaction and hence should be considered during the designing of photoreactor. The photoreactors configurations of a photocatalyst reactor must obey the following rules [43]:

a) emission spectrum of light, the transmitted spectrum of the reactor window and the absorption spectrum of the photocatalyst must all be matched to considered a suitable photocatalyst reactor.
b) There must be a correlation in the depth of the incident ray and the reactor.
c) The photoreactor configurations must match with the distribution of the photocatalyst.

The following subsection discussed the few parameters during the configuration of the photoreactor.

7.3.1 Source of Light Irradiation

The purpose of photoreactors configurations is to have a straightforward design that is thoughtful to achieve an effective outcome. However, the design of a photocatalytic reactor and its configurations is not easy as compared to others. The fact that a provision to supply sufficient irradiation to the mass of the photocatalyst in the system makes it complicated. Two types of light irradiation are used in photoreactors: (a) natural solar irradiation (sun) and (b) artificial solar irradiation (lamps). Photoreactors based on

Table 7.1 Characteristics of light sources [47].

Light source	Type	Nominal power [W]	η, %
UV lamp	UV lamp, λ ¼ 365 nm	32	25
BL strip	Blue light LED strip	5	50
WL strip	White light LED strip	5	50
BL	Blue light LED	20	50
WL	White light LED	6	50

the solar simulator, mainly for home purposes, is consequently configured in need of a lamp source [44] and a parabolic reflector [45, 46] to establish a suitable photoreactor. A low-pressure mercury lamp (LPML) is the most popular and oldest solar light lamp used in photoreactors. There are different light sources that can engage in photoreactor configurations. Still, these light sources have limitations that vary in wavelength and can undergo photodegrading of organic pollutants with the help of a photocatalyst [47]. Table 7.1 shows the various light sources with different wavelength and η% (electricity into light conversion efficiency). Therefore, it is also essential to know how to use the specific wavelength solar radiation effectively in the photoreactor under various operating conditions [48]. The band gap of the photocatalyst and feasibility of the utilization of solar light of specific spectrum should also be kept in mind during the fabrication of artificial lamp in the photoreactor. Under artificial solar simulated light irradiation, most of the photoreactions are carried out under UV and/or visible part of electromagnetic radiation.

7.3.2 Geometry of Photoreactor

Physical geometry is important to photoreactors because it defines the photon accumulation performance of the overall photochemical system [49]. The high transparent material, shape of the light source and the determination of the pipe diameter should be considered as an important factor while designing the photoreactor. The geometry of the photoreactor should be designed in order to collect the maximum emitted light. For efficient utilization, the diameter (D) of the photoreactor is designed 4 times to the specific depth of heterogeneous radiation. For example, if the penetration of incident light is 1-3 cm, then the pipe diameter should be 4 cm greater. Most importantly, the pipe diameter should not be too large and thus internal diameter should be 4 cm. Given all these measurements compiled to an effective and easy photoreactor [43].

7.3.3 Light Source Placement and Distribution

Photo-irradiation can be performed via external illumination [47, 50] or internal illumination [51]. External illumination is carried out by focusing the lens, an IR filter, thermostated bath for cooling and an interference filter or simple Pyrex glass to the reaction cell [52]. External photo-irradiation can be achieved by using one or more LEDs or lamps. Mercury lamps are a popular example of external photo-irradiation. External light irradiating photoreactors are generally used for visible-light-driven photocatalysis reactions. Figure 7.2 exhibits two different configurations for external illumination. However, in internal light irradiation photoreactors (also called as immersion lamp photoreactors), a lamp is fitted inside the quartz jacket and immersed in the reaction solution. These photoreactors consist of three types of cooling configurations: (a) lamp is secured in the additional jacket to circulate the water, (b) the reactor

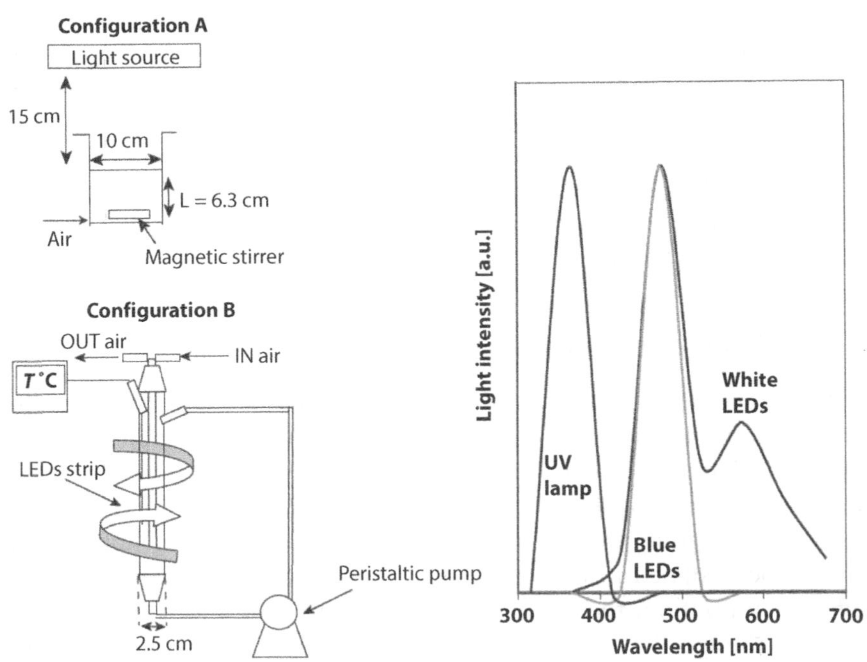

Figure 7.2 Reactor configurations (A and B) and emission spectra of the different light sources. Reproduced with permission from Ref. [47].

exhibit an additional jacket to circulate the water and (c) the entire photoreactor system with the light source is placed in the water bath [52]. The used light source can be UV-lamp, visible lamp, blue LEDs or white LEDs (Figure 7.2).

7.3.4 Photoreactor Materials

Since light sources are a critical element to be considered, there are materials and dimensions that can also favour these light sources. Certain materials with specific dimensions have been given preferences to be able to captivate as much radiation as possible. Quartz material has been appointed in photoreactors due to having excellent UV transmittance, good thermal and chemical resistance. Its high excellent UV transmittance is prompted by high iron content. However, the high cost of quartz had made it to be less considered toward an easy built of a photoreactor. Low-iron borosilicate glass is also an alternate material that has the same properties as quartz [53]. Ultraviolet solarization is the process whereby changes in valance of metal ions from a material, preference glass, undergo a transformation during the UV absorption. Based on these radical changes, an addition of 0.1 % silicon (Si) can be added to the melting material (glass) to reduce the melting process and endure the lifespan of the material. Several additional materials, such as aluminium reflectors or glass transmittance, are also introduced in the photoreactor to provide high reflective in the ultraviolet spectrum of solar radiation at a wavelength range of 300–400 nm [53, 54]. Fluorinated, acrylic polymers and several other types of glass material in photoreactor also enhance the ability to absorb as much solar radiation [55]. The fluoropolymers materials have been chosen to be the best due to their high UV and chemical stability. To achieve the minimum operating pressure, the fluorinated wall thickness should be increased, which will also reduce the UV transmittance from the system and make it compatible [56].

The use of photocatalyst during the photoreactor configurations on its reaction can cause wind-down fouling. Wind-down fouling is when the photocatalyst is being stuck or sticking on the photoreactor, therefore to manage or avoids this, an additional metal oxide, preferably TiO_2 which will have a role on the reduction of radiation flux entering the photocatalytic photoreactor is recommended [43, 53]. The selectivity of the material during the design and configuration of the photoreactor is vital for the specific operation of the reactor. The materials must be translucent in order to emit a larger absorbent resulting in size limits, sealing problems and the possibility of breakage [55].

7.4 Types of Photoreactors

Photocatalysis is influenced and affected by several factors, and the reactor type/configuration is one of the many. Therefore, to maximize the efficiency of photocatalytic reactions, a reactor with suitable configurations has to be selected. A good photoreactor must possess the following features and advantages: large surface area per volume unit reactor, unconstrained, mass transfer within the system, permits the photocatalyst to reabsorb photons reflected from the surface and ability to provide UV and/or solar radiation. Photoreactors, such as membrane, slurry, rotating drum, closed-loop step, and annular photoreactors, are some of the reactors that have been developed for use in photocatalysis. An understanding of these reactors is important in the selection a photoreactor for a particular reaction and very useful in scaling-up of the photoreactors. The following section explains the types of various photoreactors.

7.4.1 Slurry Photoreactors

Slurry photoreactors are mostly preferred for the systems with three phases, for an example; a CO_2 reduction process where CO_2 is in gaseous phase, reducing agent in liquid phase and a photocatalyst material in solid phase [57]. For a photocatalyst to remain suspended in slurry reactors there has to be a continuous agitation, or injection of gas to purge the system. To improve the purity of the slurry photoreactor system, air is usually removed from the reactor before any reaction can be started. Slurry photoreactors are also known as fluidized bed photoreactors. These photoreactors can be operated in two arrangements, i.e. batch or continuous process [58]. The difference between batch and continuous slurry photoreactors is that, in batch operation the purging of the system with a gas that will not react with the components of the system is only ran for a limited time i.e. at the beginning of the process and when the purging step is complete, the reactor is tightly closed, then illuminated for a reaction to begin [57, 59].

The configuration of slurry photoreactors has the following benefits: (1) the fluidized particles provide a high surface to volume ratio; (2) the reactor is able to retain fluidized particles, therefore a need to have an external process to separate solid material from the fluid is eliminated; (3) the highly unstable fluid and particles flow reduces mass-transfer limitation. The merits and demerits of slurry photoreactors are clearly populated in Table 7.2. The important factors that determine the reaction rate in slurry photoreactors are; (1) light intensity on the photocatalyst surface;

Table 7.2 Merits and demerits of slurry photoreactor.

Merits	Demerits
Simple design	Reduced penetration depths due to liquid and solid particle absorbing light
Easy to scale up	Large inactive surface sites of the photocatalyst
No mass transfer limitation	Reduced photoactivity due to vigorous stirring
High degradation rate	Solids particles inside the reactor foul the light source
Large surface area for chemical reactions	Separation of photocatalyst from the water is required

(2) adsorption potential of the components in the reactor; (3) the quantum efficiency of the photocatalyst [57].

7.4.2 Photocatalytic Membrane Photoreactors

One of the biggest and common obstacles of using photocatalysis in wastewater treatment is the difficulty of separating the photocatalyst from the treated water for recycling. As a result, a recovery step is usually required to separate and recover the photocatalyst in slurries for reusing. Membrane technology has been successfully used for separation purposes. Thus, the idea to couple photocatalysis with membrane technology has been considered [60]. The combination of membrane technique with photocatalysis makes systems that is known as photocatalytic membrane reactor (PMR). In simple terms, PMR is a reactor system where an effective photocatalysis and recovery of the photocatalyst occur simultaneously using the membrane (Figure 7.3) [61]. The driving force in PMRs can be pressure, concentration or partial pressure difference.

PMRs are environmental friendly equipment as they allow chemical reactions to occur and still allow retention of non-degraded, and undesired molecules in the membrane for further degradation and the retention of the photocatalyst for reuse [62]. The advantages of PMRs mentioned above are beneficial in improving photocatalytic processes in terms of stability, controllability and efficiency, recollection of photocatalyst for further use in consecutive reactions, beneficial in energy saving as well as in reduction of number of units for complete degradation of pollutants because

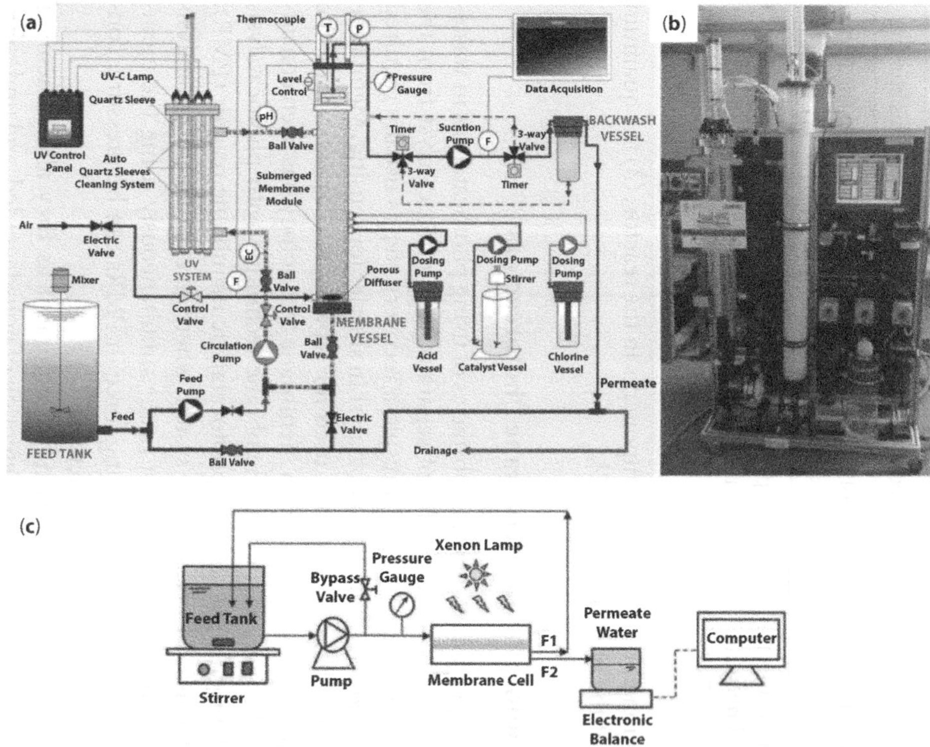

Figure 7.3 Schematic diagram (a) and front side (b) of the pilot scale continuous photocatalytic membrane reactor system with photocatalyst suspended in the feed mixture. (c) Photocatalytic membrane reaction system with photocatalyst immobilized in/on the membrane [61]. Reproduced with permission from [64] (a), and [61] (b).

additional systems for separation purposes, such as coagulation, sedimentation, and flocculation, will not be required [60].

PMRs are normally put into different categories and there are different approaches to classifying PMRs. The common classification of PMRs are based on photocatalyst configuration and on the location of the source of light [63]. On the basis of photocatalyst configurations, PMRs can be categorised into two configurations, namely, (1) PMRs with photocatalyst suspended in the feed mixture (Figure 7.3a); and (2) supported PMRs with photocatalyst immobilized in/on the membrane (photocatalytic membrane) (Figure 7.3b). PMRs with photocatalyst suspended in the feed mixture can be further classified into two subcategories; combined-type PMRs, where photocatalysis and separation system take place in one unit and separated-type PMRs, whereby photocatalysis and membrane separation process take place in two different coupled units, which are the photoreactor and the membrane vessel [62]. Most studies have indicated that PMRs with suspended photocatalyst always attain higher efficiency when compared to PMRs with photocatalytic membrane [63]. The larger surface area in PMRs with suspended photocatalyst ensures an adequate contact between the photocatalyst surface and the target pollutant. Due to the benefits mentioned above, PMRs with suspended photocatalyst have been widely studied and largely used. The disadvantage of PMRs with suspended photocatalyst are: (1) scattering of light by suspended photocatalyst particles; (2) membrane fouling as a result of photocatalyst particles deposition on the surface of the membrane [62].

However, PMRs with photocatalytic membranes has some advantages over PMRs with suspended photocatalyst such easy recovery, regeneration and reusability of the photocatalyst. Furthermore, the use of photocatalytic membranes prevents the challenge of light scattering that occurs in PMRs with suspended photocatalyst. PMRs with photocatalytic membrane give flexibility in the design of the process since the photocatalytic activities can take place on both the feed and permeate side of the membrane [65]. However, the limitation with PMRs with photocatalytic membrane is that the degradation of the pollutant happens within the membrane pores or its surface, as a result in order to initiate photocatalytic reactions, the membrane must be irradiated directly. Due to the latter limitation, it is important that membranes of higher stability under the action of hydroxyl radical and irradiation must be used. The types of membranes that are stable are usually very costly. Moreover the immobilized photocatalyst (photocatalytic membrane) results in mass-transfer limitation, which subsequently decreases the effectiveness of pollutant degradation process [63]. The need to change the entire membrane whenever the photocatalyst loses its activity is also a big obstacle of this technology.

On the other hand, on the basis of location of light sources, PMRs can be categorized as follows: (1) PMRs with light source inside/above membrane unit; (2) PMRs with light source inside/above the feed tank; (3) PMRs with light source inside/above an additional unit positioned between the membrane unit and the feed tank [60, 66].

7.4.3 Rotating Drum Photoreactors

A rotating drum is one of the innovative technologies that have gained considerable interest in photochemistry over the few years [67]. The motivation to apply this technology in photocatalysis is to improve and optimize photocatalytic reactions in terms of improving light distribution, surface area to volume ratio and providing mixing inside the photoreactor. Rotating drum photoreactor is a reactor design where a light source/UV lamp is installed inside a drum along the axis and material acting as photocatalyst is coated and immobilized outside the glass-drum [68]. The photocatalyst coated glass drum is partially immersed in contaminated water and rotated with a motor. To maximize the efficiency of rotating drum photoreactors, multiple drums are normally used in one system (usually 5 rotating drums). The drums are usually placed horizontally and parallel to each other, so that the light from the sun outside and/or the artificial UV light from the lamp inside the drum can reach the polluted water. This improves the efficiency of the photocatalyst and the effectiveness of the photocatalytic process in the destruction of the pollutants [69]. The outstanding advantages of introducing rotation in photoreactors are; (1) enhanced mass transfer rate in the system due to the continuous contact between the pollutant and surface of the drum reactor; (2) improved photocatalytic efficiency due to controlled periodic illumination; (3) continuous movement in the system due to rotating drums [68]. Moreover the continuous controlled periodic illumination is important in reducing the build-up of the intermediates and reduce the occurrence of undesirable side-reactions [70]. The other important benefits of rotating drum photoreactors is that, since the water film on the surface of the drum is thin, penetration of light to the photocatalyst surface happens seamlessly [69, 71]. Furthermore the photoreactor system is exposed to the atmosphere to receive solar light and the oxygen from air, the O_2 molecules diffuse rapidly to the photocatalyst surface and cause the formation of $^{\bullet}OH$ and $^{\bullet}OOH$ radicals, which accelerate the photodegradation of water pollutants [72].

7.4.4 Microphotoreactors

Microphotoreactors are devices in which photocatalytic reactions take place in small tubes or micro channels arranged parallel to each other and the tubes usually have an internal diameter of 10 to 50 μm [73, 74]. Microphotoreactors were developed to solve the limitations of conventional photoreactors and improve the effectiveness photocatalytic reactions [75]. Microreactors technology has been successfully utilized in organic chemistry field, and due to its desirous features in photochemistry, such as light penetration, even for concentrated solutions, microphotoreactors were designed. Microphotoreactors exhibit several advantages on photocatalysis, such as better light penetration through the entire reactor depth, higher spatial illumination homogeneity, lower mass-transfer constraints due to the high surface to volume ratio that exists, small amount of photocatalyst can be used, and reduced irradiation time [73, 75]. One of the obvious drawbacks of microphotoreactors is, not being able to scaling–up the process. Therefore if there is a desire to increase the production rate in microphotoreactors, increasing the number of tubes in reactor is one of the solutions [76]. Owing to the fact that the all the tubes are required to have the exact same features, numbering-up of microphotoreactors can be time-consuming and costly [77].

7.4.5 Annular Photoreactor (APR)

An annular photoreactor or tubular continuous reactor is a slurry photo reactor where the lamp is installed inside a vessel in such a way that the suspended photocatalyst or the material to be illuminated surrounds the light source, this arrangement gives the reactor vessel as U-shaped or annulus shape [78] (Figure 7.4). The simple schematic diagram of this configuration is displayed in Figure 7.4 [79]. A pump is configured in the photoreactor to circulate the feed between the reservoir and reactor. These photoreactors have been designed to solve issues and the limitation of the conventional slurry photoreactors [80]. APR allows for continuous mode of operation, which leads to the enhanced performance of the photoreactor in photocatalysis [81]. In this reactor, the light source is positioned inside vessel, which allows the utilization of the irradiation efficiently because the area in which the light directly beams on is very high. Moreover the U-shape that is created by the position of the light source shortens the distance travelled by the light radiation thus making it easier for light to be reflected back to the surface [81].

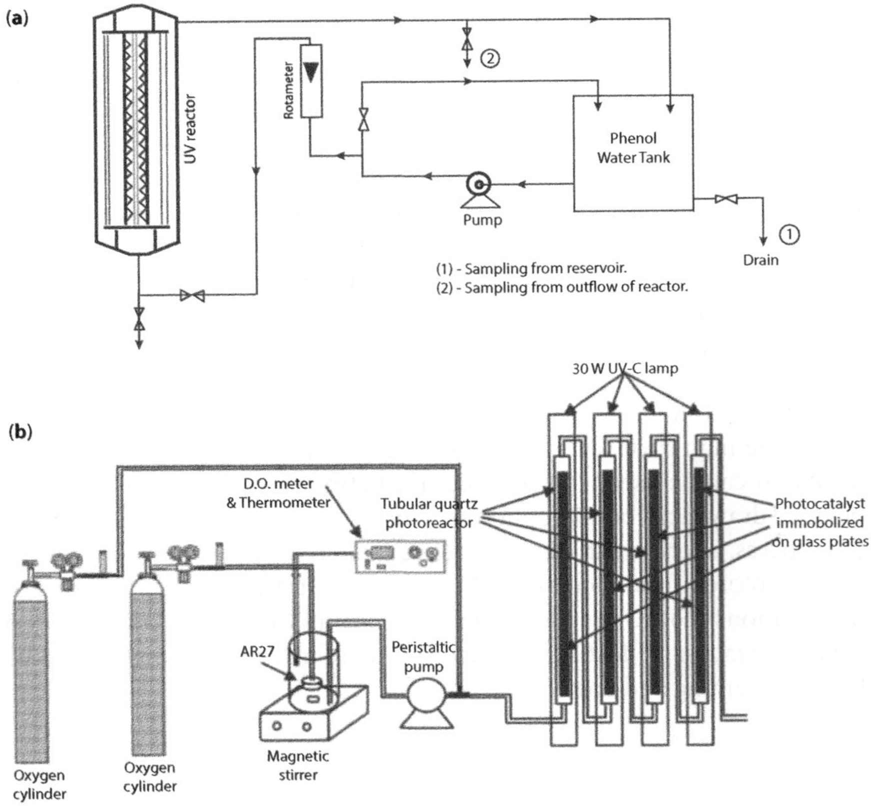

Figure 7.4 Schematic diagram of (a) annular photoreactor [79], and (b) tubular continuous-flow photoreactor with immobilised photocatalyst [82]. Reproduced with permission from [79] (a) [82] (b).

7.4.6 Closed-Loop Step Photoreactors

A closed-loop photoreactor refers to a process whereby no material cross the border of the system boundary once the photocatalytic reaction has started [83]. It is also known as a thin-film cascade photoreactor. Any components participating in the reactions can only enter the system at the beginning of the process and only be withdrawn out of the system when the reaction is complete. Closed-loop photoreactors consist of a flat thin-film where photocatalyst can be deposited, a tank that contains an aqueous solution and a pump for the movement of the liquid. Secondary or external units for separating photocatalyst from aqueous solution is not required in closed-loop systems [83, 84].

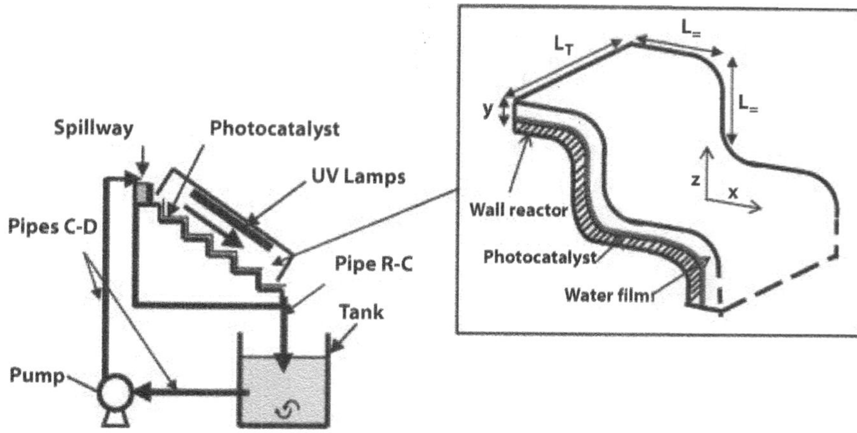

Figure 7.5 A simplified diagram of a closed-loop step photoreactor. Reproduced with permission from ref. [84].

Close loop photoreactors are improved by introducing the step geometry in the system (Figure 7.5). A step geometry provides a more surface for chemical reactions to take place when compared to flat films [84]. The circulation of the liquid in closed-loop photoreactors enhances the mass transfer, but the incorporation of the step further improves the mass transfer rate due to the breaks of the flow that is created by the presence of the steps [85]. Closed-loop step photoreactors offer the following benefits in photocatalysis: illumination zone of higher surface, the thin film of liquid absorbs little UV irradiation, which makes the photo-activation of the photocatalyst to be high, simple configuration and perfect mixing, and easy to scale up [85].

7.5 Photocatalytic Water Purification Using Photoreactors

Photocatalyst within the batch and flow photoreactors can be used by two methods: suspension/dispersed and immobilization/retained forms (Figure 7.6). In suspension approach, photocatalyst is suspended in the liquid phase with the aid of magnetic stirring, toroidal or non-toroidal agitation stimulated by gas molecules. The suspension of photocatalyst in the liquid phase is considered as a slurry system. In immobilization approach, the photocatalyst is bonded to inert support/surfaces via a physical or a chemical bonding [86].

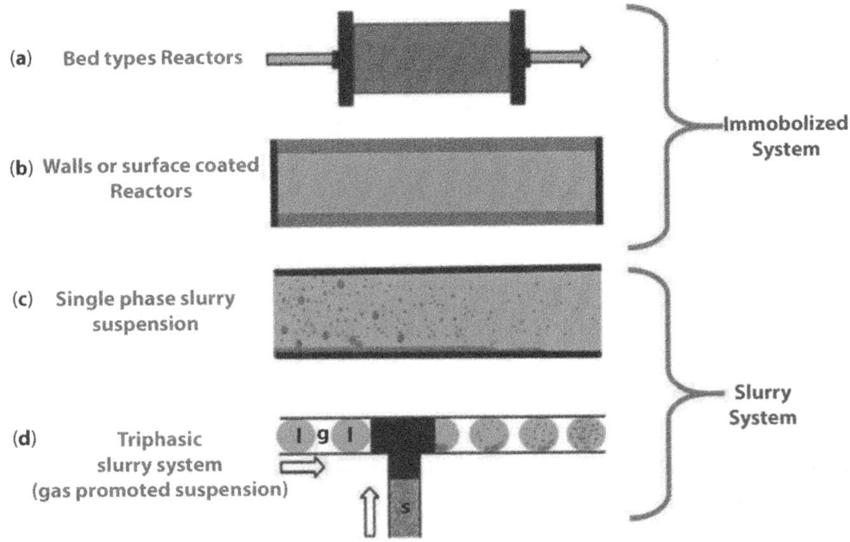

Figure 7.6 Approaches used for the effective implementation of photocatalysts in flow reactors. Reproduced with permission from ref. [86].

The high light absorption and effective reactive oxygen species interaction with pollutants are noticed in photoreactor with suspended/dispersed catalyst, but photocatalyst reuse is obstructed by the laborious recovery of nanocatalyst after completion of the process. Moreover, the toxicity of nanocatalysts and their consequences on the environment are still not completely determined. Other side, photoreactor with immobilized/retained forms of photocatalysts offers reusability of catalyst and high performance. The fluidized-bed photocatalytic reactor is an example of a photoreactor with immobilized/retained forms of the photocatalyst. In this reactor, the light source can be adjusted at the reactor walls or the center of the reactor, considering either LED lights or tubular lamps. For example, Fang *et al.* fabricated TiO_2 nanotube pillared graphene-based macrostructures for photodegradation of bisphenol A (BPA) in a fluidized bed reactor [87]. The composite macrostructures removed 86% of the BPA solution (0.05 mg/L) in the continuous flow system, whereas 97% removal of the BPA solution (5.0 mg/L) over 30 min was achieved in the batch reaction.

Furthermore, immobilization of photocatalyst on a solid, fix support enable long operation time and minimum amount of catalyst loss. Among various supports, polymeric membranes, such as polyvinylidene fluoride (PVDF) and polyethersulfone (PES) have been commonly used to enhance the mass transfer of contaminants to the photocatalytic sites. For instance,

Fischer *et al.* immobilised TiO_2 nanoparticles on PES membrane by dip coating and ultrasound treatment for dye photodegradation. The produced TiO_2/PES membrane showed almost 100% degradation of methylene blue (13 mg/L) in 40 minutes for nine consecutive cycles [89]. The degradation of toxic components can be explained on the basis of its breakdown into smaller products. Sometimes, it has also been noticed that the degraded products are more toxic and carcinogenic than the parent component. Therefore, the study of degraded products and maximum mineralization of toxic components is essential. Li *et al.* exhibited the usage of double cylindrical shell photoreactor with an immobilized monolayer of silica gel beads coated with TiO_2 [90]. Fouad *et al.* described the use of W-TiO_2 coating on inclined stainless-steel plates to form thin-film fixed-bed reactors, which was irradiated with two metal halide lamps placed above the reactor [91]. Moreover, Stephan *et al.* developed a thin-film cascade photoreactor comprised of fibres steps coated by TiO_2, where polluted water circulated in the closed-loop [84]. Alternatively, Sun *et al.* proposed a continuous magnetic aggregation bed photocatalytic recycle reactor for dye treatment (Figure 7.7a) [88]. They used a magnetic photocatalyst ($CoFe_2O_4$-Ag_2O),

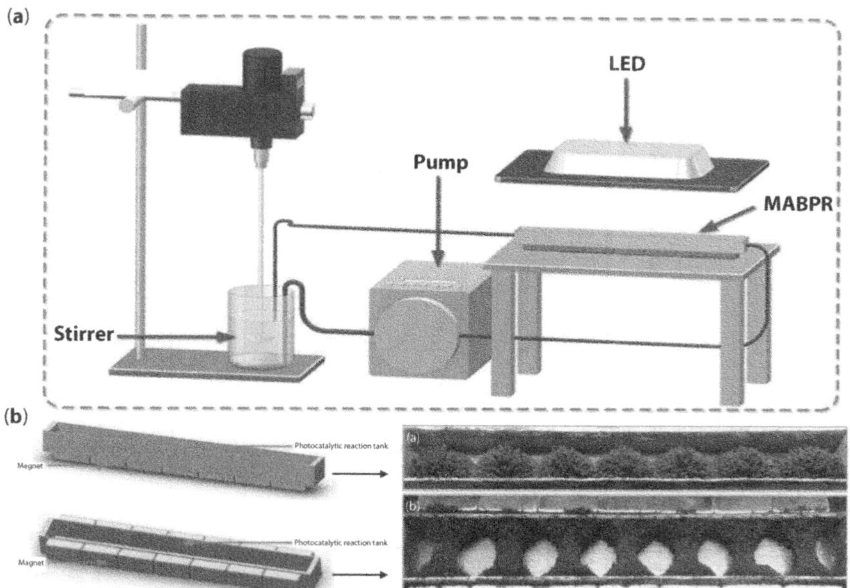

Figure 7.7 (a) Illustration of the assembled continuous magnetic aggregation bed photocatalytic reactor. (b) Different magnet arrangements in photoreactor: bottom and lateral set-up. Reproduced with permission from ref. [88].

Table 7.3 Some recent examples of different photoreactors and their performance.

Reactor	Light source	Pollutant	Photocatalyst	Efficiency	Reaction time	Reuse cycles	Ref.
Micro–meso-reactor	1700 W Xe lamp	Cr(VI)	TiO_2	100%	60 min	10	[64]
Submerged coated plate reactor	400 W metal-halide lamp	Multiple bacteria	Ru-WO_3/ZrO_2	100%	240 min	4	[93]
Coated plate reactor	400 W metal-halide lamp	Chlorpyrifos	ZrV_2O_7/graphene nanoplatelets	96.8%	90 min	5	[94]
Fluidized bed reactor	350W Xe lamp	Bisphenol A	TiO_2 nanotubes/ graphene	100%	30 min	5	[87]
Polymer membrane reactor	Solar lamp	Methylene blue	TiO_2	100%	40 min	9	[89]
Continuous flow slurry reactor	36W Philips PL-L sunlamp UV tubes	Methylene blue	TiO_2 pellet	98%	60 min	-	[95]
Magnetic aggregation bed reactor	600 W LED lamp	Methyl orange	$CoFe_2O_4$-Ag_2O	92%	60 min	3	[88]
Photocatalytic PVDF membrane	500 W Xe lamp	Fluvastatin	$ZnIn_2S_4$	99.75%	180 min	6	[61]
Submerged membrane reactor	500 W Xe lamp	p-nitro-phenol	Fe-ZnS/g-C_3N_4	93.5%	300 min	-	[96]
Immobilized double-cylindrical-shell photoreactor	UV black light lamp	Rhodamine B	TiO_2-coated silica gel beads	90.4%	12 h	5	[90]

which was anchored to the bottom of the reactor via an external magnet (using either at the bottom or lateral arrangement) (Figure 7.7b). The magnetic aggregation in this reactor increased the mass transfer and irradiation surface area. It did not require extra energy input for photocatalyst recovery and mass of photocatalyst remained same over the course of the reaction. However, the performance of magnetic reactor is lower than those of many other reported reactors, whereas it has potential to scale-up for real water treatment. Moreover, continuous slurry reactors with gas promoted suspension are also designed for water treatment. For instance, Rakhshaee *et al.* fabricated a plug flow tubular photoreactor in which catalyst (nano α-Fe_2O_3) were suspended in the liquid phase using air bubbles [92]. It was tested for photocatalytic degradation of different dyes (malachite green, methylene blue, rhodamine B, and bismarck brown Y). Table 7.3 presented some examples of photoreactors and their performance for the treatment of water pollutants.

7.6 Challenges for Effective Photoreactors

Slurry reactors including both solid-liquid and solid-liquid-gas system, are highly beneficial as compared to immobilized reactors due to better photocatalytic performance. The higher catalytic activity of the slurry reactor is attributed to the available high total surface area of photocatalysts per unit volume. As a result, the slurry reactor is very cost-effective in terms of operation and fabrication. Still, it suffers from the drawback of downstream separation of photocatalyst from solution after completion of the reaction. To avoid downstream separation, researchers are inspired to develop/switch toward continuous flow slurry photoreactors without any immobilization. Additionally, recent research efforts are also dedicated to switching the mechanically obtained suspension by air or other gases promoted suspension. This change is essential to match the demand for microreactors, which requires low photocatalyst loading. However, limited research has been done on immobilized reactor systems to improve the irradiation source and required length of scale-up reactors. There are still many challenges, which obstruct the full-scale application of photoreactor with immobilised photocatalyst. Due to inherent kinetics and mass transfer limitations of the photocatalytic process, the processing capacities of the reactor are usually restricted. To achieve enhanced photocatalytic efficiencies, the higher residence time is needed, which can further slow down the complete process. Thus, utilisation of immobilized reactors at a large scale for real water treatment is still challenging.

In order to develop an effective photoreactor system, microreactors have gained attention due to their compact design, high surface to volume ratio and better mass transfer compared to the conventionally used large-scale photoreactors [97]. In addition, they offer low catalyst loading, uniform light distribution and improved heat transfers [98].

Furthermore, more studies are needed to investigate the decrease performance in consecutive cycles, intermediate products during photocatalytic degradation, reduction in photocatalyst leaching and applicability of photoreactor in real water matrices (treated municipal wastewaters, river waters, and polluted industrial water) or multi-component pollutants.

7.7 Conclusion

Heterogeneous photocatalyst has been proven the best remediation for the organic pollutants in wastewater with the help of solar radiation. Researches have been accomplished to analyze the efficiency and mechanism of photocatalytic degradation. However, the research that focuses on the configuration and set up of photochemical reactors is limited. There are several types of photoreactors, which employed for the numerous photochemical reactions. For the photoreactors to be operational, they must uphold an effective geometry, preferred light source, materials that have higher reflective property and a thermostat for temperature control. The role of photocatalytic materials and the geometry of the photoreactor have been proved to provide effective photodegradation. Various photoreactor designs such as slurry photoreactor, rotating drum photoreactor, membrane photoreactor, closed-loop step photoreactor, annular photoreactor, mirco-photoreactor and magnetic bed photoreactor have been developed for water treatment. Photocatalyst within the batch and flow photoreactors can be used by two methods: suspension/dispersed and immobilization/retained forms. Continuous slurry flow reactors and microphotoreactor are highly advantageous compared to immobilized reactors due to better photocatalytic performance. Furthermore, there is a need to develop photoreactors that are effective for real water matrices and in the multi-pollutants system. More research efforts are required to make photocatalysis technology more commercially viable and environmentally sustainable than conventional processes.

References

1. Ahmed, S.N. and Haider, W., Heterogeneous photocatalysis and its potential applications in water and wastewater treatment: A review. *Nanotechnology*, 29, 342001, 2018.
2. Al-Mamun, M., Kader, S., Islam, M., Khan, M., Photocatalytic activity improvement and application of UV-TiO2 photocatalysis in textile wastewater treatment: A review. *J. Environ. Chem. Eng.*, 7, 103248, 2019.
3. Gusain, R., Kumar, N., Ray, S.S., Recent advances in carbon nanomaterial-based adsorbents for water purification. *Coord. Chem. Rev.*, 405, 213111, 2020.
4. Mukwevho, N., Fosso-Kankeu, E., Waanders, F., Kumar, N., Ray, S.S., Yangkou Mbianda, X., Photocatalytic activity of Gd2O2CO3·ZnO·CuO nanocomposite used for the degradation of phenanthrene. *SN Appl. Sci.*, 1, 10, 2018.
5. Kumar, N., Mittal, H., Alhassan, S.M., Ray, S.S., Bionanocomposite hydrogel for the adsorption of dye and reusability of generated waste for the photodegradation of ciprofloxacin: A demonstration of the circularity concept for water purification. *ACS Sustainable Chem. Eng.*, 6, 17011–25, 2018.
6. Ray, S.S., Gusain, R., Kumar, N., *Carbon nanomaterial-based adsorbents for water purification: Fundamentals and applications*, Elsevier, Amsterdam, 2020.
7. Moroz, P., Boddy, A., Zamkov, M., Challenges and prospects of photocatalytic applications utilizing semiconductor nanocrystals. *Front. Chem.*, 6, 353, 2018.
8. Adnan, M.A.M., Julkapli, N.M., Abd Hamid, S.B., Review on ZnO hybrid photocatalyst: Impact on photocatalytic activities of water pollutant degradation. *Rev. Inorg. Chem.*, 36, 77–104, 2016.
9. You, J., Guo, Y., Guo, R., Liu, X., A review of visible light-active photocatalysts for water disinfection: Features and prospects. *Chem. Eng. J.*, 373, 624–41, 2019.
10. Kumar, N., Kumar, S., Gusain, R., Manyala, N., Eslava, S., Ray, S.S., Polypyrrole-promoted rGO–MoS2 nanocomposites for enhanced photocatalytic conversion of CO2 and H2O to CO, CH4, and H2 products. *ACS Appl. Energy Mater.*, 3, 9897–909, 2020.
11. Kumar, N., Sinha Ray, S., Ngila, J.C., Ionic liquid-assisted synthesis of Ag/Ag2Te nanocrystals via a hydrothermal route for enhanced photocatalytic performance. *New J. Chem.*, 41, 14618–26, 2017.
12. Yu, C., Wu, Z., Liu, R., Dionysiou, D.D., Yang, K., Wang, C. *et al.*, Novel fluorinated Bi2MoO6 nanocrystals for efficient photocatalytic removal of water organic pollutants under different light source illumination. *Appl. Catal. B: Environ.*, 209, 1–11, 2017.
13. Kim, D.J. and Jo, W.-K., Sustainable treatment of harmful dyeing industry pollutants using SrZnTiO3/g-C3N4 heterostructure with a light source-dependent charge transfer mechanism. *Appl. Catal. B: Environ.*, 242, 171–7, 2019.

14. Ama, O.M., Kumar, N., Adams, F.V., Ray, S.S., Efficient and cost-effective photoelectrochemical degradation of dyes in wastewater over an exfoliated graphite-MoO3 nanocomposite electrode. *Electrocatalysis.*, 9, 623–31, 2018.
15. Lin, X., Wang, X., Zhou, Q., Wen, C., Su, S., Xiang, J. *et al.*, Magnetically recyclable MoS2/Fe3O4 hybrid composite as visible light responsive photocatalyst with enhanced photocatalytic performance. *ACS Sustainable Chem. Eng.*, 7, 1673–82, 2018.
16. Zhang, Y., Chen, J., Tang, H., Xiao, Y., Qiu, S., Li, S. *et al.*, Hierarchically-structured SiO2-Ag@ TiO2 hollow spheres with excellent photocatalytic activity and recyclability. *J. Hazard. Mater.*, 354, 17–26, 2018.
17. Kumar, N., Mittal, H., Reddy, L., Nair, P., Ngila, J.C., Parashar, V., Morphogenesis of ZnO nanostructures: Role of acetate (COOH–) and nitrate (NO3–) ligand donors from zinc salt precursors in synthesis and morphology dependent photocatalytic properties. *RSC Adv.*, 5, 38801–9, 2015.
18. Sacco, O., Vaiano, V., Sannino, D., Main parameters influencing the design of photocatalytic reactors for wastewater treatment: A mini review. *J. Chem. Technol. Biotechnol.*, 95, 2608–18, 2020.
19. Mazierski, P., Bajorowicz, B., Grabowska, E., Zaleska-Medynska, A., Photoreactor design aspects and modeling of light, in: *Heterogeneous Photocatalysis*, pp. 211–48, Springer, Florida, 2015.
20. Malato, S., Blanco, J., Maldonado, M., Fernandez-Ibanez, P., Padilla, D., Pereira, M. *et al.*, Engineering of solar photocatalytic collectors. *Sol. Energy*, 77, 513–24, 2004.
21. Cassano, A.E., Martin, C.A., Brandi, R.J., Alfano, O.M., Photoreactor analysis and design: Fundamentals and applications. *Ind. Eng. Chem. Res.*, 34, 2155–201, 1995.
22. Ameta, R. and Ameta, S.C., *Photocatalysis: Principles and applications*, CRC Press, Florida, 2016.
23. Ntakadzeni, M., Anku, W.W., Kumar, N., Govender, P.P., Reddy, L., PEGylated MoS2 nanosheets: A dual functional photocatalyst for photodegradation of organic dyes and photoreduction of chromium from aqueous solution. *Bull. Chem. React. Eng. Catal.*, 142–152, 11, 2019.
24. Fujishima, A. and Honda, K., Electrochemical photolysis of water at a semiconductor electrode. *Nature*, 238, 37–8, 1972.
25. Long, Z., Li, Q., Wei, T., Zhang, G., Ren, Z., Historical development and prospects of photocatalysts for pollutant removal in water. *J. Hazard. Mater.*, 395, 122599, 2020.
26. Antonopoulou, M., Kosma, C., Albanis, T., Konstantinou, I., An overview of homogeneous and heterogeneous photocatalysis applications for the removal of pharmaceutical compounds from real or synthetic hospital wastewaters under lab or pilot scale. *Sci. Total Environ.*, 765, 144163, 2020.
27. Parrino, F. and Palmisano, G., Highlights on recent developments of heterogeneous and homogeneous photocatalysis, 2021. Multidisciplinary Digital Publishing Institute. *Molecules*, 26, 1, 23, 2021.

28. Li, X., Yu, J., Jiang, C., Principle and surface science of photocatalysis. *Interface Sci. Technol.*, 31, 1–38, 2020.
29. Iervolino, G., Zammit, I., Vaiano, V., Rizzo, L., Limitations and prospects for wastewater treatment by UV and visible-light-active heterogeneous photocatalysis: A critical review, in: *Heterogeneous Photocatalysis*, pp. 225–64, 2020.
30. Gisbertz, S. and Pieber, B., Heterogeneous photocatalysis in organic synthesis. *ChemPhotoChem*, 4, 456–475, 2020.
31. Umukoro, E.H., Kumar, N., Ngila, J.C., Arotiba, O.A., Expanded graphite supported p-n MoS2-SnO2 heterojunction nanocomposite electrode for enhanced photo-electrocatalytic degradation of a pharmaceutical pollutant. *J. Electroanal. Chem.*, 827, 193–203, 2018.
32. Mukwevho, N., Gusain, R., Fosso-Kankeu, E., Kumar, N., Waanders, F., Ray, S.S., Removal of naphthalene from simulated wastewater through adsorption-photodegradation by ZnO/Ag/GO nanocomposite. *J. Ind. Eng. Chem.*, 81, 393–404, 2020.
33. Gusain, R., Kumar, N., Opoku, F., Govender, P.P., Ray, S.S., MoS2 nanosheet/ZnS composites for the visible-light-assisted photocatalytic degradation of oxytetracycline. *ACS Appl. Nano Mater.*, 4, 4721–4734, 2021.
34. Mukwevho, N., Fosso-Kankeu, E., Waanders, F., Bunt, J., Kumar, N., Visible light-excitable ZnO/2D graphitic-C3N4 heterostructure for the photodegradation of naphthalene, *Desalin. Water Treat.*, 163, 286–296, 2019.
35. Naik, A.P., Mittal, H., Wadi, V.S., Sane, L., Raj, A., Alhassan, S.M. *et al.*, Super porous TiO2 photocatalyst: Tailoring the agglomerate porosity into robust structural mesoporosity with enhanced surface area for efficient remediation of azo dye polluted wastewater. *J. Environ. Manage.*, 258, 110029, 2020.
36. Maswanganyi, S., Gusain, R., Kumar, N., Fosso-Kankeu, E., Waanders, F.B., Ray, S.S., Bismuth molybdate nanoplates supported on reduced graphene oxide: An effective nanocomposite for the removal of naphthalene via adsorption–photodegradation. *ACS Omega*, 6, 16783–94, 2021.
37. Loddo, V., Bellardita, M., Camera-Roda, G., Parrino, F., Palmisano, L., Heterogeneous photocatalysis: A promising advanced oxidation process, in: *Current Trends and Future Developments on (bio-) Membranes*, pp. 1–43, Elsevier, Amsterdam, 2018.
38. Gusain, R., Kumar, N., Ray, S.S., Factors influencing the photocatalytic activity of photocatalysts in wastewater treatment, in: *Photocatalysts in Advanced Oxidation Processes for Wastewater Treatment*, pp. 229–70, 2020.
39. Aziz, N.A.A., Palaniandy, P., Aziz, H.A., Dahlan, I., Review of the mechanism and operational factors influencing the degradation process of contaminants in heterogenous photocatalysis. *J. Chem. Res.*, 40, 704–12, 2016.
40. Zhang, J., Tian, B., Wang, L., Xing, M., Lei, J., Mechanism of photocatalysis, in: *Photocatalysis*, pp. 1–15, Springer, New York, 2018.
41. Shen, Y.-S. and Wang, D.-K., Development of photoreactor design equation for the treatment of dye wastewater by UV/H2O2 process. *J. Hazard. Mater.*, 89, 267–77, 2002.

42. Daisey, J.M., Lewandowski, C.G., Zorz, M., A photoreactor for investigations of the degradation of particle-bound polycyclic aromatic hydrocarbons under simulated atmospheric conditions. *Environ. Sci. Technol.*, 16, 857–61, 1982.
43. Wei, Q., Yang, Y., Hou, J., Liu, H., Cao, F., Zhao, L., Direct solar photocatalytic hydrogen generation with CPC photoreactors: System development. *Sol. Energy*, 153, 215–23, 2017.
44. Arslan, I., Balcioglu, I.A., Bahnemann, D.W., Heterogeneous photocatalytic treatment of simulated dyehouse effluents using novel TiO2-photocatalysts. *Appl. Catal. B: Environ.*, 26, 193–206, 2000.
45. Vidal, A. and Luengo, M.M., Inactivation of titanium dioxide by sulphur: Photocatalytic degradation of Vapam®. *Appl. Catal. B: Environ.*, 32, 1–9, 2001.
46. Oliveira, H.G., Nery, D.C., Longo, C., Effect of applied potential on photocatalytic phenol degradation using nanocrystalline TiO2 electrodes. *Appl. Catal. B: Environ.*, 93, 205–11, 2010.
47. Vaiano, V., Sacco, O., Stoller, M., Chianese, A., Ciambelli, P., Sannino, D., Influence of the photoreactor configuration and of different light sources in the photocatalytic treatment of highly polluted wastewater. *Int. J. Chem. Reactor Eng.*, 12, 63–75, 2014.
48. Spasiano, D., Marotta, R., Malato, S., Fernandez-Ibañez, P., Di Somma, I., Solar photocatalysis: Materials, reactors, some commercial, and pre-industrialized applications. A comprehensive approach. *Appl. Catal. B: Environ.*, 170-171, 90–123, 2015.
49. Alfano, O., Bahnemann, D., Cassano, A., Dillert, R., Goslich, R., Photocatalysis in water environments using artificial and solar light. *Catal. Today*, 58, 199–230, 2000.
50. Górska, P., Zaleska, A., Kowalska, E., Klimczuk, T., Sobczak, J.W., Skwarek, E. *et al.*, TiO2 photoactivity in vis and UV light: The influence of calcination temperature and surface properties. *Appl. Catal. B: Environ.*, 84, 440–7, 2008.
51. Kowalska, E., Janczarek, M., Hupka, J., Grynkiewicz, M., H2O2/UV enhanced degradation of pesticides in wastewater. *Water Sci. Technol.*, 49, 261–6, 2004.
52. Kowalska, E. and Rau, S., Photoreactors for wastewater treatment: A review. *Recent Pat. Eng.*, 4, 242–66, 2010.
53. Blanco, J., Malato, S., Fernández, P., Vidal, A., Morales, A., Trincado, P. *et al.*, Compound parabolic concentrator technology development to commercial solar detoxification applications. *Sol. Energy*, 67, 317–30, 1999.
54. Lazar, M.A., Varghese, S., Nair, S.S., Photocatalytic water treatment by titanium dioxide: Recent updates. *Catalysts*, 2, 572–601, 2012.
55. Lu, M., *Photocatalysis and water purification: From fundamentals to recent applications*, John Wiley & Sons, New York, 2013.
56. Enteria, N. and Akbarzadeh, A., *Solar energy sciences and engineering applications*, CRC Press, Florida, 2013.

57. Khan, A.A. and Tahir, M., Recent advancements in engineering approach towards design of photo-reactors for selective photocatalytic CO2 reduction to renewable fuels. *J. CO2 Util.*, 29, 205–39, 2019.
58. Akyol, A. and Bayramoglu, M., The degradation of an azo dye in a batch slurry photocatalytic reactor. *Chem. Eng. Process.: Process Intensif.*, 47, 2150–6, 2008.
59. Damodar, R.A., You, S.-J., Ou, S.-H., Coupling of membrane separation with photocatalytic slurry reactor for advanced dye wastewater treatment. *Sep. Purif. Technol.*, 76, 64–71, 2010.
60. Zheng, X., Shen, Z.-P., Shi, L., Cheng, R., Yuan, D.-H., Photocatalytic membrane reactors (PMRs) in water treatment: Configurations and influencing factors. *Catalysts*, 7, 224, 2017.
61. Liu, T., Wang, L., Liu, X., Sun, C., Lv, Y., Miao, R. *et al.*, Dynamic photocatalytic membrane coated with ZnIn2S4 for enhanced photocatalytic performance and antifouling property. *Chem. Eng. J.*, 379, 122379, 2020.
62. Molinari, R., Palmisano, L., Loddo, V., Mozia, S., Morawski, A., Photocatalytic membrane reactors: Configurations, performance and applications in water treatment and chemical production, in: *Handbook of Membrane Reactors*, pp. 808–45, Elsevier, Amsterdam, 2013.
63. Mozia, S., Morawski, A., Molinari, R., Palmisano, L., Loddo, V., Photocatalytic membrane reactors: Fundamentals, membrane materials and operational issues, in: *Handbook of Membrane Reactors*, pp. 236–95, Elsevier, Amsterdam, 2013.
64. Plakas, K.V., Sarasidis, V.C., Patsios, S.I., Lambropoulou, D.A., Karabelas, A.J., Novel pilot scale continuous photocatalytic membrane reactor for removal of organic micropollutants from water. *Chem. Eng. J.*, 304, 335–43, 2016.
65. Molinari, R., Argurio, P., Szymański, K., Darowna, D., Mozia, S., Photocatalytic membrane reactors for wastewater treatment, in: *Current Trends and Future Developments on (Bio-) Membranes*, pp. 83–116, 2020.
66. Mozia, S., Photocatalytic membrane reactors (PMRs) in water and wastewater treatment. A review. *Sep. Purif. Technol.*, 73, 71–91, 2010.
67. Durán, A., Monteagudo, J., San Martín, I., Merino, S., Photocatalytic degradation of aniline using an autonomous rotating drum reactor with both solar and UV-C artificial radiation. *J. Environ. Manage.*, 210, 122–30, 2018.
68. Meunier, S.M., Gamage, J., Duvnjak, Z., Zhang, Z., Design and characterization of a novel rotating corrugated drum reactor for wastewater treatment. *Int. J. Photoenergy*, 2010, Article ID 146743, 2010.
69. Boiarkina, I., Pedron, S., Patterson, D.A., An experimental and modelling investigation of the effect of the flow regime on the photocatalytic degradation of methylene blue on a thin film coated ultraviolet irradiated spinning disc reactor. *Appl. Catal. B: Environ.*, 110, 14–24, 2011.
70. Tanthapanichakoon, W., Charinpanitkul, T., Yamamoto, T., Sano, N., Thongprachan, N., A novel rotary drum filtering photoreactor for wastewater

treatment using titanium dioxide nanoparticles. *2007 2nd IEEE International Conference on Nano/Micro Engineered and Molecular Systems*, IEEE, pp. 438–42, 2007.
71. Hamill, N.A., Weatherley, L.R., Hardacre, C., Use of a batch rotating photocatalytic contactor for the degradation of organic pollutants in wastewater. *Appl. Catal. B: Environ.*, 30, 49–60, 2001.
72. Dionysiou, D.D., Balasubramanian, G., Suidan, M.T., Khodadoust, A.P., Baudin, I., Laîné, J.-M., Rotating disk photocatalytic reactor: Development, characterization, and evaluation for the destruction of organic pollutants in water. *Water Res.*, 34, 2927–40, 2000.
73. Matsushita, Y., Ohba, N., Kumada, S., Sakeda, K., Suzuki, T., Ichimura, T., Photocatalytic reactions in microreactors. *Chem. Eng. J.*, 135, S303–S8, 2008.
74. Palm, G., *Contribution to the development of a microphotoreactor for the production of pharmaceutical compounds*, Semantic scholar, WA, 2017.
75. Eskandarloo, H. and Badiei, A., Fabrication of an inexpensive and high efficiency microphotoreactor using CO2 laser technique for photocatalytic water treatment applications. *Environ. Technol.*, 36, 1063–73, 2015.
76. Aillet, T., Loubière, K., Prat, L., Dechy-Cabaret, O., Impact of the diffusion limitation in microphotoreactors. *AIChE J.*, 61, 1284–99, 2015.
77. Drhova, M., Hejda, S., Kristal, J., Kluson, P., Performance of continuous micro photo reactor–comparison with batch process. *Proc. Eng.*, 42, 1365–72, 2012.
78. Sarkar, S., Bhattacharjee, C., Sarkar, S., Studies on the performance of annular photo reactor (APR) for pharmaceutical wastewater treatment. *J. Water Process Eng.*, 19, 26–34, 2017.
79. Bhattacharjee, S., Chakraborty, S., Mandol, K., Liu, L., Choi, H., Bhattacharjee, C., Optimization of process parameters during photocatalytic degradation of phenol in UV annular reactor. *Desalin. Water Treat.*, 54, 2270–9, 2015.
80. Peres, J.C.G., Tambani, P.C., Teixeira, A.C.S.C., Guardani, R., Vianna Jr, A.d.S., Study of an annular photoreactor with tangential inlet and outlet. II. The UV/H2O2 reactive flow. *Chem. Eng. Technol.*, 42, 316–26, 2019.
81. Chong, M.N., Lei, S., Jin, B., Saint, C., Chow, C.W., Optimisation of an annular photoreactor process for degradation of Congo Red using a newly synthesized titania impregnated kaolinite nano-photocatalyst. *Sep. Purif. Technol.*, 67, 355–63, 2009.
82. Behnajady, M.A., Modirshahla, N., Daneshvar, N., Rabbani, M., Photocatalytic degradation of an azo dye in a tubular continuous-flow photoreactor with immobilized TiO2 on glass plates. *Chem. Eng. J.*, 127, 167–76, 2007.
83. Yahiat, S., Fourcade, F., Brosillon, S., Amrane, A., Removal of antibiotics by an integrated process coupling photocatalysis and biological treatment–case of tetracycline and tylosin. *Int. Biodeterior. Biodegrad.*, 65, 997–1003, 2011.
84. Stephan, B., Ludovic, L., Dominique, W., Modelling of a falling thin film deposited photocatalytic step reactor for water purification: Pesticide treatment. *Chem. Eng. J.*, 169, 216–25, 2011.

85. Li, Y., Ma, Y., Li, K., Chen, S., Yue, D., Photocatalytic reactor as a bridge to Link the commercialization of photocatalyst in water and air purification. *Catalysts,* 12, 724, 2022.
86. Xu, C., Ravi Anusuyadevi, P., Aymonier, C., Luque, R., Marre, S., Nanostructured materials for photocatalysis. *Chem. Soc. Rev.*, 48, 3868–902, 2019.
87. Fang, Z., Hu, Y., Cheng, J., Chen, Y., Continuous removal of trace bisphenol A from water by high efficacy TiO2 nanotube pillared graphene-based macrostructures in a photocatalytically fluidized bed. *Chem. Eng. J.*, 372, 581–9, 2019.
88. Sun, F., He, J., Wu, P., Zeng, Q., Liu, C., Jiang, W., Magnetic photocatalyst CoFe2O4-Ag2O with magnetic aggregation bed photocatalytic reactor for continuous photodegradation of methyl orange. *Chem. Eng. J.*, 397, 125397, 2020.
89. Fischer, K., Schulz, P., Atanasov, I., Abdul Latif, A., Thomas, I., Kühnert, M. *et al.*, Synthesis of high crystalline TiO2 nanoparticles on a polymer membrane to degrade pollutants from water. *Catalysts*, 8, 376, 2018.
90. Li, D., Zheng, H., Wang, Q., Wang, X., Jiang, W., Zhang, Z. *et al.*, A novel double-cylindrical-shell photoreactor immobilized with monolayer TiO2-coated silica gel beads for photocatalytic degradation of Rhodamine B and Methyl Orange in aqueous solution. *Sep. Purif. Technol.*, 123, 130–8, 2014.
91. Fouad, K., Gar Alalm, M., Bassyouni, M., Saleh, M.Y., A novel photocatalytic reactor for the extended reuse of W–TiO2 in the degradation of sulfamethazine. *Chemosphere*, 257, 127270, 2020.
92. Rakhshaee, R. and Darvazeh, J., Studying role of air bubbles on suspension of hematite particles with three size ranges in plug flow reactor to improve dyes photo degradation compared to conventional packed bed photo reactors. *J. Hazard. Mater.*, 356, 61–72, 2018.
93. Fouad, M., Gar Alalm, M., El-Etriby, H.K., Boffito, D.C., Ookawara, S., Ohno, T. *et al.*, Visible-light-driven photocatalytic disinfection of raw surface waters (300–5000 CFU/mL) using reusable coated Ru/WO3/ZrO2. *J. Hazard. Mater.*, 402, 123514, 2021.
94. Samy, M., Ibrahim, M.G., Gar Alalm, M., Fujii, M., Diab, K.E., ElKady, M., Innovative photocatalytic reactor for the degradation of chlorpyrifos using a coated composite of ZrV2O7 and graphene nano-platelets. *Chem. Eng. J.*, 395, 124974, 2020.
95. McCullagh, C., Robertson, P.K.J., Adams, M., Pollard, P.M., Mohammed, A., Development of a slurry continuous flow reactor for photocatalytic treatment of industrial wastewater. *J. Photochem. Photobiol. A: Chem.*, 211, 42–6, 2010.
96. Wang, Q., Wang, P., Xu, P., Hu, L., Wang, X., Qu, J. *et al.*, Submerged membrane photocatalytic reactor for advanced treatment of p-nitrophenol wastewater through visible-light-driven photo-Fenton reactions. *Sep. Purif. Technol.*, 256, 117783, 2021.

97. Pieber, B., Shalom, M., Antonietti, M., Seeberger, P.H., Gilmore, K., Continuous Heterogeneous Photocatalysis in Serial Micro-Batch Reactors. *Angew. Chem. Int. Ed.*, 57, 9976–9, 2018.
98. Elvira, K.S., i Solvas, X.C., Wootton, R.C.R., deMello, A.J., The past, present and potential for microfluidic reactor technology in chemical synthesis. *Nat. Chem.*, 5, 905–15, 2013.

8

Design of Photoreactors for Effective Dye Degradation

Rajashree Sahoo[1]* and Arpan Kumar Nayak[2]†

[1]*Department of Physics, School of Applied Sciences, KIIT Deemed to be University, Bhubaneswar, Odisha, India*
[2]*Department of Physics, School of Advanced Sciences, Vellore Institute of Technology, Vellore, India*

Abstract

Sky rocketing industrialization and modernization has brought hazardous effluents into the water, which deteriorates aquatic environment and causes threat to the living organisms. To overcome these problems, an expensive, self-sustainable and environmental friendly heterogeneous photocatalysts have been developed. The different photoreactors are developed like solar/artificial light photoreactors for the decomposition of organic contaminants. Therefore, this chapter highlights the construction and design of different photoreactors, the parameters affecting reactor design for the dye degradation and its efficiency. These photoreactors also may be used in the field of biological, chemical, and pharmaceutical industries or laboratories. The basic design of photoreactor modeling will be helpful for designing the instrument for biologist, physicists, and chemists. These applications give instruction for construction of a cost-effective reactor with optimized parameters for the analysis of degradation data.

Keywords: Effluents, heterogeneous, solar/artificial light, photocatalyst, photoreactors, dye degradation, cost-effective, optimized parameters

Abbreviations

AOPs	Advanced Oxidation process
LPML	Low-pressure mercury lamp

**Corresponding author*: rajashreesahoo92@gmail.com
†*Corresponding author*: aknayakju@gmail.com; arpankumar.nayak@vit.ac.in

Elvis Fosso-Kankeu, Sadanand Pandey, and Suprakas Sinha Ray (eds.) Photoreactors in Advanced Oxidation Processes: The Future of Wastewater Treatment, (247–276) © 2023 Scrivener Publishing LLC

VUV	Vacuum ultraviolet
FL	Fluorescent lamp
MPML	Medium-pressure mercury lamp
XeL	Xenon lamp
Xe2*L	Xe2* excilamp
KrCl*L	KrCl* excilamp
W-MDEL	Tungsten-triggered microwave discharge electrodless lamp
ML	Mercury lamp
DMDT	Methoxychlor
Bl	Black light tube
BlB	Black light blue
BLFL	Black light fluorescent lamp
MVL	Metal vapor lamp
TL	Tubular lamp
BSA	Benzenesulfonic acid
THL	Tungsten-halogen lamp
LEDs	Light-emitting diodes
OLEDs	Organic light-emitting diodes
PHOLEDs	Phosphorescent organic light-emitting diodes
WL	Tungsten lamp
LFFS	Laminar falling film slurry
LSPP	Linear source parallel plane emission model
TFM	Two-flux model
ZRM	Zero reflectance model
TFS	Thin-film, slurry (reactor)
CFD	Computational fluid dynamics
PIV	Particle image velocimetry
CDCR	Cocurrent downflow contactor reactor
PBTPR	Pulsed baffled tubular photochemical reactor
CPC	Compound parabolic collector
PRC	Parabolic round sunlight concentrator
3-NBSA	3-nitrobenzenesulfonic acid
2, 5-ADSA	2,5-anilinedisulfonic acid
DB	Sodium benzene sulfonate
DBS	Sodium dodecylbenzene sulfonate
MB	Methylene blue

8.1 Introduction

Rapid industrialization and urbanization is increasing rapidly with the result of large amount of organic contaminants enters into the water.

To remove these pollutants from wastewater, enormous efforts have given. Different water treatment processes like flocculation, trickling filters, and electrodialysis are developed [1–3], but these are unable for complete removal of organic contaminants. So, to overcome these problems, advanced oxidation processes have been developed. In this process, the oxidation of bigger molecules is converted into harmless molecules likes CO_2 and H_2O [4, 5]. It generates hydroxyl radical, which is a strong oxidizing agent that helps in the degradation of pollutants into harmless products. This process is cheap, ecofriendly, and efficient for the degradation of organic pollutants. The degradation efficiency depends on the energy required to break the chemical bond and also the presence of dissolved molecular oxygen. Artificial light sources required much more energy for the decomposition of organic contaminants for a specific concentration.

Proper light sources (natural/artificial), amount of catalyst, pH, and concentration of wastewater and also an oxidation system play a major role in wastewater treatment efficiency [6]. UV exposure is an important parameter for an opaque environment. The efficient interaction between the photocatalyst and pollutants with high oxygen (O_2) uptake at gas-liquid interface is an alternative requirement for high transfer rate in practical applications. In this regard, it is a challenge for the design of photoreactor for wastewater treatment.

8.1.1 Mechanisms and Theory of AOP

Ozone

Ozone acts as a disinfectant in water and is used in the wastewater treatment due to its strong oxidizing nature. It reacts with double bonds or triple bonds, such as C≡C, C=N, C=C or N=N [7]. When direct ozonation becomes not effective in some cases, ozone decomposes in water at pH medium to form a large number of hydroxyl radicals. The decomposition of ozone follows these equations

$$O_3 + H_2O \rightarrow HO_3^+ + OH^- \quad (8.1)$$

$$HO_3^+ + OH \rightarrow 2HO_2 \quad (8.2)$$

$$HO_2 + O_3 \rightarrow OH + 2O_2 \quad (8.3)$$

Hydrogen Peroxide

It is one of the most powerful oxidizing agents able to oxidize different types of chemicals, such as aldehyde, azo compound, phenols, cyanides, and metals. It decomposes in the presence of iron catalyst to form large number of hydroxyl radicals. It is generally known as Fenton's reaction because it was discovered by Fenton in 1894. So, the Fenton's reaction is

$$Fe^{2+} + H_2O_2 \rightarrow Fe^{3+} + OH + OH \tag{8.4}$$

$$Fe^{3+} + H_2O_2 \rightarrow Fe^{2+} + OOH + H^{+} \tag{8.5}$$

It also decomposes in presence of ozone

$$2O_3 + H_2O_2 \rightarrow 2\ OH + 3O_2 \tag{8.6}$$

UV/H_2O_2

Advanced oxidation based on UV/H_2O_2 is very efficient in the decomposition of aqueous effluents. H_2O_2 has a weak molar absorption coefficient, which increases as the wavelength decreases in 200 to 300 nm wavelength range. The photolysis of H_2O_2 produces hydroxyl radicals.

$$H_2\ O_{2+}\ hv \rightarrow 2OH \tag{8.7}$$

$$OH + H_2\ O_2\ /\mathrm{HO}_2 \rightarrow \mathrm{HO}_2/O_2 + H_2\mathrm{O} \tag{8.8}$$

$$HO_2 + H_2O_2 \rightarrow OH + H_2O + O_2 \tag{8.9}$$

$$HO_2 + HO_2 \rightarrow H_2O_2 + O_2 \tag{8.10}$$

UV/Ozone

Ultraviolet photolysis of ozone produces hydroxyl radicals rapidly. So, ozonation in the presence of UV irradiation has become the most usable AOPs for the degradation of organic compounds like acid, alcohols and trihalomethanes, etc. It strongly absorbs UV irradiation at a wavelength of 254 nm [8].

8.1.2 Design of Photoreactors

8.1.2.1 Source of Irradiation

Two types of light sources are required for the photoreaction, such as artificial light source (lamps) and natural light source (sun) [9]. Low pressure mercury lamp (LPML) which is known as gas discharge lamp is used as a light source. In this lamp, tungsten filament is located inside each side of the lamp, a jacket filled with inert gas like argon, and also the Hg vapor under negative pressure. So, these lamps are known as vacuum ultraviolet (VUV). Hg is excited by the glow discharge line spectrum within 185 and 253.7 nm wavelength ranges. It is similar to common fluorescent lamp (FL) except its jacket is made up of quartz. It helps in the transmission of ultraviolet rays. In the case of FL, the jacket is covered with luminophore which absorbs radiation at 253.7 nm and emit radiation in a long wavelength range. To avoid the impact of environmental temperature on lamp capacity, often, mercury is replaced by amalgam. The other lamp also often used, i.e., middle pressure mercury lamp (MPML) called polychromatic lamp. This light emitted from 200 nm to IR range. Here, the pressure inside the quartz tube is 1300 hpa. It is working under high temperature. LED technology is the prominent technique for the future as they are more energy efficient than the conventional technique. However, continuous sources of light are most stable than the light sources.

Usually, photocatalytic reactions occurred in the presence of UV light. This light consists of four different ranges, such as VUV, UVC, UVB, and UVA. Far, mid, and near UV are also often used instead of VUV/UVC, UVB, and UVA.

8.1.2.2 Wavelength/Lamp Selection

A larger wavelength (l > 400 nm) of solar irradiation is used in solar disinfection study. But the photo-killing mechanism is not clear yet as it includes a variety of microbial and mixed spectrum of UV-A and solar irradiation. Also, in the cell destruction mechanism, UV-A irradiation is used in mixed light spectrum. In solar disinfection study, in the presence 6 hours of sun light exposure, the pathogens in the drinking water (in PET bottles) are found to be inactivated. So, significant research and developments are to be conducted for the disinfection using TiO_2 as catalysts.

UV irradiation is necessary for photocatalysis, photolysis, and also for the destruction of microorganisms. Previously, UV/visible light irradiation are used for wastewater management [10–13]. Basically, the wavelength is necessary for the advanced oxidation processes. So, LPMLs (254.7 nm) are generally used for disinfection, H_2O_2 photolysis, MPMLs, HPMLs, FLs. Xenon lamps are used for photocatalysis on the surface of common TiO_2 photocatalyst, which requires UVA/vis activation (< 400 nm). Tungsten lamps, halogen lamps, and LEDs are also used for photocatalysis

Table 8.1 Categories, sources, and application of electromagnetic radiation.

Light source	Wavelength (nm)	Different light	Application
UVC	100 - 280	XeL [14], W-MDEL [15], KrCl*L [16], TUV [17], LMPL [18], ML [19]	UV: germicidal [16, 18, 20]; UV/H_2O_2: germicidal [18], river water decolorization [21], phenol removal [22], DMDT [23], pesticide removal [24]; phenol degradation: UV/O_3 [19]
UVB	280 - 315	XeL [14, 25], FL [26], ML [19]	Pesticide photolysis: UV [25]
UVA	315 - 400	HPML [27, 28], FL [26, 29], Bl [30], BlB [31, 32], XeL [14], MVL [33], BLFL [34], MPML [35], ML [19], TL [36]	UV/TiO_2: inorganic and organic pollutants [37–39], dyehouse Effluents [29], phenol [14, 28], 4-CP [28], 4-NP [28], salicylic acid [34], chloroform [40], chloroorganic pesticides [41–43] Cyanide [44], BSA [26], dodecane [45]; $Fe^{+3}/H_2O_2/$UV-vis:fenitrothion [46]
Visible (VIS)	380 - 760	XeL [14], THL [47], LED [48], OLEDs [49], PHOLEDs [49], FL [26, 32], WL [36, 50], MVL [33]	Model treatment of wastewater by: Vis/ modified-TiO_2 [51]: - nonmetal/TiO_2: C [52, 53], S [54], N [55–57], F [58], B [59], I [60] - metal/TiO_2: Fe [32], Pt [61, 62], Au [63, 64], Ag [41]
IR-A	760 - 1400	IR lamp [65]	IR-A, IR-A/vis, IR-A/ UV decomposition of: B-Carotene, Chlorophyll a [65]
IR-C	3000 - 1000 000	LWIRE [66], Heater [65]	IR-C: decomposition of: B-Carotene, Chlorophyll a [65]

on the surfaces of novel photocatalysts with the ability of working under visible-light irradiation, e.g., modified TiO_2, Pt/WO_3, Au/CeO_2.

Some light sources, its wavelength, and different applications of electromagnetic radiations are explained in Table 8.1.

8.1.3 Placement of Light Source and Light Distribution

Irradiation of light can be carried out in two ways, such as external illumination and internal illumination. In case of external illumination, the lamps are placed perpendicular or parallel to side walls of the reactor [67–69]. Point radiation occurs. Light passes through collimating lens like fused silica, silicate glass, and quartz. In case of internal illumination, lamps are placed inside the reactor (immersed in liquid) [68, 70, 71]. The UV light with the required intensity can be achieved using neutral density filters or stainless-steel screens with different mesh sizes [69, 72]. The reaction cell, which is usually made up of quartz for UV transmission or a cell made of glass equipped with quartz window, is used. When temperature inside the cell increases, the required cell cooling is used (where the cell is placed in thermostated bath filled with water). The other resources of cell cooling through the flow of air [73], direct lamp cooling by compressed air, and also application of cold mirror [74]. It is very important to major the intensity of light with the flat surface of front wall of the reactor. Even a small change of reactor position creates large change in the light intensity and its distribution. So, the position of irradiated cell should be fixed using the cell holder like the one in Figure 8.1.

In Figure 8.2, it has been shown that some test tubes are placed at some distance from the lamp. The advantage of this setup is simultaneous irradiation of light from the source. The experimental setup has been made in a wooden box with black color, a dark room, also the reactor coated with black adhesive tape, aluminium foil [75], or closing the reactor with a mirror cover to avoid the effect of stray light.

Mostly, external illumination used for the testing of new photocatalysts under visible light irradiation is shown in Figure 8.1. Sometimes, for separation of UV light, cutoff filter is used on the way of incident light which is insufficient, and also, there is a possibility of scattered UV rays can reach reaction mixture. So, the reaction tubes are fixed with the extra tube holders provided with irradiation window with cut-off filter. Due to stiffness of the sample holder, the temperature of cooling water inside the holder can increase. So, a new setup is developed with circulating cooling water and well separated UV radiation shown in Figure 8.3. This reactor consists of mainly two sections, i.e., A & B. In section A, xenon lamp (1) which is the

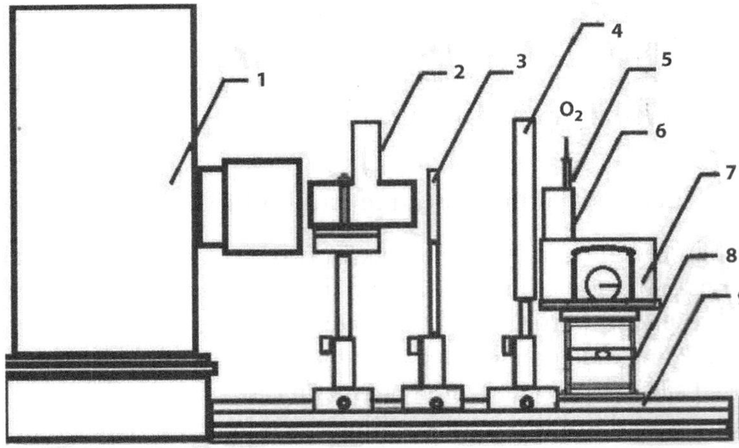

Figure 8.1 Schematic diagram of experimental set-up with point irradiation [9]. 1—UV/VIS lamp (XeL), 2—IR water filter, 3—UV cutoff filter, 4—diaphragm, 5—oxygen supply (needle), 6—stand for reaction cuvette, 7—magnetic stirrer, 8—laboratory lift, 9—optical bench.

light source passes through water bath (8). A UV cutoff filter (3) is used to remove IR and UV radiation. For the circulation of water in both section and removal of UV light in section B, a separating wall has been developed (in part A). There is a hole (9) in the bottom of partition wall between two parts. In the section B, there are two test-tubes (12) which are placed behind the filter at the same distance from the light source. Their content is stirred (13) and cooled by circulating water.

In internal illumination, the UV/visible lamp is placed inside the quartz tube filled with water. For cooling the lamp, three types of methods can be applied. (i) The lamp is placed inside the quartz tube, which is filled with cooling water as shown in Figure 8.4. (ii) The reactor contains a jacket in which water is circulated [76]. (iii) The whole reactor containing immersed lamp is placed in a water bath [77].

In heterogeneous photocatalytic reactor, there are several models like LSSE-LSPP model for laminar falling film slurry (LFFS) photoreactor, two-flux model (TFM) and zero reflectance models (ZRM) have been developed [78, 79]. A novel, thin-film, slurry "fountain" photocatalytic reactor has been developed for wastewater treatment. There are two models like flat fountain model and a parabolic fountain model have been proposed [80].

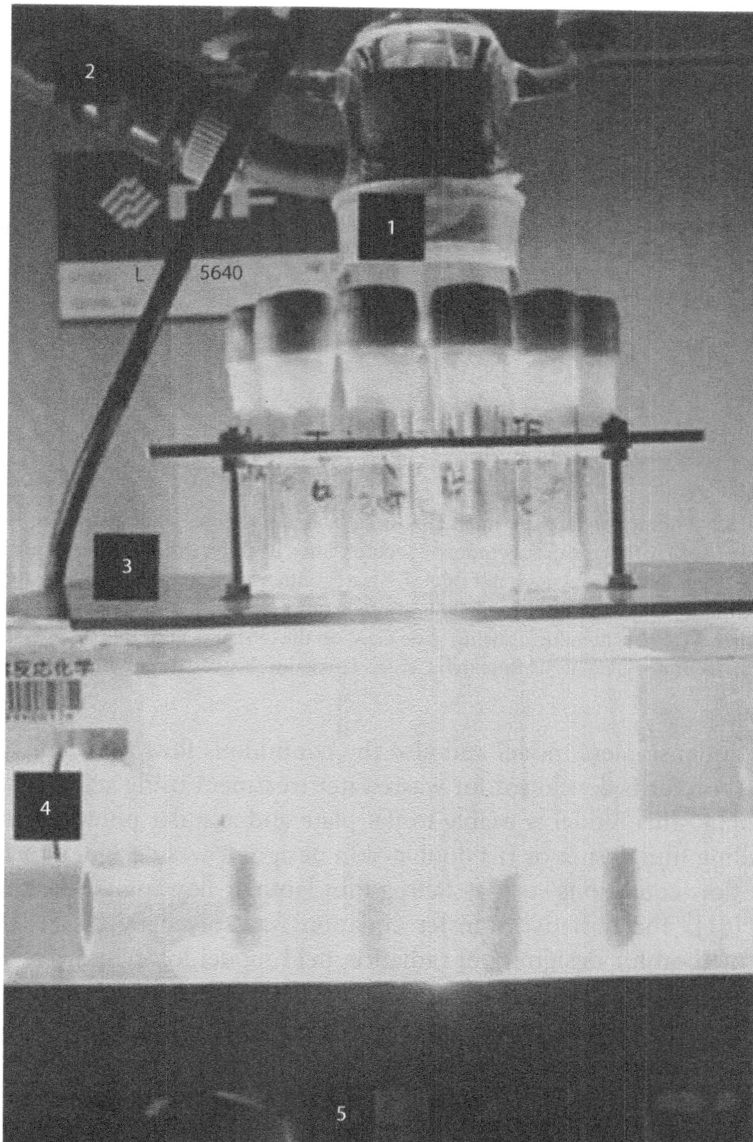

Figure 8.2 Experimental set-up photograph with 12 simultaneously. irradiated testing tubes: 1—UV/VIS lamp (MPML), 2—thermostated lamp cooling water, 3—testing tubes holder, 4—thermostated water bath, 5—magnetic stirrer [Photography was taken in Ohtani's laboratory, Catalysis Research Center, Hokkaido University] [9].

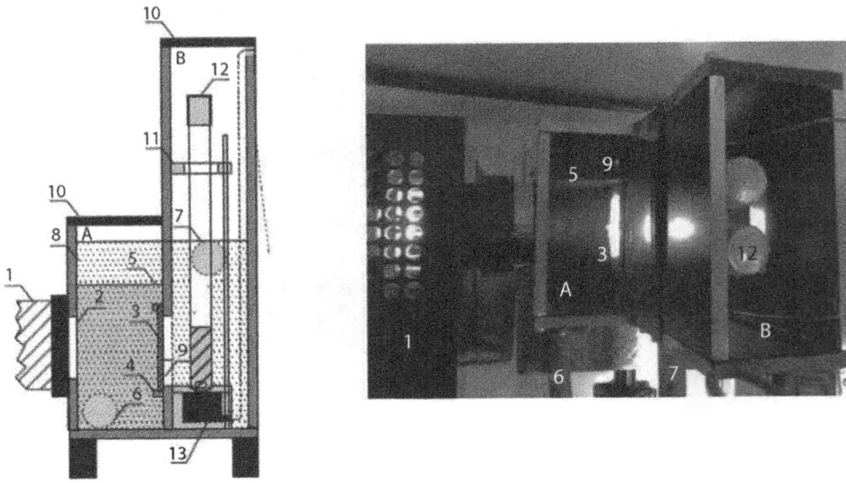

Figure 8.3 Photocatalytic experimental setup tests under visible-light irradiation: 1—UV/vis lamp (XeL), 2—pyrex glass window, 3—cutoff filter, 4—filter holder, 5—separating wall for water circulation, 6—water inlet, 7—water outlet, 8—water bath, 9—hole for circulating water, 10—reactor covers, 11—sample tubes holder, 12—sample tubes (two), 13—stirrers (two) schematic drawing (side view, on the left) and a photograph taken in Ohtani's laboratory, Catalysis Research Center, Hokkaido University (on the right) [9].

A dimensionless model and also the continuous flow thin film, slurry (TFS) reactor is developed for wastewater treatment using solar light and UV light. This model is usable to flat plate and annular photoreactors of (i) falling film design or (ii) double-skin design. It was designed for three ideal flow conditions such as failing film laminar flow, plug flow and slit flow [81]. The radiative transfer equation was solved by discrete ordinate method for designing of radiation field model for cylindrical slurry reactor [82]. Monte Carlo modeling for photocatalysis was developed. This model includes photogeneration, interfacial transfer of holes and electrons, charge kinetics, recombination to simulate the photocatalytic reaction. Imoberdorf *et al.* made a complete model of the radiation field. He explained the UV-radiation interaction with TiO_2-coated silica-fused sphere beds for a fixed bed photocatalytic reactor [83]. Four submodels like light emission model, light absorption model, reaction kinetic model, and fluid dynamic model are applied for the modeling of photocatalytic reactors [84]. In general, modeling is used meticulously for the analysis of the radiation field in the photoreactor resulting a numerical treatment to form a solution and also form a complex set of integrodifferential equations.

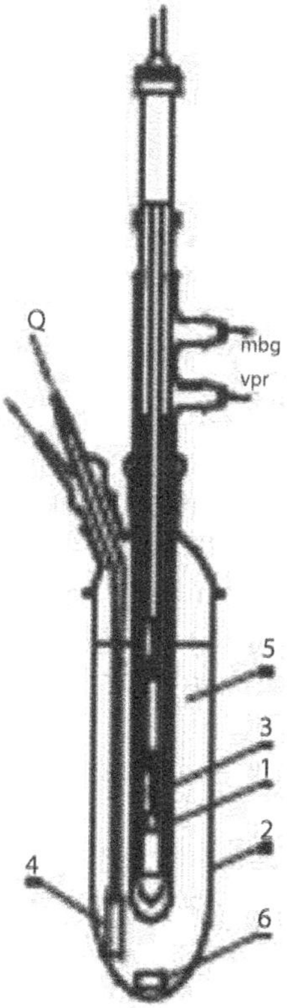

Figure 8.4 Schematic diagram of Heraeus photoreactor: 1—UV/vis lamp (MPML) inside quartz tube filled with cooling water, 2—glass reactor, 3—UV lamp cooling system, 4—porous frit for sparging gas, 5—liquid phase, 6—magnetic stirrer [9].

The experimental data were fitted by the models of salicylic acid during photocatalytic oxidation in a pilot-scale LFFS photocatalytic reactor, with a pilot plant "fountain" photocatalytic reactor [82], degradation of 4-chlorophenol in a cylindrical slurry reactor [85], for hydrodynamic measurement via CFD (computational fluid dynamics) with visualization data by PIV (particle image velocimetry). However, for long reaction duration (>120 min), realistic reactors performances are very different than expected by theoretical calculation. Romero *et al.* made annular, tubular reactor inside a recycle for three dissimilar modes: (a) perfectly mixed, (b) operating at pseudo-steady-state, and (c) operating at transient state, and only the third one, after consideration of deposition and agglomeration effects, fitted to the experimental results. It is believed that insufficient amount of investigation on photoreactor modeling, and thus a deficiency of readily applicable mathematical models to reactor design and scale-up, is the reason of impeding application of photocatalytical method on an industrial scale for wastewater treatment. Fortunately, a large number of modeling of different photoreactors has been developed by Cassano and Alfano group [86–88].

8.2 Different Photoreactors Are Used for Wastewater Treatment

- ❖ Construction: The reactor system consists of two parts, such as spiral [89] and specially design reactor [90].
- ❖ Shape: There are different shapes of reactor like cylindrical [90–93], vertical, annular round-shaped, horizontal, rectangular, flat, round flask, tubular [94, 95], round bottom, and flat-surfaced reactor [96].
- ❖ Material: glass, Pyrex, stainless steel, quartz [97], and Teflon reactor [98].
- ❖ Type of mixing: recirculating [99], completely mixed batch (CMBR) reactor [100], concurrent downflow contactor [98]
- ❖ Cooling system: jacket cooling, bubble reactor, double jacket [99], water jacketed reactor, oxygen-permeable membrane
- ❖ Flow: cocurrent downflow contactor reactor (CDCR) [100], fountain, laminar falling film slurry (LFFS), thin film pulsed baffled tubular photochemical (PBTPR) [101] and thin-film reactor
- ❖ Operating mode: batch, step [102] and semi-batch reactor
- ❖ Producer: Rayox [103–105], Rayoney, Heraeus and Rayonet [106] reactor
- ❖ Light source: parabolic round concentrator [107], solar simulator, and vis-LED reactor
- ❖ Irradiation mode: collimated beam apparatus [108], double-wall, and multilamp [109] reactor
- ❖ Photocatalyst arrangement: with thin film, fixed bed and heterogeneous reactor, membrane [110], suspension, packed-bed, TiO_2-immobilized, slurry
- ❖ Experiment scale: photocatalytic cell, bench scale, laboratory scale, and pilot-size reactor
- ❖ Multifunctional (cooperating with other processes): ultrasonic and sonuv reactor.

8.2.1 Some Typical Photoreactors Used for Wastewater Treatment Are Described Below

The photodegradation experiment was performed in a quartz cuvette. The intensity of light can be measured by power meter [111], radiometer, photomultipliers, or chemical actinometry. Various gases like Ar, air, N_2, O_2 was entered by syringes or Pasteur pipettes. Sometimes there is no need for gas, but oxygen is necessary for reaction. Oxygen is introduced from surrounding atmosphere when reactor is open. Stirring of the solution was done by the magnetic stirrers and the reaction temperature was controlled using the IR filters, which was filled with stationary or circulating water. The "cylindrical reactor" is commonly used for testing of organic photo-decomposition and the UV lamp is either dipped in the liquid or placed outside the reactor. Usually, the reactors having lesser volume have external source of light while centrically dipped lamp is used for larger volumes of liquid. The water jacket or thermostate is used to cool the reactor. Mechanic stirrers with blade producing axial and upwardly directed flow have been used for larger volumes. The simple home-made photoreactor consists of a glass beaker with a magnetic stirring setup and light is placed above the beaker [112].

There is negligible difference between a cylindrical reactor and other photoreactors used for the laboratory testing. Depending on the construction of the reactor, "horizontal reactor" and "reactor with flat surface," in which the light is fixed above or below liquid surface have been reported. It has also been reported that more advanced reactor designs have higher mass transfer efficiency. One example is a pilot scale cocurrent down flow contactor reactor (CDCR) which is operated in closed loop recycle mode. The equipment like cuvette, quartz reactor, glass reactor, batch reactor, etc. used in homogeneous and heterogeneous photocatalysis usually does not differ from each other. There are only a few examples of photoreactors, which emphasized the application of heterogeneous photocatalyst, e.g., photocatalytic membrane reactor, reactor with packing, packed-bead reactor. In this type of reactor, a lamp is placed outside the photoreactor. Earlier, it was reported that, a system with a dipped lamp can be used as a photocatalyst coated on the photoreactor walls and on packed glass beads.

Commercial application of artificial sources of light bears high costs. So enormous efforts have been given on the solar irradiation for wastewater treatment. Laboratory photoreactors usually use several cut-off or band-pass filters or solar simulator. Direct solar radiation is not used for laboratory experiments, because of low repeatability of experimental data, due to large changes in solar radiation. Moreover, the intensity of solar emission is quite low, and thus involves very long irradiation. The photoreactors based on homemade solar simulator are composed of a parabolic reflector [113] and a lamp (or set of lamps). The front of the light source can be also covered with a light homogenizer to obtain uniform light output. In such experiments, in the UV/vis lamps (xenon arc) there is fluorescent metallic vapor discharge and thus only distribution of light

Table 8.2 Recommended dose of photocatalyst for various compounds and photoreactors.

Dye name	C_0/mM	Volume/ mL	Photoreactor// light intensity/ mW cm^{-2}	Kind of TiO_2 photocatalyst	Ref.
3-NBSA 2,5-ADSA	0.04		vertical & horizontal//5.6 × 10^{-9} Einstcm^{-2} s^{-1}	P25 or anatase	[26]
Cr (VI)		100	flat reactor inside Solarbox	P25	[134]
DBS	0.1	3000	PRC (25–70 suns) and a round flask photoreactor, circulation	P25	[117]
BS	0.1	3000	PRC (25-70 suns) and a round flask photoreactor, Circulation	P25	[107]
MB	0.025		PRC (25–70 suns) and a round flask photoreactor, circulation	P25	[107]
4-NP	0.144		semi-batch with two circular glass plates//	P25	[28]

(homogeneous), but not its spectrum, is the same as in the case of natural solar radiation. More recently solar simulators [114] with the same radiation as the Sun spectrum at the Earth surface have been also applied. The solar simulators are either "homemade" equipped with xenon lamp and special glass filters (e.g., Oriel, AM0, and AM1.0) or with solar simulator lamp or commercial ones. The photochemical systems based on direct solar radiation are composed of focusing/concentrating solar collectors and a photoreactor. The Compound Parabolic Collector (CPC) and the Parabolic Sunlight Round Concentrator (PRC) are the most often applied collectors for wastewater treatment [115–118]. The CPC photoreactor is composed of several collectors, one tank and one pump. Each collector consists of a few Pyrex tubes, filled with flowing treated wastewater, connected in series, and mounted on a fixed platform. The CPC photoreactor has been used as a wastewater treatment. The PRC is composed of one reactor on which solar light is concentrated by a parabolic round mirror (geometric concentration of sunlight equivalent to 25 or 70 suns in dependence on mirror area and aperture diameter), one pump and one tank. The usage of direct, nonconcentrated solar radiation has been also reported, e.g. a photoreactor consists of only photocatalyst deposited on a glass wool mat spread over a wide area on the ground. Such system is very simple, i.e., the wastewater is poured onto mats. Moreover, it is very cheap and thus highly recommended in developing countries for water purification. Another application of photocatalysis under direct solar radiation has been reported for tomato hydroponic culture system [119]. This system allows decomposition of organic compounds with simultaneous recovery of nutrients using porous ceramic plates coated with titania photocatalyst in recycling system.

8.2.2 Homogenous and Heterogenous Systems

There are two types of photochemical processes i.e. homogeneous or heterogeneous systems. Before irradiation, hydrogen peroxide is usually added once. At high concentration, hydrogen peroxide acts as a trapping agent for radicals

$$\text{According to equation: } HO\cdot + H_2O_2 = HOO\cdot + H_2O$$

At low H_2O_2 concentration, absorption of UV radiation becomes insignificant causing impaired generation of hydroxyl radicals.

There is two types of reaction design are used i.e.

- Fixed-bed or packed-bed photoreactors,
- Suspension or slurry photoreactors.

8.2.3 Heterogenous Photocatalyst Arrangement

In heterogeneous photocatalytic arrangement, the catalysts are mixed by magnetic or mechanical stirring. Slurry type reactors are efficient due to large surface area of photocatalysts and high mass transfer rates [120, 121]. Photocatalysts having small size basically create two practical problems such as i) particles can adsorb on the elements of photoreaction system and ii) a post-treatment photocatalyst recovery stage involving sedimentation or microfiltration must be done [122]. Such post treatment is not usable at industrial scale as it would add to the capital and operating costs of the treatment process. In addition, the penetration depth of the light into the slurry is limited.

1) suspended systems where photocatalyst can be:- immobilised on the surface of larger particles, e.g. silica, silica gel microspheres [123], activated carbon [124], microporous zeolites (NaY, Na-mordenite), mesoporous moleculra sieves (MCM-41), glass microspheres recovered from fly ashes, glass Raschig rings, sand, in the form of larger particles: granulated, e.g. VP AEROPERL® P-25/20 is a granulated form of well know TiO_2, Degussa P25 [125], spheres, e.g. titania spheres synthesized through polystyrene spheres templating,
2) Fixed systems with photocatalyst immobilised on:- elements of irradiation system, e.g. quartz protection tube of UV lamps, - photoreactor walls, - additional elements, e.g. glass beads packed in the reactor (PBR), mesh, cloth [126], membrane, plates [127], films [128] and foils. The support is made of stainless steel, alloys (Ti-4V-6Al), titanium, glass, polymers, cotton, ceramic. Pyrex glass is used as a support of allows cutting off light shorter than 300 nm thereby eliminating the direct photolysis of the organic compound. The photocatalyst can be also incorporated into the support structure, e.g. titanium-containing zeolites. It is important to note, that activity and reaction mechanism can change after photocatalyst immobilization as follow.

8.2.4 Amount of Photocatalyst

The required amount of photocatalyst used in the experiment depends on the organic compound, geometry and working conditions of a photoreactor. With increase of photocatalyst amount the reaction rate increases, due to absorption of more photons by photocatalyst and larger adsorption of organic compounds on the photocatalyst [129]. In other case, large amount of photocatalyst will decrease the reaction rate [129]. It is due to the "screening effect" of excess particles, which masks part of photosensitive surface decreasing in penetration depth of light. Chen and Ray developed a model for predicting the optimal photocatalyst amount in aqueous solutions for different photosystems [130]. Using this model, the optimal photocatalyst dose can be determined by using these parameters like the light absorption coefficient of catalyst and influence of light intensity before photocatalysis experiment.

8.3 Photoreactors Designed to Work Under Visible-Light Irradiation Toward Wastewater Treatment

Sutisna *et al.* made a prototype of flat panel photoreactor for the degradation of MB dye under solar light exposure. It was concluded that two parameters, i.e., ratio of reactor volume to total volume and amount of catalyst was necessary for designing of a photoreactor. A hollow rectangular panel is made up of transparent glass to improve the light absorption by the catalyst shown in Figure 8.5. Photons reflected by the mirror fall on the catalyst surface and the illuminated catalyst surface are improved. The catalysts (a) were added into the reactor panel and kept inside by installing a filter at the bottom edge of the reactor panel. A control channel (b) was placed at the center of the reactor panel to keep the waste solution from overflowing, which would cause flow resistance by the catalysts. The dye degradation efficiency was obtained 98% after 48 hours of solar light exposure [131] was observed in Figure 8.6(b). Wastewater was flowed to the reactor panel using a pipe (c) with tiny channels to emit water. A test solution was supplied between the catalysts, going out of the reactor panel through a pipe at the bottom's edge (d). The reactor panel having two sizes: $48 \times 56 \times 1$ cm^3 (small) and $68 \times 76 \times 1$ cm^3 (large). An image of the reactor panel is shown in Figure 8.5.

A solar photoreactor is a system that consists of a reactor panel and a water circulator. Wastewater is circulated from a container (e) through the

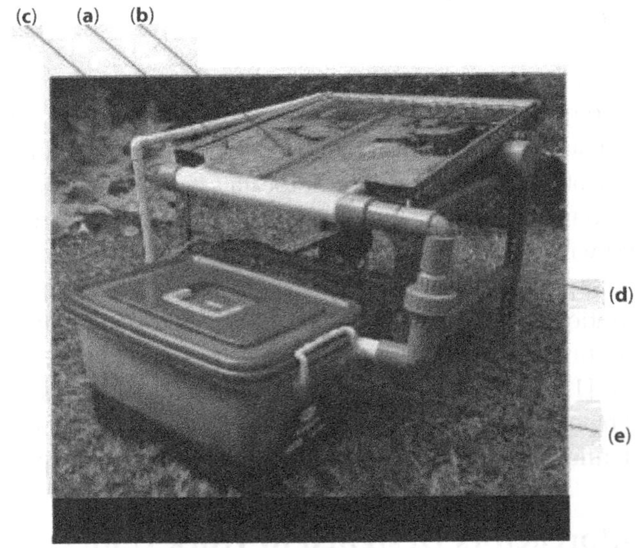

Figure 8.5 Installation of the reactor panel in the solar photoreactor [131].

reactor panel using an electric pump. In the reactor panel, the wastewater will come in contact with the catalysts creating a photocatalytic reaction.

The performance of the FP photoreactor was also improved by arranging several reactor panels in series shown in Figure 8.7(a). Using four panels,

Figure 8.6 MB photodegradation experiment using the proposed photoreactor: (a) initial condition, (b) after 48 hours of irradiation [131].

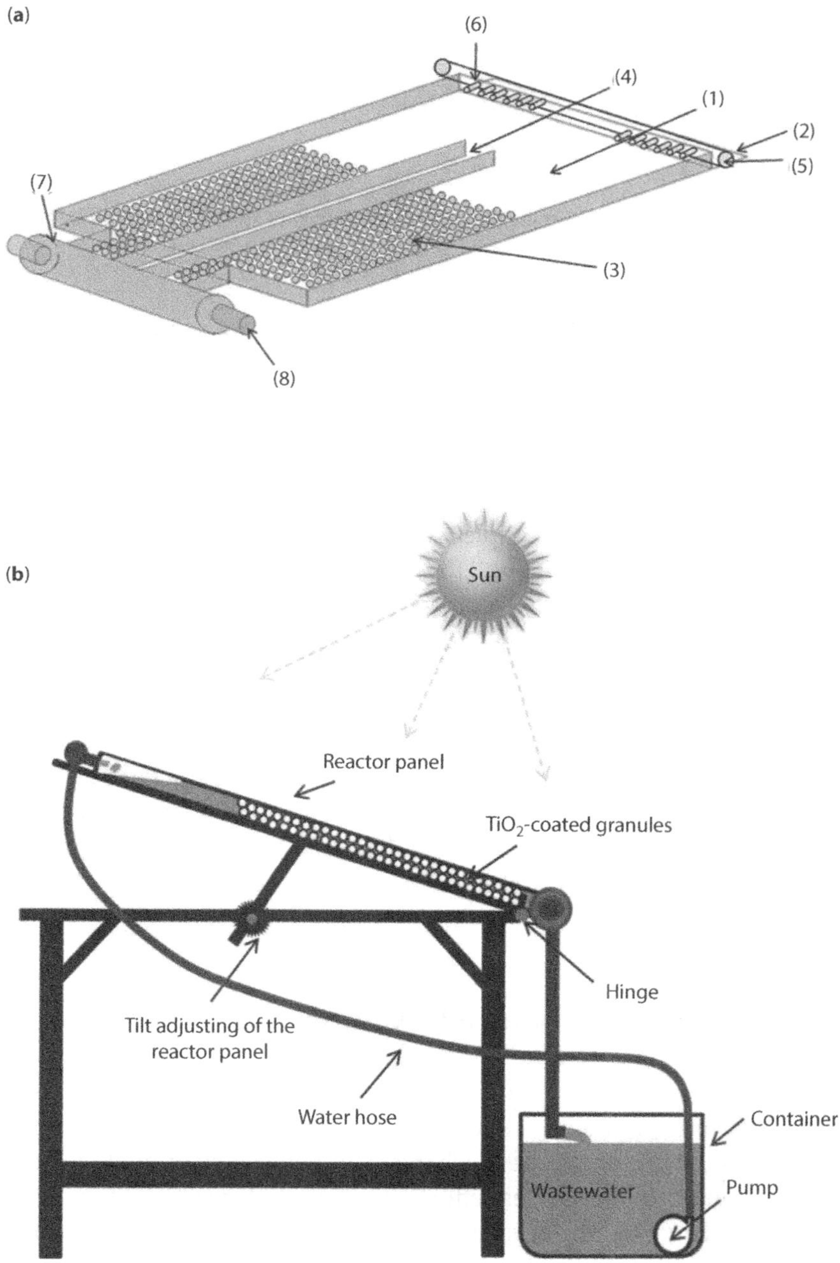

Figure 8.7 (a) The design of the reactor panel as main part of FP photoreactor and (b) Installation of the reactor panel in a solar photoreactor.

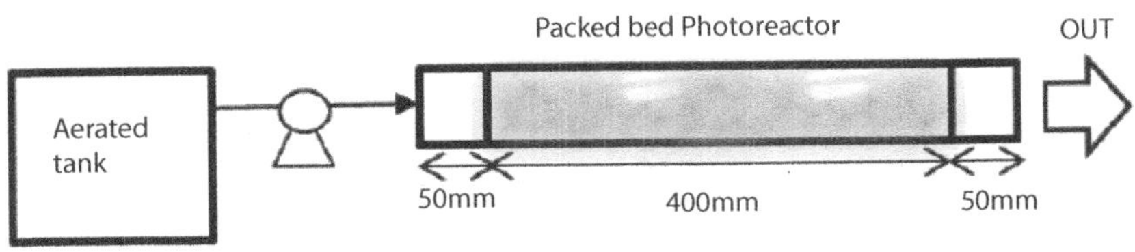

Figure 8.8 Schematic picture of the packed bed photoreactor working in continuous mode.

we observed that the complete decomposition of the same MB solution can be achieved within 10 hours shown in Figure 8.7(b). The proposed FP photoreactor is a very promising alternative for use in decomposing recalcitrant organic pollutants in wastewater [132].

Sacco *et al.* developed a packed bed photoreactor for methylene blue dye decomposition using N-doped TiO_2 particles under visible light exposure shown in Figure 8.8. The complete dye decomposition was achieved after 120 minutes of light irradiation [133].

8.3.1 Limitations of the Currently Employed Photoreactors and Future Scope

The reactors having annular geometry have disadvantages like the incomplete irradiation of the photocatalyst along the annular thickness.

The disadvantage of thin film rotating disk reactor is scale up can only be achieved by widening the rotating disk on a horizontal plane and also when sun light is used as the light source it is not possible to stack the rotating disk.

8.4 Current and Future Developments

Current technology has focused on the decomposition of organic contaminants using reusable photocatalysts under solar radiation. Some of the studies explain about the mechanism and efficiency of photochemical degradation but there is a less effort given on photoreactor setups. These reactors design starts from a beaker under daylight irradiation, a laboratory cuvette with point irradiation, a cylindrical reactor with an immersed lamp, and interference filters for photocatalyst activity testing and also more complex photoreactors show the technological problems of photocatalyst arrangement, gas phase distribution, mass transfer limitation, optimal irradiation conditions, and connection of photochemical reaction with other processes enhancing organic compound decomposition. The photoreactor design is also highly dependent on kind of treated wastewater. Mainly, the focused has given on low-cost photoreactors and efficient photocatalysts for the decomposition of contaminants under solar light radiation.

References

1. Arques, A., Amat, A.M., Garcia-Ripoll, A., Vicente, R., Detoxification and/or increase of the biodegradability of aqueous solutions of dimethoate by means of solar photocatalysis. *J. Hazard. Mater.*, 146, 447–452, 2007.
2. Ahmed, S., Rasul, M.G., Martens, W.N., Brown, R.J., Hashib, M.A., Heterogeneous photocatalytic degradation of phenols in wastewater: A review on current status and developments. *Desalination*, 261, 3–18, 2010.
3. Gupta, V.K., Ali, I., Saleh, T.A., Chemical treatment technologies for waste-water recycling—An overview. *RSC Adv.*, 2, 6380, 2012.
4. Glaze, W.H., Kang, J.W., Chapin, D.H., The chemistry of water treatment processes involbing ozone, hydrogen peroxide, and ultraviolet radiation. *Ozone Sci. Eng.*, 9, 335, 1987.
5. Al-Kdasi, A., Idris, A., Saed, K., Guan, C.T., Treatment of textile wastewater by advanced oxidation processes: A review. *Global Nest Int. J.*, 6, 222, 2004.
6. Sahoo, R., Mundamajhi, A., Das, S.K., Growth of ZnO nanoparticles prepared from cost effective laboratory grade ZnO powder and their application in UV photocatalytic dye decomposition. *J. Mater. Sci. Mater. Electron.*, 30, 4541–4547, 2019.
7. Rostam, A.B. and Taghizadeh, M., Advanced oxidation processes integrated by membrane reactors and bioreactors for various wastewater treatments: A critical review. *J. Environ. Chem. Eng.*, 2020.
8. Cuerda-Correa, E.M., Alexandre-Franco, M.F., Fernández-González, C., Advanced oxidation processes for the removal of antibiotics from water. An overview. *Water*, 2019.
9. Kowalska, E. and Rau, S., Photoreactors for wastewater treatment: A review. *Recent Pat. Eng.*, 4, 242–266, 2010.
10. Artuna-Kowalska, E. and Hupka, J., H_2O_2/UV/air oxidation of organic contaminants in the gas-sparged cyclone reactor. *Cent. Eur. J. Public Health*, 8, Suppl: 88-9, 2000.
11. Kowalska, E., Janczarek, M., Hupka, J., Grynkiewicz, M., H_2O_2/UV enhanced degradation of pesticides in wastewater. *Water Sci.*, 49, 261–6, 2004.
12. Hoffmann, M.R., Martin, S.T., Choi, W.Y., Bahnemann, D.W., Environmental applications of semiconductor photocatalysis. *Chem. Rev.*, 95, 69–96, 1995.

13. Zaleska, A., Hupka, J., Wiergowski, M., Biziuk, M., Photocatalytic degradation of lindane, p,p'-DDT and methoxychlor in an aqueous environment. *J. Photochem. Photobiol. A Chem.*, 135, 213–20, 2000.
14. Byrne, J.A., Eggins, B.R., Brown, N.M.D., McKinney, B., Rouse, M., Immobilisation of TiO2 powder for the treatment of polluted water. *Appl. Catal. B Environ.*, 17, 25–36, 1998.
15. Horikoshi, S., Tsuchida, A., Sakai, H., Abe, M., Sato, S., Serpone, N., Microwave discharge electrodeless lamps (MDEL) Part IV. Novel self-ignition system incorporating metallic microwave condensing cones to activate MDELs in photochemical reactions. *Photochem. Photobiol. Sci.*, 8, 1618–25, 2009.
16. Wang, D., Oppenlander, T., El-Din, M.G., Bolton, J.R., Comparison of the disinfection effects of vacuum-UV (VUV) and UV light on bacillus subtilis spores in aqueous suspensions at 172, 222 and 254 nm. *Photochem. Photobiol.*, 86, 176–81, 2010.
17. Philips Lighting, Keep your pond clean and healthy with Philips TUV lamps. Available at: http://www.lih.de/images/kataloge/ katalog_79.pdf.
18. Labas, M.D., Brandi, R.J., Zalazar, C.S., Cassano, A.E., Water disinfection with UVC radiation and H_2O_2. A comparative study. *Photochem. Photobiol. Sci.*, 8, 670–6, 2009.
19. Villasenor, J., Reyes, P., Pecchi, G., Catalytic and photocatalytic ozonation of phenol on MnO_2 supported catalysts. *Catal. Today*, 76, 121–31, 2002.
20. Taghipour, F., Ultraviolet and ionizing radiation for microorganism inactivation. *Water Res.*, 38, 3940–8, 2004.
21. Martin, C.A., Alfano, O.M., Cassano, A.E., Decolorization of water for domestic supply employing UV radiation and hydrogen peroxide. *Catal. Today*, 60, 119–27, 2000.
22. Artuna-Kowalska, E. and Hupka, J., H2O2/UV/air oxidation of organic contaminants in the gas-sparged cyclone reactor. *Cent. Eur. J. Public Health*, 8, Suppl: 88-9, 2000.
23. Kowalska, E., Zaleska, A., Wiergowski, M., Hupka, J., Biziuk, M., H_2O_2/UV oxidation of methoxychlor. *Proceedings of the Twelfth International Conference on Photochemical Conversion and Storage of Solar Energy*, Berlin, Germany, August 9-14, 1998.
24. Kowalska, E., Janczarek, M., Hupka, J., Grynkiewicz, M., H_2O_2/UV enhanced degradation of pesticides in wastewater. *Water Sci. Technol.*, 49, 261–6, 2004.
25. Villaverde, J., Maqueda, C., Undabeytia, T., Morillo, E., Effect of various cyclodextrins on photodegradation of a hydrophobic herbicide in aqueous suspensions of different soil colloidal components. *Chemosphere*, 69, 575–84, 2007.
26. Rachel, A., Sarakha, M., Subrahmanyam, M., Boule, P., Comparison of several titanium dioxides for the photocatalytic degradation of benzenesulfonic acids. *Appl. Catal. B Environ.*, 37, 293–300, 2002.

27. Hussein, F.H., Halbus, A.F., Hassan, H.A.K., Hussein, W.A.K., Photocatalytic degradation of bismarck brown G using irradiated ZnO in aqueous solutions. *Eur. J. Chem.*, 7, 540–4, 2010.
28. Chen, D. and Ray, A.K., Photocatalytic kinetics of phenol and its derivatives over UV irradiated TiO_2. *Appl. Catal. B. Environ.*, 23, 143–57, 1999.
29. Arslan, I., Balcioglu, I.A., Bahnemann, D.W., Heterogeneous photocatalytic treatment of simulated dyehouse effluents using novel TiO_2-photocatalysts. *Appl. Catal. B Environ.*, 26, 193–206, 2000.
30. Arabatzis, I.M., Stergiopoulos, T., Andreeva, D., Kitova, S., Neophytides, S.G., Falaras, P., Characterization and photocatalytic activity of Au/TiO_2 thin films for azo-dye degradation. *J. Catal.*, 220, 127–35, 2003.
31. Arabatzis, I.M., Stergiopoulos, T., Bernard, M.C., Labou, D., Neophytides, S.G., Falaras, P., Silver-modified titanium dioxide thin films for efficient photodegradation of methyl orange. *Appl. Catal. B Environ.*, 42, 187–201, 2003.
32. Teoh, W.Y., Amal, R., Madler, L., Pratsinis, S.E., Flame sprayed visible light-active Fe-TiO_2 for photomineralisation of oxalic acid. *Catal. Today*, 120, 203–13, 2007.
33. Wilke, K. and Breuer, H.D., The influence of transition metal doping on the physical and photocatalytic properties of titania. *J. Photochem. Photobiol. A*, 121, 49–53, 1999.
34. Davydov, L., Reddy, E.P., France, P., Smirniotis, P.G., Sonophotocatalytic destruction of organic contaminants in aqueous systems on TiO_2 powders. *Appl. Catal. B Environ.*, 32, 95–105, 2001.
35. Ilisz, I., Laszlo, Z., Dombi, A., Investigation of the photodecomposition of phenol in near-UV-irradiated aqueous TiO_2 suspensions. I: Effect of charge-trapping species on the degradation kinetics. 180, 25–33, 1999.
36. Romero, R.L., Alfano, O.M., Cassano, A.E., Photocatalytic reactor employing titanium dioxide: from a theoretical model to realistic experimental results. *Ind. Eng. Chem. Res.*, 48, 10456–66, 2009.
37. Serpone, N. and Pelizzetti, E., *Photocatalysis. Fundamentals and Applications*, Wiley, New York, Chichester, Brisbane, Toronto, Singapore, 1989.
38. Hoffmann, M.R., Martin, S.T., Choi, W.Y., Bahnemann, D.W., Environmental applications of semiconductor photocatalysis. *Chem. Rev.*, 95, 69–96, 1995.
39. Pelizzetti, E. and Minero, C., Mechanism of the photooxidative degradation of organic pollutants over titanium dioxide particles. *Electrochim. Acta*, 38, 47–55, 1993.
40. Kormann, C., Bahnemann, D.W., Hoffmann, M.R., Photolysis of chloroform and other organic molecules in aqueous titanium dioxide suspensions. *Environ. Sci. Technol.*, 25, 494–500, 1991.
41. Zielinska, A., Kowalska, E., Sobczak, J.W. *et al.*, Silver-doped TiO_2 prepared by microemulsion method: Surface properties, bio- and photoactivity. *Sep. Purif. Technol.*, 72, 309–18, 2010.
42. Hupka, J., Zaleska, A., Kowalska, E., Bokotko, R., Tyszkiewicz, H., Biziuk, M., Photocatalytic degradation of almost non-soluble organics in gas-sparged

reactor. *Proceedings of the EF Conference Environmental Technology for Oil Pollution*, Jurata, Poland, August 29-September 3, 1999.

43. Zaleska, A., Hupka, J., Wiergowski, M., Biziuk, M., Photocatalytic degradation of lindane, p,p'-DDT and methoxychlor in an aqueous environment. *J. Photochem. Photobiol. A Chem.*, 135, 213–20, 2000.
44. Dabrowski, B., Zaleska, A., Janczarek, M., Hupka, J., Miller, J.D., Photooxidation of dissolved cyanide using TiO_2 catalyst. *J. Photochem. Photobiol. A Chem.*, 151, 201–5, 2002.
45. Pelizzetti, E., Minero, C., Maurino, V., Hidaka, H., Serpone, N., Terzian, R., Photocatalytic degradation of dodecane and of some dodecyl derivatives. *Ann. Chim.*, 80, 81–7, 1990.
46. Derbalah, A.S., Nakatani, N., Sakugawa, H., Photocatalytic removal of fenitrothion in pure and natural waters by photo-Fenton reaction. *Chemosphere*, 57, 635–44, 2004.
47. Pal, B. and Sharon, M., Preparation of iron oxide thin film by metal organic deposition from Fe(III)-acetylacetonate: A study of photocatalytic properties. *Thin Solid Films*, 379, 83–8, 2000.
48. Subagio, D.P., Srinivasan, M., Lim, M., Lim, T.T., Photocatalytic degradation of bisphenol-A by nitrogen-doped TiO_2 hollow sphere in a vis-LED photoreactor. *Appl. Catal. B Environ.*, 95, 414–22, 2010.
49. Fan, C., Chen, Y.H., Jiang, Z.Q. *et al.*, Diarylmethylene-bridged triphenylamine derivatives encapsulated with fluorene: Very high T-g host materials for efficient blue and green phosphorescent OLEDs. *J. Mater. Chem.*, 20, 3232–7, 2010.
50. Chatterjee, D. and Mahata, A., Demineralization of organic pollutants on the dye modified TiO_2 semiconductor particulate system using visible light. *Appl. Catal. B Environ.*, 33, 119–25, 2001.
51. Zaleska, A., Doped-TiO2: A review. *Recent Pat. Eng.*, 2, 157–64, 2008.
52. Zabek, P., Eberl, J., Kisch, H., On the origin of visible light activity in carbon-modified titania. *Photochem. Photobiol. Sci.*, 8, 264–9, 2009.
53. Gorska, P., Zaleska, A., Kowalska, E. *et al.*, TiO2 photoactivity in vis and UV light: The influence of calcination temperature and surface properties. *Appl. Catal. B Environ.*, 84, 440–7, 2008.
54. Ohno, T., Akiyoshi, M., Umebayashi, T., Asai, K., Mitsui, T., Matsumura, M., Preparation of S-doped TiO_2 photocatalysts and their photocatalytic activities under visible light. *Appl. Catal. A Gen.*, 265, 115–21, 2004.
55. Asahi, R., Morikawa, T., Ohwaki, T., Aoki, K., Taga, Y., Visible-light photocatalysis in nitrogen-doped titanium oxides. *Science*, 293, 269–71, 2001.
56. Sakthivel, S., Janczarek, M., Kisch, H., Visible light activity and photoelectrochemical properties of nitrogen-doped TiO2. *J. Phys. Chem. B*, 108, 19384–7, 2004.
57. Mitoraj, D. and Kisch, H., The nature of nitrogen-modified titanium dioxide photocatalysts active in visible light. *Angew. Chem. Int. Ed.*, 47, 9975–8, 2008.

58. Huang, D.G., Liao, S.J., Liu, J.M., Dang, Z., Petrik, L., Preparation of visible-light responsive N-F-codoped TiO_2 photocatalyst by a solgel-solvothermal method. *J. Photochem. Photobiol. A Chem.*, 184, 282–8, 2006.
59. Zaleska, A., Sobczak, J.W., Grabowska, E., Hupka, J., Preparation and photocatalytic activity of boron-modified TiO_2 under UV and visible light. *Appl. Catal. B Environ.*, 78, 92–100, 2008.
60. He, Z.Q., Xu, X., Song, S. *et al.*, A visible light-driven titanium dioxid photocatalyst codoped with lanthanum and iodine: An application in the degradation of oxalic acid. *J. Phys. Chem. C*, 112, 16431–7, 2008.
61. Kisch, H., Burgeth, G., Macyk, W., Visible light photocatalysis by a titania transition metal complex. *Adv. Inorg Chem.*, 56, 241–59, 2004.
62. Kowalska, E., Remita, H., Colbeau-Justin, C., Hupka, J., Belloni, J., Modification of titanium dioxide with platinum ions and clusters: Application in photocatalysis. *J. Phys. Chem. B*, 112, 1124–31, 2008.
63. Kowalska, E., Abe, R., Ohtani, B., Visible light-induced photocatalytic reaction of gold-modified titanium(IV) oxide particles: Action spectrum analysis. *Chem. Commun.*, 241–3, 2009.
64. Kowalska, E., Mahaney, O.O.P., Abe, R., Ohtani, B., Visible-lightinduced photocatalysis through surface plasmon excitation of gold on titania surfaces. *Phys. Chem. Chem. Phys.*, 12, 2344–55, 2010.
65. Han, C.S. and Itoh, K., *Studies on a Heating Characteristics of Far IR Rays. Part 2. Decomposition Rate of Coloring Matter*, vol. 16, pp. 277–81, Memoirs of the Faculty of Agriculture, Hokkaido University, 1989.
66. Maxey, C.D., Ahmed, M.U., Jones, C.L. *et al.*, Growth of long wavelength infrared MCT emitters on conductive substrates. *J. Electron. Mater.*, 30, 723–7, 2001.
67. Gorska, P., Zaleska, A., Kowalska, E. *et al.*, TiO_2 photoactivity in vis and UV light: The influence of calcination temperature and surface properties. *Appl. Catal. B Environ.*, 84, 440–7, 2008.
68. Hupka, J., Zaleska, A., Janczarek, M., Kowalska, E., Gorska, P., Aranowski, R., UV/VIS light- enhanced photocatalysis for water treatment and protection, in: *Soil and Water Pollution Monitoring, Protection and Remediation, Nato Science Series: IV. Earth and Environmental Sciences*, I. Twardowska, H.E. Allen, M.M. Häggblom, S. Stefaniak, (Eds.), pp. 151–66, Springer, 2005.
69. Zhang, Z., Wang, C.-C., Zakaria, R., Ying, J.Y., Role of particle size in nanocrystalline TiO_2-based photocatalysts. *J. Phys. Chem. B*, 102, 10871–8, 1998.
70. Alnaizy, R. and Akgerman, A., Advanced oxidation of phenolic compounds. *Adv. Environ. Res.*, 4, 233–44, 2000.
71. Kowalska, E., Chrzanowska, M., Janczarek, M., Blaszke, A., Hupka, J., Characterization and treatment of industrial pesticide wastewater. *Proceedings of the REWAS'04 - Global Symposium on Recycling, Waste Treatment and Clean Technology*, Madrid, Spain, September 26-29, 2004.

72. Yan, X., Abe, R., Ohno, T., Toyofuku, M., Ohtani, B., Action spectrum analyses of photoinduced superhydrophilicity of titania thin films on glass plates. *Thin Solid Films*, 516, 5872–6, 2008.
73. Mitoraj, D., Origin of visible light activity in nitrogen modified titanium dioxide: Mechanistic studies on urea modified titanium dioxides, in: *Suedwestdeutscher Verlag fuer Hochschulschriften*, 2010.
74. Amano, F., Nogami, K., Tanaka, M., Ohtani, B., Correlation between surface area and photocatalytic activity for acetaldehyde decomposition over bismuth tungstate particles with a hierarchical structure. *Langmuir*, 2010.
75. Zyoud, A.H., Zaatar, N., Saadeddin, I. *et al.*, CdS-sensitized TiO_2 in phenazopyridine photo-degradation: Catalyst efficiency, stability and feasibility assessment. *J. Hazard. Mater.*, 173, 318–25, 2010.
76. Vel Leitner, N.K. and Dore, M., Role de l'oxygene dissous dans le mecanisme de decomposition de l'acide formique en solution aqueuse par irradiation UV en presence de peroxyde d'hydrogene. *J. Chim. Phys.*, 91, 503–18, 1994.
77. Spangenberg, D., Moller, U., Kleinermanns, K., Photooxidation of exhaust pollutants 4. Photocatalytic and thermal decomposition of trichloroacetic acid: Degradation efficiencies and products. *Chemosphere*, 33, 43–9, 1996.
78. Brucato, A. and Rizzuti, L., Simplified modeling of radiant fields in heterogeneous photoreactors. 1. Case of zero reflectance. *Ind. Eng. Chem. Res.*, 36, 4740–7, 1997.
79. Brucato, A. and Rizzuti, L., Simplified modeling of radiant fields in heterogeneous photoreactors. 2. Limiting "two-flux" model for the case of reflectance greater than zero. *Ind. Eng. Chem. Res.*, 36, 4748–55, 1997.
80. Puma, G.L. and Yue, P.L., A novel fountain photocatalytic reactor for water treatment and purification: Modeling and design. *Ind. Eng. Chem. Res.*, 40, 5162–9, 2001.
81. Puma, G.L. and Yue, P.L., Modelling and design of thin-film slurry photocatalytic reactors for water purification. *Chem. Eng. Sci.*, 58, 2269–81, 2003.
82. Satuf, M.L., Brandi, R.J., Cassano, A.E., Alfano, O.M., Photocatalytic degradation of 4-chlorophenol: A kinetic study. *Appl. Catal. B Environ.*, 82, 37–49, 2008.
83. Imoberdorf, G.E., Alfano, O.M., Cassano, A.E., Irazoqui, H.A., Monte Carlo model of UV-radiation interaction with TiO_2-coated spheres. *AICHE J.*, 53, 2688–703, 2007.
84. Puma, G.L. and Yue, P.L., A laminar falling film slurry photocatalytic reactor. Part II-experimental validation of the model. *Chem. Eng. Sci.*, 53, 3007–21, 1998.
85. Satuf, M.L., Brandi, R.J., Cassano, A.E., Alfano, O.M., Photocatalytic degradation of 4-chlorophenol: A kinetic study. *Appl. Catal. B Environ.*, 82, 37–49, 2008.
86. Brandi, R.J., Alfano, O.M., Cassano, A.E., Modeling of radiation absorption in a flat plate photocatalytic reactor. *Chem. Eng. Sci.*, 51, 3169–74, 1996.

87. Brandi, R.J., Alfano, O.M., Cassano, A.E., Rigorous model and experimental verification of the radiation field in a flat-plate solar collector simulator employed for photocatalytic reactions. *Chem. Eng. Sci.*, 54, 2817–27, 1999.
88. Satuf, M.L., Brandi, R.J., Cassano, A.E., Alfano, O.M., Modeling of a flat plate, slurry reactor for the photocatalytic degradation of 4-chlorophenol. *Int. J. Chem. React. Eng.*, 5, A59, 2007.
89. Coleman, H.M., Vimonses, V., Leslie, G., Amal, R., Degradation of 1,4-dioxane in water using TiO_2 based photocatalytic and H_2O_2/UV processes. *J. Hazard. Mater.*, 146, 496–501, 2007.
90. Ballari, M.D., Alfano, O.O., Cassano, A.E., Photocatalytic degradation of dichloroacetic acid. A kinetic study with a mechanistically based reaction model. *Ind. Eng. Chem. Res.*, 48, 1847–58, 2009.
91. Yeber, M.C., Rodriguez, J., Freer, J., Duran, N., Mansilla, H.D., Photocatalytic degradation of cellulose bleaching effluent by supported TiO_2 and ZnO. *Chemosphere*, 41, 1193–7, 2000.
92. Marugan, J., van Grieken, R., Cassano, A.E., Alfano, O.M., Scaling-up of slurry reactors for the photocatalytic oxidation of cyanide with TiO_2 and silica-supported TiO_2 suspensions. *Catal. Today*, 144, 87–93, 2009.
93. Bezzon, M., Carminati, M., Lagrasta, C., Bellobono, I.R., Tozzi, P.M., Influence of concentrations of iron(III) thiocyanate and hydrogen peroxide on photomineralization rate of phenol onto photocatalytic membranes immobilizing titanium dioxide. *Fresenius Environ. Bull.*, 12, 1551–7, 2003.
94. Dijkstra, M.F.J., Buwalda, H., de Jong, A.W.F., Michorius, A., Winkelman, J.G.M., Beenackers, A.A.C.M., Experimental comparison of three reactor designs for photocatalytic water purification. *Chem. Eng. Sci.*, 56, 547–55, 2001.
95. Hsien, Y.H., Chang, C.F., Chen, Y.H., Cheng, S.F., Photodegradation of aromatic pollutants in water over TiO_2 supported on molecular sieves. *Appl. Catal. B Environ.*, 31, 241–9, 2001.
96. Chatterjee, D. and Mahata, A., Visible light induced photodegradation of organic pollutants on dye adsorbed TiO_2 surface. 153, 199–204, 2002.
97. Grzechulska, J., Hamerski, M., Morawski, A.W., Photocatalytic decomposition of oil in water. *Water Res.*, 34, 1638–44, 2000.
98. Chu, W.H., Gao, N.Y., Deng, Y., Performance of a combination process of UV/H_2O_2/micro-aeration for oxidation of dichloroacetic acid in drinking water. *Clean*, 37, 233–8, 2009.
99. Vidal, A. and Luengo, M.A.M., Inactivation of titanium dioxide by sulphur: Photocatalytic degradation of Vapam. *Appl. Catal. B Environ.*, 32, 1–9, 2001.
100. Chu, W.H., Gao, N.Y., Deng, Y., Performance of a combination process of UV/H_2O_2/micro-aeration for oxidation of dichloroacetic acid in drinking water. *Clean*, 37, 233–8, 2009.
101. Gao, P., Ching, W.H., Herrmann, M., Chan, C.K., Yue, P.L., Photooxidation of a model pollutant in an oscillatory flow reactor with baffles. *Chem. Eng. Sci.*, 58, 1013–20, 2003.

102. Thu, H.B., Karkmaz, M., Puzenat, E., Guillard, C., Herrmann, J.M., From the fundamentals of photocatalysis to its applications in environment protection and in solar purification of water in arid countries. *Res. Chem. Intermediat.*, 31, 449–61, 2005.
103. Cater, S.R., Stefan, M.I., Bolton, J.R., Safarzadeh-Amiri, A., UV/H_2O_2 treatment of methyl tert-butyl ether in contaminated waters. *Environ. Sci. Technol.*, 34, 659–62, 2000.
104. Stefan, M.I. and Bolton, J.R., Mechanism of the degradation of 1, 4-dioxane in dilute aqueous solution using the UV hydrogen peroxide process. *Environ. Sci. Technol.*, 32, 1588–95, 1998.
105. Stefan, M.I., Hoy, A.R., Bolton, J.R., Kinetics and mechanism of the degradation and mineralization of acetone in dilute aqueous solution sensitized by the UV photolysis of hydrogen peroxide. *Environ. Sci. Technol.*, 30, 2382–90, 1996.
106. Chu, W. and Tsui, S.M., Photo-sensitization of diazo disperse dye in aqueous acetone. *Chemosphere*, 39, 1667–77, 1999.
107. Zhang, T.Y., Oyama, T., Horikoshi, S., Hidaka, H., Zhao, J.C., Serpone, N., Photocatalyzed N-demethylation and degradation of methylene blue in titania dispersions exposed to concentrated sunlight. *Sol. Energy Mater. Sol. C*, 73, 287–303, 2002.
108. Pereira, V.J., Linden, K.G., Weinberg, H.S., Evaluation of UV irradiation for photolytic and oxidative degradation of pharmaceutical compounds in water. *Water Res.*, 41, 4413–23, 2007.
109. Mokrini, A., Ousse, D., Esplugas, E., Oxidation of aromatic compounds with UV radiation/ozone/hydrogen peroxide. *Water Sci. Technol.*, 35, 95–102, 1997.
110. Molinari, R., Mungari, M., Drioli, E. *et al.*, Study on a photocatalytic membrane reactor for water purification. *Catal. Today*, 55, 71–8, 2000.
111. Tahiri, H., Aitichou, Y., Herrmann, J.-M., Photocatalytic degradation of chlorobenzoic isomers in aqueous suspensions of neat and modified titania. *J. Photochem. Photobiol. A Chem.*, 114, 219–26, 1998.
112. Kimura, I., Kase, T., Taguchi, Y., Tanaka, M., Preparation of titania/silica composite microspheres by sol-gel process in reverse suspension. *Mater. Res. Bull.*, 38, 585–97, 2003.
113. Oliveira, H.G., Nery, D.C., Longo, C., Effect of applied potential on photocatalytic phenol degradation using nanocrystallinie TiO_2 electrodes. *Appl. Catal. B Environ.*, 93, 205–11, 2010.
114. Rodriguez, M., Malato, S., Pulgarin, C. *et al.*, Optimizing the solar photo-fenton process in the treatment of contaminated water. *Determination intrinsic kinetic constants scale-up. Sol. Energy*, 79, 360–8, 2005.
115. Malato, S., Blanco, J., Richter, C., Curco, D., Gimenez, J., Low concentrating CPC collectors for photocatalytic water detoxification: Comparison with a medium concentrating solar collector. *Water Sci. Technol.*, 35, 157–64, 1997.

116. Malato, S., Blanco, J., Caceres, J., Fernandez-Alba, A.R., Aguera, A., Rodriguez, A., Photocatalytic treatment of water-soluble pesticides by photo-fenton and TiO_2 using solar energy. *Catal. Today*, 76, 209–20, 2002.
117. Zhang, T.Y., Oyama, T., Horikoshi, S., Zhao, J.C., Serpone, N., Hidaka, H., Photocatalytic decomposition of the sodium dodecylbenzene sulfonate surfactant in aqueous titania suspensions exposed to highly concentrated solar radiation and effects of additives. *Appl. Catal. B Environ.*, 42, 13–24, 2003.
118. Minero, C., Pelizzetti, E., Malato, S., Blanco, J., Large solar plant photocatalytic water decontamination: Degradation of pentachlorophenol. *Chemosphere*, 26, 2103–19, 1993.
119. Hashimoto, K., Irie, H., Fujishima, A., TiO_2 photocatalysis: A historical overview and future prospects. *Jpn. J. Appl. Phys. 1*, 44, 8269–85, 2005.
120. Zhang, T.Y., Oyama, T., Horikoshi, S., Zhao, J.C., Serpone, N., Hidaka, H., Photocatalytic decomposition of the sodium dodecylbenzene sulfonate surfactant in aqueous titania suspensions exposed to highly concentrated solar radiation and effects of additives. *Appl. Catal. B Environ.*, 42, 13–24, 2003.
121. McMurray, T.A., Byrne, J.A., Dunlop, P.S.M., Winkelman, J.G.M., Eggins, B.R., McAdams, E.T., Intrinsic kinetics of photocatalytic oxidation of formic and oxalic acid on immobilised TiO_2 films. *Appl. Catal. A Gen.*, 262, 105–10, 2004.
122. Wu, X.H., Jiang, Z.H., Liu, H.L., Li, X.D., Hu, X.U., TiO_2 ceramic films prepared by micro-plasma oxidation method for photodegradation of rhodamine B. *Mater. Chem. Phys.*, 80, 39–43, 2003.
123. Kimura, I., Kase, T., Taguchi, Y., Tanaka, M., Preparation of titania/silica composite microspheres by sol-gel process in reverse suspension. *Mater. Res. Bull.*, 38, 585–97, 2003.
124. Tryba, B., Morawski, A.W., Inagaki, M., Application of TiO_2-mounted activated carbon to the removal of phenol from water. *Appl. Catal. B Environ.*, 41, 427–33, 2003.
125. Malato, S., Blanco, J., Campos, A. *et al.*, Effect of operating parameters on the testing of new industrial titania catalysts at solar pilot plant scale. *Appl. Catal. B Environ.*, 42, 349–57, 2003.
126. A. Tabatabaie-Raissi, N.Z. Muradov, E. Martin, Apparatus for high flux photocatalytic pollution control using rotating fluidized bed reactor. US200030085111, 2003.
127. Wu, X.H., Jiang, Z.H., Liu, H.L., Li, X.D., Hu, X.U., TiO_2 ceramic films prepared by micro-plasma oxidation method for photodegradation of rhodamine B. *Mater. Chem. Phys.*, 80, 39–43, 2003.
128. Pasqui, D., Atrei, A., Barbucci, R., The immobilization of titania nanoparticles on hyaluronan films and their photocatalytic properties. *Nanotechnology*, 20, 015703, 2009.
129. Zhang, T.Y., Oyama, T., Horikoshi, S., Zhao, J.C., Serpone, N., Hidaka, H., Photocatalytic decomposition of the sodium dodecylbenzene sulfonate

surfactant in aqueous titania suspensions exposed to highly concentrated solar radiation and effects of additives. *Appl. Catal. B Environ.*, 42, 13–24, 2003.
130. Chen, D.W. and Ray, A.K., Photocatalytic kinetics of phenol and its derivatives over UV irradiated TiO_2. *Appl. Catal. B Environ.*, 23, 143–57, 1999.
131. Sutisna, R., Wibowo, M., Murniati, E., Khairurrijal, R., Abdullah, M., Novel solar photocatalytic reactor for wastewater treatment. *Iop Conf. Ser: Mater. Sci. Eng.*, 214, 2017.
132. Sutisna, R., Wibowo, M., Khairurrijal, E., Abdullah, M., Prototype of a flat-panel photoreactor using TiO_2 nanoparticles coated on transparent granules for the degradation of Methylene Blue under solar illumination. *Sus. Environ. Res.*, 27, 4, 172–180, 2017.
133. Sacco, O., Sannino, D., Vaiano, V., Packed bed photoreactor for the removal of water pollutants using visible light emitting diodes. *Appl. Sci.*, 9, 472, 2019.
134. Curco, D., Gimenez, J., Addardak, A., Cervera-March, S., Esplugas, S., Effects of radiation absorption and catalyst concentration on the photocatalytic degradation of pollutants. *Catal. Today.*, 76, 177–188, 2002.

9

Simulation of Photocatalytic Reactors

John Akach[1,2]*, John Kabuba[1] and Aoyi Ochieng[3]

[1]Department of Chemical Engineering, Vaal University of Technology, Vanderbijlpark, South Africa
[2]Department of Chemical and Process Engineering, Technical University of Kenya, Nairobi, Kenya
[3]Botswana International University of Science and Technology, Palapye, Botswana

Abstract
Recently, the removal of recalcitrant chemicals in wastewater by photocatalysis has attracted significant attention due to its economic and environmental benefits. Despite numerous laboratory-scale studies and a few pilot-scale applications, commercialization of photocatalysis is yet to be realized. This can be attributed in part to inadequate tools for the design and optimization of photocatalytic reactors. In this respect, several studies have focused on the modeling and simulation of photocatalytic reactors. It has been observed that the distribution of light has a strong influence on photoreactor performance. Consequently, several techniques have been developed for the analysis of light distribution, including the P1, six flux, discrete ordinates, and Monte Carlo methods. In this chapter, these methods will be discussed, as well as other issues affecting reactor simulation, such as the boundary conditions, catalyst optical properties, and validation. Also, the relationship between light absorption and kinetics will be evaluated.

Keywords: Photocatalysis, modeling, simulation, light distribution, kinetics

Abbreviations

CFD	Computational Fluid Dynamics
CPC	Compound Parabolic Concentrator

**Corresponding author*: johnakach@gmail.com

Elvis Fosso-Kankeu, Sadanand Pandey, and Suprakas Sinha Ray (eds.) Photoreactors in Advanced Oxidation Processes: The Future of Wastewater Treatment, (277–304) © 2023 Scrivener Publishing LLC

DR	Diffuse reflectance phase function
HG	Henyey-Greenstein phase function
ISO	Isotropic phase function
LED	Light Emitted Diode
LSDE	Line Source Diffuse Emission
LSSE	Line Source Specular Emission
LVREA	Local Volumetric Rate of Energy Absorption
RTE	Radiation Transport Equation
SSDE	Surface Source Diffuse Emission
SSSE	Surface Source Specular Emission
UV	Ultraviolet
VREA	Volumetric Rate of Energy Absorption
VSE	Volume Source Emission

9.1 Introduction

Photocatalytic reactors are multiphase contacting devices, which bring together the solid catalyst, liquid substrates and sometimes air bubbles in the case of air fluidized reactors [1]. These reactors are complex due to the interplay of hydrodynamics and light distribution. In order to design, analyse, optimize and scale up such reactors, appropriate mathematical models need to be developed. These models should account for the distribution of pollutants, catalysts, photons and bubbles. Photoreactor modeling is usually divided into three elements: hydrodynamics, light distribution and reaction kinetics [2]. Hydrodynamics modeling shows the distribution of catalyst particles in the reactor. Then, light distribution modeling is carried out to analyse the light distribution in the photoreactor. Reaction kinetics modeling is then carried out to determine the rate of photocatalysis. Most photoreactors employ nanocatalysts, which are usually dispersed in the reactor using either air or liquid fluidization. This results in a well-mixed reactor in which the influence of hydrodynamics on photocatalysis is insignificant. Therefore, in most cases photocatalytic reactor modeling consists of an analysis of the light distribution and reaction kinetics only.

The field of photocatalytic reactor modeling and simulation has experienced a gradual growth in recent years. A few review papers on different aspects of photocatalytic reactor modeling have been published. For example, Pareek *et al.* [3] reviewed the methods commonly used to evaluate the light distribution in photocatalytic reactors. Boyjoo *et al.* [2] carried out a review of different aspects of photocatalytic reactor modeling using computational fluid dynamics (CFD) such as hydrodynamics, light distribution and reaction rate modeling. Since these publications, considerable developments have been made in the field of photocatalytic reactor modeling. Numerous studies have been published on the modeling of light distribution in different photocatalytic reactors using simulation methods, such as six-flux and Monte Carlo methods. These studies have reported novel methods of establishing the catalyst optical properties and validating the models. This chapter will review these recent developments, with respect to the modeling of light distribution and reaction kinetics in photocatalysis reactors.

9.2 Modeling of Light Distribution

9.2.1 Light Distribution

In a photoreactor, the participation of photons has to be accounted for by modeling the light distribution for a comprehensive analysis of reactor performance. An analysis of the light distribution in the reactor is crucial as it is the basis of reactor design parameters, such as efficiency parameters, intrinsic kinetics, and the local volumetric rate of energy absorption (LVREA) [4]. In a photoreactor, light photons interact with catalysts, bubbles, and reactor walls. This results in a highly inhomogeneous distribution of light in the photoreactor [5]. Experimental determination of the light distribution in a photoreactor is not feasible due to time and cost constraints. Moreover, any instrument used to measure the local light intensities inside the reactor would distort the radiation field, thus making it impossible to accurately measure the light intensity profile in the reactor [6]. Therefore, light distribution is normally established using experimentally validated simulation [7].

The modeling of light distribution is usually carried out by solving the radiation transport equation (RTE) [8]:

$$\frac{dI_\lambda(s,\Omega)}{ds} = -\kappa_\lambda I_\lambda(s,\Omega) - \sigma_\lambda I_\lambda(s,\Omega) + \frac{1}{4\pi}\sigma_\lambda \int_0^{4\pi} p(\Omega' \to \Omega) I_\lambda(s,\Omega') d\Omega' \qquad (9.1)$$

where I_λ is a ray of light of wavelength λ travelling in direction Ω and distance s; $p(\Omega' \to \Omega)$ is the scattering phase function, which defines the probability that light from direction Ω' will be scattered to direction Ω; κ_λ and σ_λ are the absorption and scattering coefficients of the medium, respectively.

An analytical solution to the RTE is impossible due to the presence of scattering terms on the right-hand side of the RTE. Therefore, the equation is usually solved using numerical methods such as the rigorous Monte Carlo method [4, 7] and the discrete ordinates method [9]. Computationally simpler methods, such as the P1 method [10] and the six-flux method [11], have also been employed to solve the RTE.

9.2.2 Light Distribution Methods

Two-flux method

This is the simplest light distribution method, which considers multiple scattering effects. The method assumes that light is scattered only in the forward and backward directions. This transforms the RTE into two first-order differential equations, which can then be solved analytically. The two-flux method is only appropriate for reactors, which may be considered to be enclosed within two infinite parallel walls. It has been applied by Puma and Brucato [12] to simulate the light distribution in an annular UV reactor with nanophase TiO_2 catalysts.

P1 method

The P1 method assumes that the direction of light propagation is unimportant i.e. that it is fully isotropic. This assumption transforms the RTE into a second-order partial differential equation, which can then be solved analytically [13]. In a typical reactor, the propagation of collimated light from such sources as sunlight may be strongly directional, especially near the light source. Therefore, the P1 method has been found to be particularly inaccurate near solar-illuminated reactor boundaries [13]. The assumptions of this method also break down at low catalyst loading where there is insufficient catalyst scattering to ensure isotropic light propagation [14]. The P1 method has been used to simulate the light distribution in annular UV [14] and hybrid light reactors with nanophase TiO_2 as the catalyst [10, 13].

Six-flux method

This method improves upon the two-flux method by assuming that incoming photons are scattered randomly by the catalyst in the six cartesian directions. The original six-flux method was developed under the assumption that light scattering could be described by the diffuse reflectance scattering phase function [15]. The probability of light scattering in each of the six directions was then calculated from an exact solution using the Monte Carlo method. Recently, the six-flux method coupled with the

Henyey-Greenstein scattering phase function has been developed and employed in several studies [6]. This version of the six-flux method was more accurate as compared to those which employ the diffuse reflectance and isotropic scattering phase functions. It should be noted that the six-flux method transforms into the two-flux method when all photons are backscattered.

The accuracy of the six-flux method has been investigated by comparing its results with those of the more accurate finite volume method. It was found that, under collimated light, a slight error of 5% was observed. However, under diffuse light, this error increased to 120% [16]. Due to its limitation of light scattering in six directions, the six-flux method has been observed to be inaccurate in predicting the light distribution at high catalyst loading [6]. The six-flux method has been used to evaluate the light distribution in annular UV reactors [15] and solar illuminated tubular and compound parabolic concentrator (CPC) reactors with nanophase TiO_2 catalysts [6, 17–20].

Discrete ordinates method

In this method, the reactor is divided into a series of control volumes and directions. The RTE is then transformed into a series of algebraic equations according to the number of volume and angular discretizations [21]. The discrete ordinates method has been included in several CFD packages such as Ansys Fluent. This method has been used to simulate the light distribution in several reactors. Casado *et al.* [22] used the discrete ordinates method to simulate the light distribution in a solar illuminated CPC reactor. Moreno-SanSegundo *et al.* [23] used the discrete ordinates method in Ansys Fluent and openFOAM to simulate the light distribution in an annular UV reactor, solar CPC reactor and LED illuminated reactor. Boyjoo and co-workers simulated the light distribution in an annular UV reactor [24] and multi-annular UV reactors [9, 21] using the discrete ordinates method in Ansys Fluent. All these simulation studies were carried out in reactors with nanophase catalysts.

Monte Carlo method

In the Monte Carlo method, the paths of numerous photons are tracked from the reactor wall until they are absorbed by the catalyst. This method has been preferred over other methods for complicated reactor geometries [7]. The accuracy of the Monte Carlo method is so well established that it is often used to validate the accuracy of other less accurate methods like the six-flux method [6]. The major drawback of the Monte Carlo method is that for an accurate simulation, a high number of photons need to be tracked while considering the interactions of the photons with catalyst particles and the reactor boundary [25]. This is computationally intensive and requires a long simulation time. In one study, 60 million photons reportedly took 32 hours to track in a quad core computer [7].

The Monte Carlo method has been used to simulate the light distribution in several reactors with nanoparticulate P25 TiO_2 catalysts. These include annular UV reactors [4, 7, 25], externally illuminated annular UV reactor [26], solar illuminated tubular and surface uniform concentrators [27] as well as slurry bubble column reactors under UV light [28] and solar illumination [29]. The light distribution in a fluidized bed reactor with large TiO_2 coated spheres has also been determined using the Monte Carlo method [30].

9.2.3 Simulation Parameters

In order to solve the RTE, appropriate boundary conditions, scattering phase functions, absorption, and scattering coefficients need to be specified.

Boundary conditions

The boundary conditions for simulation of the light distribution include the spectrum of the light source and how the light interacts with the boundaries such as the lamp, lamp sleeve and reactor wall. Most light sources exhibit a wide band spectrum spanning across several wavelengths. An accurate accounting of the light spectrum is important since catalyst absorption and scattering are strong functions of the light wavelength. The most accurate simulations utilize a complete model of the light spectrum. This is common practice in Monte Carlo simulations where each photon is randomly assigned a specific wavelength according to the light spectrum [7]. The six-flux method generally assumes a single wavelength and then uses wavelength-averaged optical properties. Some finite volume methods have also employed this approach. A few finite volume methods improve upon this by dividing the light spectrum into multiple zones [21].

Before simulation, it is necessary to establish how the light interacts with the boundaries. A study by Valadés-Pelayo *et al.* [7] found that the best boundary conditions for their annular reactor consisted of a combination of total absorption of photons by the lamp and the wall. These values were revised to 30% lamp absorption and 65% wall absorption after further refinements using a better validation technique [25]. It is also important to consider how the light interacts with transparent boundaries. An accurate simulation would consider the refraction and reflection of the light across the boundaries. This is especially important for direct or collimated light sources in which refraction is crucial. Acosta-Herazo *et al.* [31] found that an error of up to 12% can result if refraction/reflection effects are not included in the simulation of the CPC absorber.

Boundary conditions also include the intensity of the incoming radiation at the reactor boundary. For reactors illuminated by a lamp, the light intensity at the lamp sleeve depends on the type of lamp. The light intensity around the lamp is usually modeled using a lamp emission model. These models are classified according to the source of photons and their direction of propagation. According to the source of photons, lamp emission models can be classified as line source, surface source and volume source models (Figure 9.1). These can be further classified as specular and diffuse emission depending on the direction of the emitted light [3].

Line source emission models assume that the lamp can be simplified as an emitting line. These models can be used to approximate the light output of long low-pressure lamps with a narrow diameter. The surface source models assume that the photons originate from the lamp surface. This is mostly true of black light lamps in which the UVA photons are emitted from the phosphor on the lamp coating. The volume source emission model treats the photons as originating from the entire volume of the lamp. This model can be used to describe the light output from arc lamps [24]. The equations for line source specular emission (LSSE), line source diffuse emission (LSDE), surface source specular emission (SSSE), surface source diffuse emission (SSDE) and volume source emission (VSE) are given in Table 9.1.

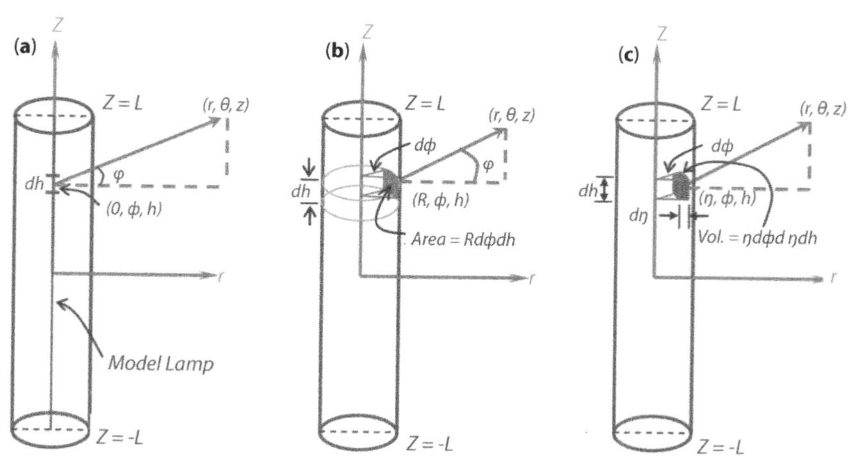

Figure 9.1 Lamp emission models (a) line source (b) surface source (c) volume source. Reprinted from [2] with permission from Elsevier.

Table 9.1 Lamp emission models [24].

Model	Equation
line source specular emission (LSSE)	$E=\frac{K_l}{4\pi}\int_{-L}^{L}\frac{dh}{(r^2+(z-h)^2)}$ $E=\frac{K_l}{4\pi r}\left[\tan^{-1}\left(\frac{z+L}{r}\right)-\tan^{-1}\left(\frac{z-L}{r}\right)\right]$
line source diffuse emission (LSDE)	$E=\frac{K_l}{4\pi}\int_{-L}^{L}\frac{rdh}{[r^2+(z-h)^2]^{3/2}}$
surface source specular emission (SSSE)	$E=\frac{K_s}{4\pi}\int_{-L}^{L}\int_{-\pi/2}^{\pi/2}\frac{R\,d\phi\,dh}{[(\mathrm{r}\cos\theta-R\cos\phi)^2+(r\sin\theta-R\sin\phi)^2+(z-h)^2]}$
surface source diffuse emission (SSDE)	$E=\frac{K_s}{4\pi}\int_{-L}^{L}\int_{-\pi/2}^{\pi/2}\frac{\cos\varphi\,R\,d\phi\,dh}{[(\mathrm{r}\cos\theta-R\cos\phi)^2+(r\sin\theta-R\sin\phi)^2+(z-h)^2]}$
volume source emission (VSE)	$E=\frac{K_V}{4\pi}\int_{-L}^{L}\int_{0}^{R}\int_{-\pi}^{\pi}\frac{\eta\,d\eta\,d\phi\,dh}{[(\mathrm{r}\cos\theta-\eta\cos\phi)^2+(\mathrm{r}\sin\theta-\eta\sin\phi)^2+(z-h)^2]}$

Scattering phase function

The scattering phase function defines the direction to which the catalyst particles scatter incoming light. For very dilute heterogeneous solutions, in which a single scattering event occurs on a spherical particle, the theoretical Mie theory can apply. However, in real photocatalytic reactors, multiple scattering events occur on agglomerated catalyst particles with non-spherical shapes. In such a situation, it is more useful to apply an empirical phase function, which describes the probability of scattering in a certain direction [32]. The phase functions, which have been employed in photocatalytic light distribution studies, are isotropic, diffuse reflecting, and Henyey-Greenstein functions.

In the isotropic phase (ISO) function, the photons have an equal probability of being scattered in any direction (Figures 9.2a–b). This phase function is expressed as:

$$p_{iso}(\theta) = \frac{1}{2} \tag{9.2}$$

where θ is the exit scattering angle.

In the diffuse reflectance (DR) phase function, the catalyst particle is assumed to reflect the light in a completely diffuse manner [33]. This phase function scatters light mainly in a backward direction (Figure 9.2a). It is expressed mathematically as [27]:

$$p_{DR}(\theta) = \frac{8}{3\pi}(\sin\theta - \theta\cos\theta) \tag{9.3}$$

The Henyey-Greenstein (HG) phase function describes a wide range of scattering phenomenon using an adjustable asymmetry factor g, which can take any value between −1 and 1 [7]:

$$p_{HG}(\theta) = \frac{1}{2}\frac{(1-g^2)}{(1+g^2-2g\cos\theta)^{3/2}} \tag{9.4}$$

Positive values of g specify forward scattering while negative values specify backward scattering (Figure 9.2b). When g is set to 0, eq. (9.4) reduces to eq. (9.2) and the HG phase function specifies isotropic scattering. By adjusting the asymmetry factor, the HG phase function can accurately describe the scattering behaviour of a wide range of catalysts in different

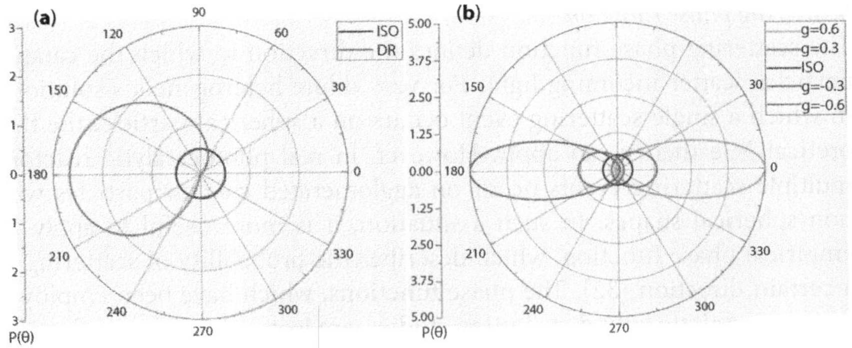

Figure 9.2 Scattering probability of (a) isotropic (ISO), diffuse reflectance (DR) and (b) Henyey-Greenstein phase functions. Reprinted from [27] with permission from Elsevier.

agglomeration states [6]. The other scattering phase functions are not able to differentiate the scattering behaviour of different catalysts, which limits their accuracy. This has been confirmed by studies, which have compared the accuracy of the different phase functions and concluded that the Henyey-Greenstein function was superior to the others [6, 7, 27, 32].

The optimal value of *g* for Aeroxide P25 TiO_2 in different reactors has been determined by comparing the experimental and simulated light intensity at specific locations in the reactor. These studies found that Aeroxide P25 TiO_2 catalyst generally exhibits forward scattering behaviour with

Table 9.2 Henyey-Greenstein asymmetry factor for Aeroxide P25 TiO_2 from different studies.

HG asymmetry factor	Catalyst loading (g/L)	Wavelength (nm)	Reference
0.4–0.7	0.2–2	295–405	[34]
0.6–0.8	0.01–0.4	345–388	[7]
0.68	0.025–0.4	345–388	[25]
0.41	0–0.1	-	[35]
0.64	0.08–0.5	-	[27]
−0.399 to 0.929	0.04–0.4	254–365	[32]
0.84	0.025–0.6	340–387	[28]
0.87	0.005–0.2	300–387	[29]

different values of *g* depending on the type of reactor, light source, catalyst loading and solution pH (Table 9.2). The study by Satuf *et al.* [34] assumed that the asymmetry factor depends only on the light wavelength. Valadés-Pelayo *et al.* [35] found that *g* had a lower value under acidic pH than under neutral pH, which showed that *g* depends on catalyst agglomeration. The majority of the studies were carried out in real reactors under UV lamp illumination with the assumption that *g* was independent of catalyst loading [7, 25, 27, 28, 32, 35]. However, Turolla *et al.* [32] found that *g* also depends on the catalyst loading such that the asymmetry factor decreased with an increase in catalyst loading resulting in backward scattering at very high catalyst loadings. Only one study [29] determined the optimum *g* under natural sunlight. They found a slight increase in *g* under sunlight as compared to UV lamp which confirmed that *g* depends on the wavelength.

Absorption and scattering coefficients

These coefficients specify the amount of light that is absorbed or scattered at specific wavelengths. The sum of the absorption and scattering coefficients of a catalyst slurry is the extinction coefficient, which can be easily measured using a spectrophotometer, as shown in Figure 9.3a. The extinction coefficient is not useful for the solution of the RTE. It is still necessary to extract either the absorption or scattering coefficient from the extinction coefficient. Cabrera *et al.* [33] were the first researchers to tackle this challenge. They used an integrating sphere to measure the diffuse transmission through a catalyst slurry (Figure 9.3b). Then, they modeled the light transmission through the sample cell using one-dimensional, one-directional RTE, which they solved using the discrete ordinates method. By comparing the experimental and simulated transmission values, they were able to obtain the absorption and scattering coefficients. In solving the RTE, they assumed the diffuse reflectance phase function.

Subsequent studies used a similar methodology but employed the isotropic phase function [8, 37, 38]. All these studies reported the absorption and scattering coefficients of several commercial brands of TiO_2. Other studies have used a similar experimental methodology to determine the optical properties of $NiFe_2O_4$ [36] and $LiVMoO_6$ [39]. These studies used the six-flux method instead of the discrete ordinates method to solve the RTE.

These previous studies assumed that the catalysts scattered light according to a predetermined phase function. Instead of assuming a phase function, Satuf *et al.* [34] set out to determine the appropriate phase function in addition to the scattering and absorption coefficients of several commercial TiO_2 catalysts. They used the Henyey-Greenstein

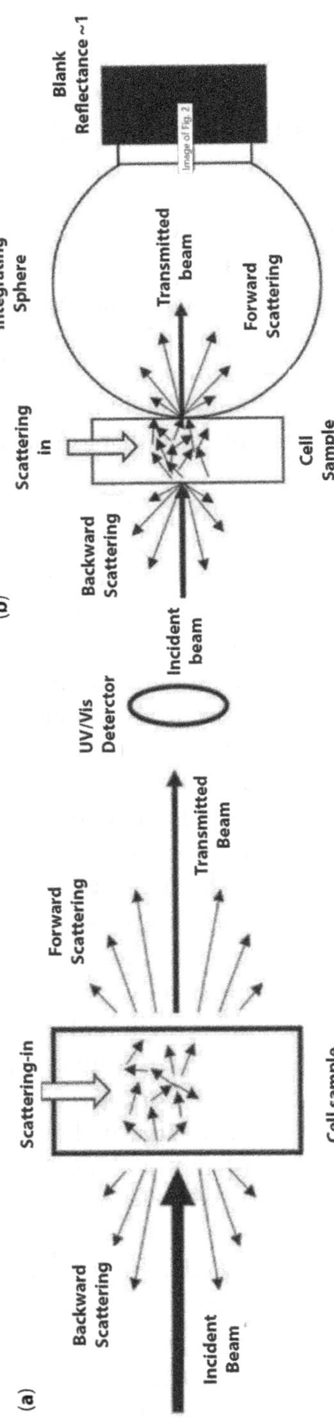

Figure 9.3 Experimental set up for (a) Extinction coefficient, (b) absorption coefficient. Reprinted from [36] with permission from Elsevier.

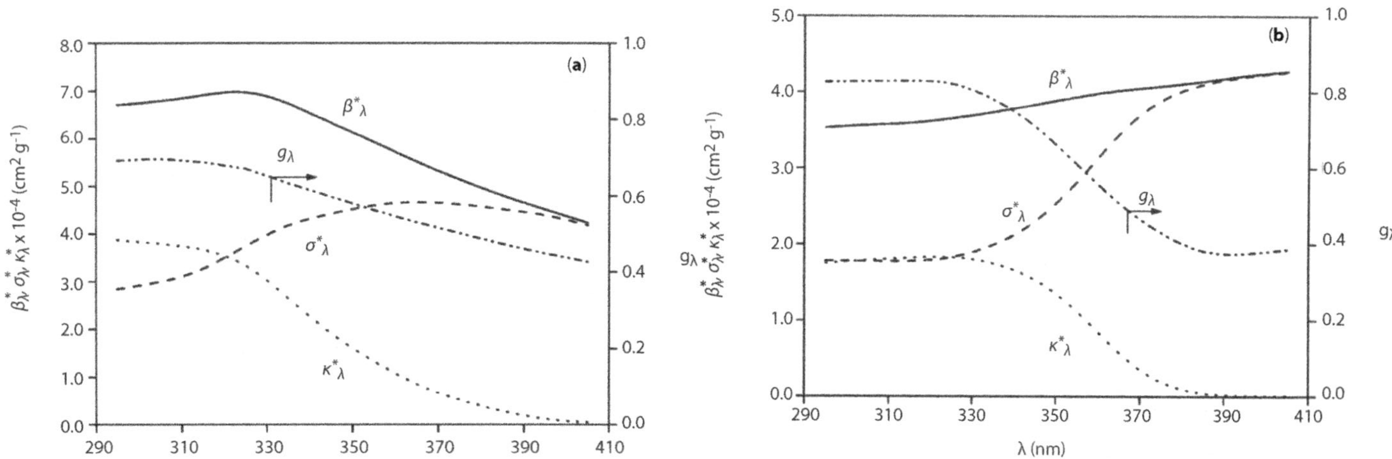

Figure 9.4 Absorption coefficient (κ_λ), scattering coefficient (σ_λ), extinction coefficients (β_λ), HG asymmetry factor (g_λ) of (a) Degussa P25 TiO_2 and (b) Aldrich TiO_2. Reprinted with permission from [34] Copyright 2005 American Chemical Society.

phase function in the RTE and computed the asymmetry factor that resulted in the least error between measured and simulated transmittance. To do this, they measured the diffuse reflectance in addition to the light transmittance through the catalyst slurry. The optical properties of two of the catalysts from this study are shown in Figure 9.4. A similar methodology was used to determine the absorption and scattering coefficients as well as the HG asymmetry factor of commercial carbon doped TiO_2 [40] and synthesized TiO_2-rGO [41]. The latter study employed the Monte Carlo method instead of the discrete ordinates method to solve the RTE.

Simpler methods have also been employed to determine absorption and scattering coefficients. Grčić and Puma *et al.* [42] measured the diffuse reflectance spectra (DRS) of the solid catalysts and the extinction coefficients of the slurry catalyst. From these two measurements, they obtained a series of simple equations from which they obtained the absorption and scattering coefficients of P25 TiO_2 and synthesized Ag@TiO_2 catalysts. Moreira *et al.* [43] obtained absorption and scattering coefficients of three commercial TiO_2 catalysts using an annular reactor. They used Monte Carlo simulation with isotropic scattering to determine the wavelength-averaged absorption and scattering coefficients that resulted in the least error between experimental and measured light intensity at the reactor wall. Turolla *et al.* [32] used an optical goniometer and CFD-based discrete ordinates method to estimate the absorption and scattering coefficients of Aeroxide P25 TiO_2 at a wavelength of 254 and 355 nm. A summary of the studies, which have reported absorption and scattering coefficients, is given in Table 9.3.

The three optical properties: absorption coefficient, scattering coefficient and scattering phase function are interrelated. Satuf *et al.* [34] showed that using different scattering phase functions resulted in remarkably different values of the absorption and scattering coefficients. Several authors have used previously reported optical properties to simulate the light distribution in their reactors. However, the catalyst agglomeration size in their reactor may not be similar to the catalyst size in the reactor that was used to generate the optical properties. This is especially true in cases where optical properties are determined under ultrasonication while the catalysts in the simulated reactor are only fluidized by wastewater recirculation. For this reason, several studies have chosen to evaluate the HG asymmetry factor specific to their reactor before simulating the light distribution [25, 27, 32].

Table 9.3 Experimental determination of absorption and scattering coefficients.

Catalyst	Experimental set up	Simulation model	Catalyst loading (g/L)	Wavelength (nm)	Reference
Aldrich/Merck/Fisher/Fluka/ Degussa P25/Hombikat TiO_2	Extinction Diffuse transmittance	DOM DR phase function	0.1–1	275–405	[33]
Aldrich/Degussa P25 TiO_2	Extinction Diffuse transmittance	DOM ISO phase function	0.025–5	315–435	[8, 37]
P25 Aeroxide/P90 Aeroxide/ P25/20 VP Aeroperl TiO_2	Extinction Diffuse transmittance	DOM ISO phase function	0.05–1.5	345–405	[38]
Commercial $NiFe_2O_4$ Synthesized $NiFe_2O_4$	Extinction Diffuse transmittance	Six-flux ISO phase function	0.1–0.4	300–850	[36]
$LiVMoO_6$	Extinction Diffuse transmittance Diffuse reflectance	Six-flux ISO phase function	0.25–0.75	280–500	[39]

(Continued)

Table 9.3 Experimental determination of absorption and scattering coefficients. (*Continued*)

Catalyst	Experimental set up	Simulation model	Catalyst loading (g/L)	Wavelength (nm)	Reference
Aldrich/Degussa P25/ Hombikat UV100 TiO_2	Extinction Diffuse transmittance Diffuse reflectance	DOM HG phase function	0.2–2	295–405	[34]
P25 Aeroxide/Kronos vlp 7000 TiO_2	Extinction Diffuse transmittance Diffuse reflectance	DOM HG phase function	0.1–0.5	P25: 400–550 vlp: 300–550	[40]
TiO_2-rGO	Extinction Diffuse transmittance Diffuse reflectance	Monte Carlo HG phase function	0.05–0.4	315–415	[41]
P25 Aeroxide TiO_2/Ag@TiO_2	DRS Extinction	-	0.1–2	300–800	[42]
Degussa P25/Aldrich/ Hombikat UV 100 TiO_2	Total transmittance Direct transmittance	Monte Carlo, ISO phase function	0–0.2	300–388 (wavelength-averaged)	[43]
P25 Aeroxide TiO_2	Angular irradiance	CFD-DOM HG phase function	0.04–0.4	254–365	[32]

9.2.4 Influence of Bubbles on Light Distribution

In air fluidized bed photoreactors, bubbles play an important role in fluidization and supply of oxygen. From the perspective of light distribution, they can be considered as inert media with a possible impact on light distribution and absorption by the catalyst. Several studies have analysed the light distribution in fluidized bed reactors in which bubbles interact with light. Using the discrete ordinates method, Trujillo *et al.* [44] modeled the light distribution on catalyst-coated flat plates immersed in a bubble column. They concluded that bubble scattering improved the light distribution on the flat plates. Motegh *et al.* [45] simulated the effect of bubbles on catalyst light absorption in a three phase reactor using a bi-directional scattering model. They found that, at normal air flow rates and typical sizes, bubbles do not significantly affect the light absorption by catalysts. Boyjoo *et al.* [9] employed the discrete ordinates method to simulate the light distribution in a multi-lamp slurry bubble column reactor. They found that the light scattered by bubbles was negligible as compared to that scattered by catalyst particles.

Akach *et al.* [29] used the Monte Carlo method to simulate photon-bubble interactions in addition to photon-catalyst interactions. Their simulations showed that air bubbles affected the transmission of direct light much more than diffuse light. Nevertheless, at the optimum catalyst loading, bubbles had a negligible effect on the light absorption by the catalyst. A similar result was obtained by simulating bubble-photon interactions in a UV illuminated photoreactor [28]. This showed that air was a good fluidization method as it achieved good catalyst mixing and provided oxygen electron acceptor without negatively impacting light absorption.

9.2.5 Validation of Light Distribution Models

In order to ensure that the light distribution model is accurate, validation is normally carried out by comparing the experimental and simulated light intensity at certain locations in the reactor. In the past, validation was carried out using actinometric techniques. In a typical set up, the reactor was surrounded by an annulus filled with potassium ferrioxalate actinometer, which was used to measure the radiation exiting the reactor wall [8]. Nowadays, light distribution models are mostly validated by measuring the forward transmitted radiation at the reactor wall using a radiometer (Figure 9.5a). This is common practice in internally illuminated annular reactors [7, 28, 30, 43]. This technique has also been employed in a solar

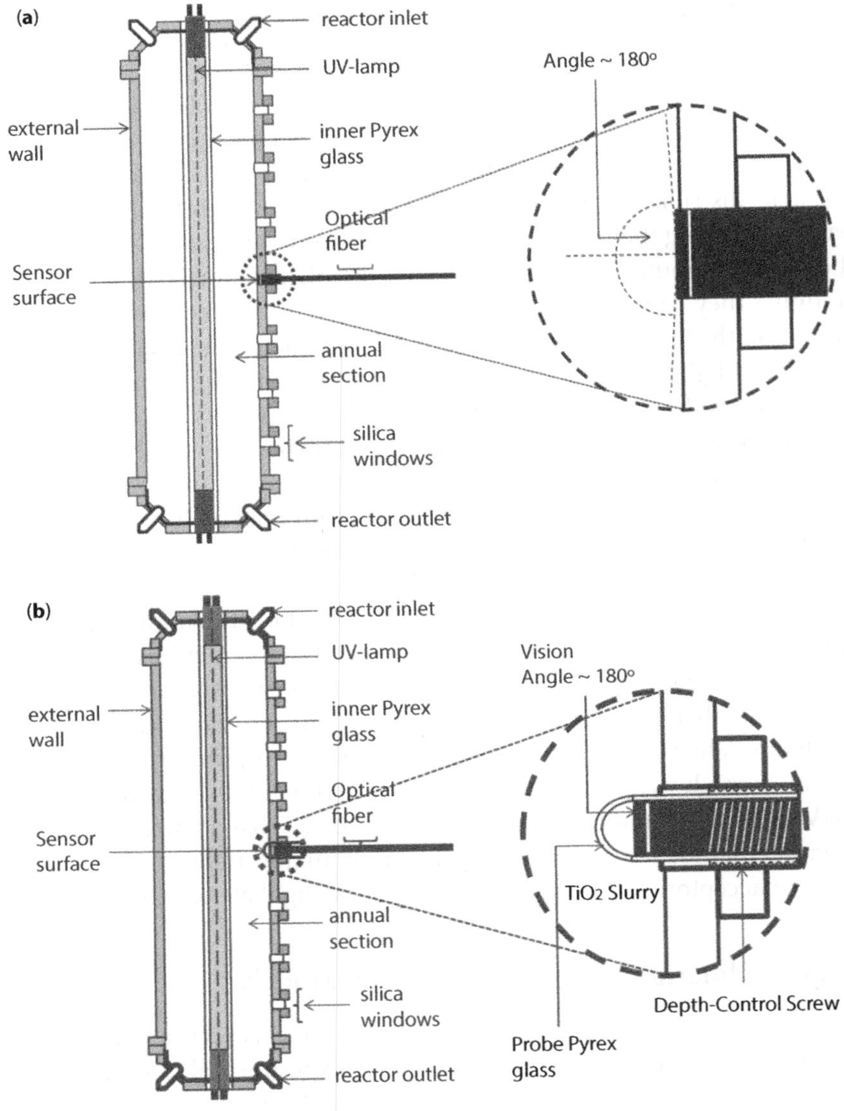

Figure 9.5 Validation methods. (a) Total transmittance. Reprinted from [7] with permission from Elsevier. (b) Radial intensity. Reprinted from [25] with permission from Elsevier. *(Continued)*

(c)

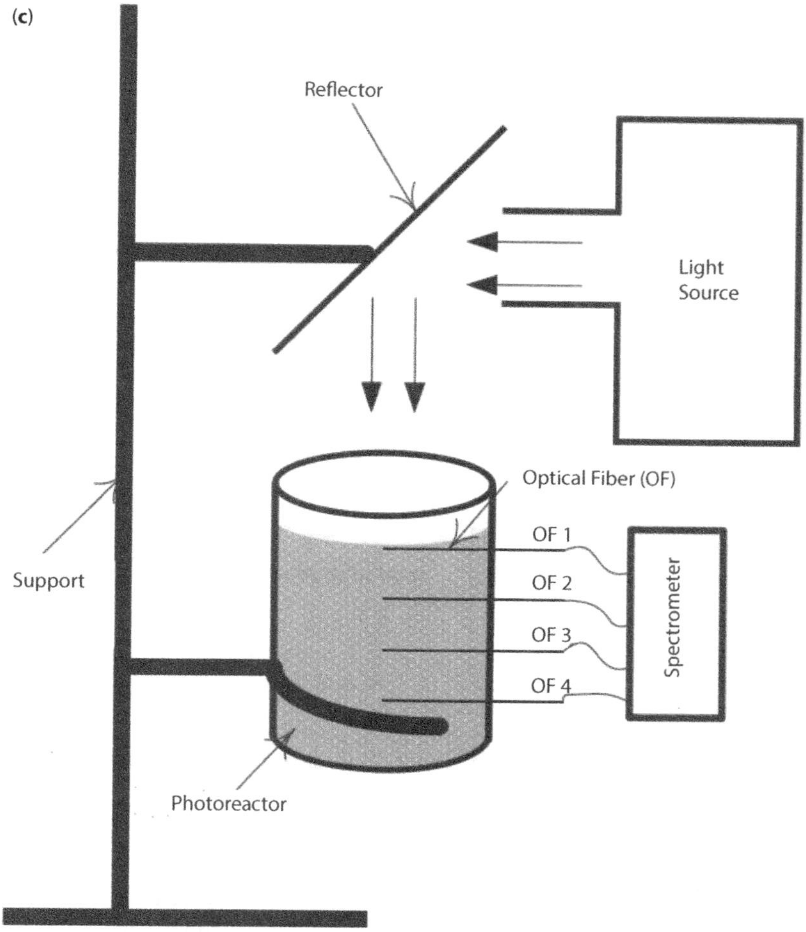

Figure 9.5 (Continued) Validation methods. (c) Optical fibres. Reprinted from [46] with permission from Elsevier. (*Continued*)

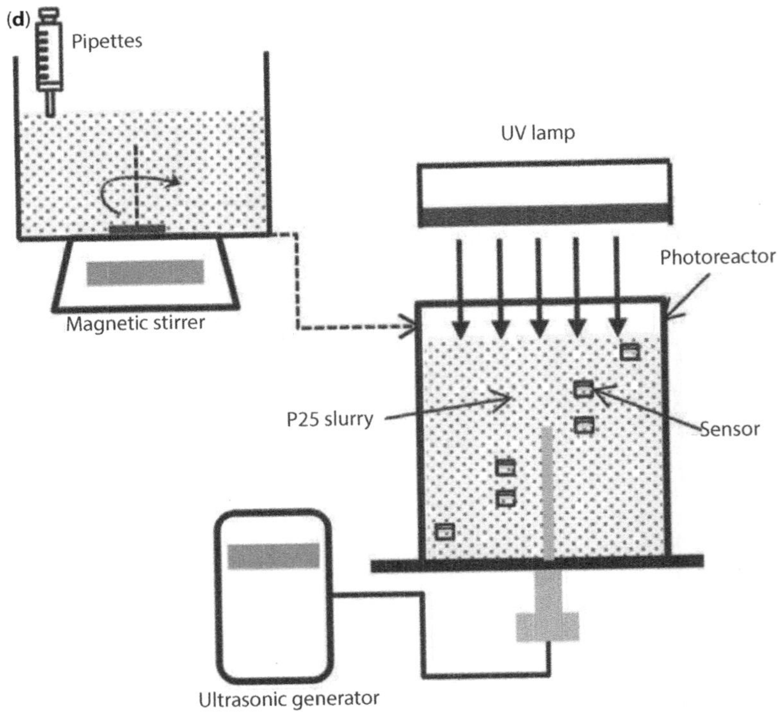

Figure 9.5 (Continued) Validation methods. (d) Underwater sensors. Reprinted from [27] with permission from Elsevier. *(Continued)*

illuminated reactor in which the forward transmitted reactor is refracted on the radiometer positioned at the opposite side of the reactor wall [29].

Validation has also been carried out by measuring the light intensity at different radial locations within the annulus using a spectroradiometric probe (Figure 9.5b). The measured radial intensity profiles provide better data for the validation of radiation gradients within the reactor as compared to measurements at the reactor wall [25]. A similar approach was employed to establish the local light intensity at different depths of a reactor illuminated from the top with a collimated light source [46] (Figure 9.5c) and UV lamp [27] (Figure 9.5d). One of the most versatile validation methods was developed by Valadés-Pelayo *et al.* [26] to measure the light intensity at various axial locations and azimuthal directions in an externally illuminated annular reactor (Figure 9.5e).

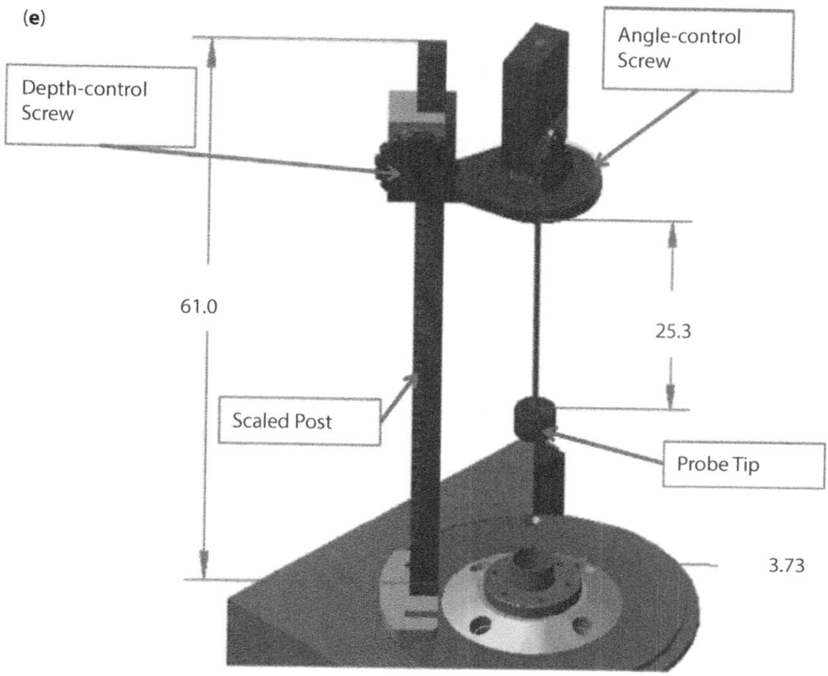

Figure 9.5 (Continued) Validation methods. (e) Probe depth/angle mechanism. Reprinted from [26] with permission from Elsevier.

9.3 Photocatalysis Kinetics

The rate of photocatalysis generally follows the Langmuir-Hinshelwood kinetics [47]:

$$r = -\frac{dC}{dt} = \frac{kKC}{1+KC} \tag{9.5}$$

where r is the rate of reaction, C is the substrate concentration at time t, k is the reaction rate constant and K is the Langmuir adsorption constant. Photocatalysis is normally carried out in wastewaters with a low substrate concentration of several mM. Therefore, $C \ll 1$ and eq. (9.5) simplifies to the pseudo first-order kinetics:

$$r = -\frac{dC}{dt} = kKC = k_{app}C \tag{9.6}$$

where k_{app} is the apparent first-order rate constant. Integrating eq. (9.6) yields:

$$\ln\frac{C_0}{C} = k_{app}t \tag{9.7}$$

In eq. (9.7), k_{app} can be obtained as the slope of a graph of ln C_0/C vs. t where C_0 is the initial substrate concentration.

The rate of photocatalysis depends on several factors such as the solution pH, temperature, dissolved oxygen concentration, catalyst type and loading, pollutant concentration, reactor diameter and light intensity [2]:

$$-r = f(pH,\ T,\ DO,\ cat,\ C,\ D,\ I) \tag{9.8}$$

In photochemical reactions, the catalyst has to absorb light in order to be activated. Consequently, the reaction rate depends on the absorbed light rather than the incident light intensity [25]. Therefore, the amount of light absorbed by the catalysts in the reactor, known as the volumetric rate of energy absorption (VREA), is a more meaningful parameter than the intensity of the light source [2]. This is due to the fact that the VREA encapsulates aspects of the light source and reactor geometry [11]. The rate of the reaction can then be expressed as a function of VREA and other terms as:

$$-r = f(pH,\ T,\ DO,\ cat,\ C,\ VREA) \tag{9.9}$$

The pseudo first-order kinetics (eq. 9.6) can be expressed using a power law dependence on the VREA [11, 15] as:

$$-r = k_{int}VREA^{\alpha}C \tag{9.10}$$

where k_{int} is the intrinsic rate constant, which depends on the dissolved oxygen, temperature, solution pH, catalyst type and loading but is independent of the reactor geometry.

The exponent α defines the extent to which absorbed light results in the successful generation and separation of electron-hole pairs. At low values of VREA, very little electron-hole recombination occurs resulting in an α of 1. As VREA increases, the rate of generation of electrons and holes outstrips the rate of photocatalysis. This increases electron-hole recombination resulting in an α value of 0.5 [15]. In a typical, optically thick photoreactor, several reaction regimes exist in the same reactor. At regions near the light source, half-order reactions (α = 0.5) are predominant due to a high value of light absorption. Deeper in the photoreactor, as light absorption reduces, full-order reactions (α = 1) dominate [2, 21]. Therefore, most photoreactors have an α value of between 0.5 and 1. For example, Li Puma *et al.* [15] reported a value of 0.82 for a 30 mm diameter reactor illuminated by a UVA lamp.

9.4 Conclusion

In this chapter, the modeling and the simulation of photocatalytic reactors have been discussed with a focus on the light distribution and reaction kinetics. The various analytical and numerical methods of solving the radiation transport equation have been discussed. These include the two-flux, P1, six-flux, discrete ordinates, and Monte Carlo methods. Several aspects of light distribution modeling, such as boundary conditions, phase function, catalyst optical properties, and validation techniques, have also been analysed. Finally, a discussion of the relationship between light absorption and photocatalytic kinetics has been carried out. Photoreactor modeling is an emerging area of research with a gradual increase in publications. Most studies so far have employed UV lamps and nanophase catalysts, particularly Aeroxide P25 TiO_2. Very little modeling has been carried out using emerging light sources, such as light emitting diodes (LEDs) and sunlight. The use of novel catalysts, such as supported catalysts, composites, and visible light active catalysts, has been few. More simulation work needs to be done in this area since the next frontier of photocatalysis research is trending toward lower cost light sources and novel catalysts.

References

1. Braham, R.J. and Harris, A.T., Review of major design and scale-up considerations for solar photocatalytic reactors. *Ind. Eng. Chem. Res.*, 48, 8890–8905, 2009.
2. Boyjoo, Y., Ang, M., Pareek, V., Some aspects of photocatalytic reactor modeling using computational fluid dynamics. *Chem. Eng. Sci.*, 101, 764–784, 2013. https://doi.org/10.1016/j.ces.2013.06.035.

3. Pareek, V., Chong, S., Tadé, M., Adesina, A.A., Light intensity distribution in heterogenous photocatalytic reactors. *Asia-Pac. J. Chem. Eng.*, 3, 171–201, 2008.
4. Moreira, J., Serrano, B., Ortiz, A., de Lasa, H., Evaluation of photon absorption in an aqueous TiO2 slurry reactor using Monte Carlo simulations and macroscopic balance. *Ind. Eng. Chem. Res.*, 49, 10524–10534, 2010.
5. Camera-Roda, G., Augugliaro, V., Cardillo, A.G., Loddo, V., Palmisano, L., Parrino, F., Santarelli, F., A reaction engineering approach to kinetic analysis of photocatalytic reactions in slurry systems. *Catal. Today*, 259, 87–96, 2016.
6. Acosta-Herazo, R., Monterroza-Romero, J., Mueses, M.Á., Machuca-Martínez, F., Li Puma, G., Coupling the six flux absorption–scattering model to the henyey–greenstein scattering phase function: Evaluation and optimization of radiation absorption in solar heterogeneous photoreactors. *Chem. Eng. J.*, 302, 86–96, 2016.
7. Valadés-Pelayo, P.J., Moreira, J., Serrano, B., de Lasa, H., Boundary conditions and phase functions in a photo-crec water-II reactor radiation field. *Chem. Eng. Sci.*, 107, 123–136, 2014.
8. Romero, R.L., Alfano, O.M., Cassano, A.E., Radiation field in an annular, slurry photocatalytic reactor. 2. Model and experiments. *Ind. Eng. Chem. Res.*, 42, 2479–2488, 2003.
9. Boyjoo, Y., Ang, M., Pareek, V., Light intensity distribution in multi-lamp photocatalytic reactors. *Chem. Eng. Sci.*, 93, 11–21, 2013.
10. Orozco, S.L., Arancibia-Bulnes, C.A., Suárez-Parra, R., Radiation absorption and degradation of an azo dye in a hybrid photocatalytic reactor. *Chem. Eng. Sci.*, 64, 2173–2185, 2009.
11. Li Puma, G., Puddu, V., Tsang, H.K., Gora, A., Toepfer, B., Photocatalytic oxidation of multicomponent mixtures of estrogens (estrone (E1), 17β-estradiol (E2), 17α-ethynylestradiol (EE2) and estriol (E3)) under UVA and UVC radiation: Photon absorption, quantum yields and rate constants independent of photon absorp. *Appl. Catal. B Environ.*, 99, 388–397, 2010.
12. Li Puma, G. and Brucato, A., Dimensionless analysis of slurry photocatalytic reactors using two-flux and six-flux radiation absorption–scattering models. *Catal. Today*, 122, 78–90, 2007. https://doi.org/10.1016/j.cattod.2007.01.027.
13. Orozco, S.L., Villafán-Vidales, H.I., Arancibia-Bulnes, C.A., Photon absorption in a hybrid slurry photocatalytic reactor: Assessment of differential approximations. *AICHE J.*, 58, 3256–3265, 2012.
14. Cuevas, S.A., Arancibia-Bulnes, C.A., Serrano, B., Radiation field in an annular photocatalytic reactor by the P1 approximation. *Int. J. Chem. React. Eng.*, 5, 1–14, 2007.
15. Li Puma, G., Khor, J.N., Brucato, A., Modeling of an annular photocatalytic reactor for water purification: Oxidation of pesticides. *Environ. Sci. Technol.*, 38, 3737–3745, 2004.

16. Ramírez-Cabrera, M.A., Valadés-Pelayo, P.J., Arancibia-Bulnes, C.A., and Ramos, E., Validity of the Six-Flux model for photoreactors. *Chem. Eng. J.*, 330, 272–280, 2017. https://doi.org/10.1016/j.cej.2017.07.120.
17. Colina-Marquez, J., Machuca-Martínez, F., Li Puma, G., Photocatalytic mineralization of commercial herbicides in a pilot-scale solar CPC reactor: Photoreactor modeling and reaction kinetics constants independent of radiation field. *Environ. Sci. Technol.*, 43, 8953–8960, 2009.
18. Colina-Márquez, J., Machuca-Martínez, F., Puma, G.L., Radiation absorption and optimization of solar photocatalytic reactors for environmental applications. *Environ. Sci. Technol.*, 44, 5112–5120, 2010.
19. Mueses, M.A., Machuca-Martinez, F., Li Puma, G., Effective quantum yield and reaction rate model for evaluation of photocatalytic degradation of water contaminants in heterogeneous pilot-scale solar photoreactors. *Chem. Eng. J.*, 215–216, 937–947, 2013.
20. Ochoa-Gutiérrez, K.S., Tabares-Aguilar, E., Mueses, M.Á., Machuca-Martínez, F., Li Puma, G., A novel prototype offset multi tubular photoreactor (OMTP) for solar photocatalytic degradation of water contaminants. *Chem. Eng. J.*, 341, 628–638, 2018.
21. Boyjoo, Y., Ang, M., Pareek, V., CFD simulation of a pilot scale slurry photocatalytic reactor and design of multiple-lamp reactors. *Chem. Eng. Sci.*, 111, 266–277, 2014. https://doi.org/10.1016/j.ces.2014.02.022.
22. Casado, C., García-Gil, Á., van Grieken, R., Marugán, J., Critical role of the light spectrum on the simulation of solar photocatalytic reactors. *Appl. Catal. B Environ.*, 252, 1–9, 2019.
23. Moreno-SanSegundo, J., Casado, C., Marugán, J., Enhanced numerical simulation of photocatalytic reactors with an improved solver for the radiative transfer equation. *Chem. Eng. J.*, 388, 124183, 2020. https://doi.org/10.1016/j.cej.2020.124183.
24. Boyjoo, Y., Ang, M., Pareek, V., Lamp emission and quartz sleeve modelling in slurry photocatalytic reactors. *Chem. Eng. Sci.*, 111, 34–40, 2014. https://doi.org/10.1016/j.ces.2014.02.023.
25. Valadés-Pelayo, P.J., Moreira del Rio, J., Solano-Flores, P., Serrano, B., de Lasa, H., Establishing photon absorption fields in a photo-CREC water II reactor using a CREC-spectroradiometric probe. *Chem. Eng. Sci.*, 116, 406–417, 2014.
26. Valadés-Pelayo, P.J., Guayaquil Sosa, F., Serrano, B., de Lasa, H., Photocatalytic reactor under different external irradiance conditions: Validation of a fully predictive radiation absorption model. *Chem. Eng. Sci.*, 126, 42–54, 2015. https://doi.org/10.1016/j.ces.2014.12.003.
27. Hou, J., Wei, Q., Yang, Y., Zhao, L., Experimental evaluation of scattering phase function and optimization of radiation absorption in solar photocatalytic reactors. *Appl. Therm. Eng.*, 127, 302–311, 2017.

28. Akach, J. and Ochieng, A., Monte Carlo simulation of the light distribution in an annular slurry bubble column photocatalytic reactor. *Chem. Eng. Res. Des.*, 129, 248–258, 2018.
29. Akach, J., Kabuba, J., Ochieng, A., Simulation of the light distribution in a solar photocatalytic bubble column reactor using the Monte Carlo method. *Ind. Eng. Chem. Res.*, 59, 17708–17719, 2020.
30. Imoberdorf, G.E., Taghipour, F., Keshmiri, M., Mohseni, M., Predictive radiation field modeling for fluidized bed photocatalytic reactors. *Chem. Eng. Sci.*, 63, 4228–4238, 2008.
31. Acosta-Herazo, R., Valadés-Pelayo, P.J., Mueses, M.A., Pinzón-Cárdenas, M.H., Arancibia-Bulnes, C., Machuca-Martínez, F., An optical and energy absorption analysis of the solar compound parabolic collector photoreactor (CPCP): The impact of the radiation distribution on its optimization. *Chem. Eng. J.*, 395, 125065, 2020. https://doi.org/10.1016/j.cej.2020.125065.
32. Turolla, A., Santoro, D., de Bruyn, J.R., Crapulli, F., Antonelli, M., Nanoparticle scattering characterization and mechanistic modelling of UV–TiO2 photocatalytic reactors using computational fluid dynamics. *Water Res.*, 88, 117–126, 2016. https://doi.org/10.1016/j.watres.2015.09.039.
33. Cabrera, M.I., Alfano, O.M., Cassano, A.E., Absorption and scattering coefficients of titanium dioxide particulate suspensions in water. *J. Phys. Chem.*, 100, 20043–20050, 1996.
34. Satuf, M.L., Brandi, R.J., Cassano, A.E., Alfano, O.M., Experimental method to evaluate the optical properties of aqueous titanium dioxide suspensions. *Ind. Eng. Chem. Res.*, 44, 6643–6649, 2005.
35. Valadés-Pelayo, P.J., Guayaquil Sosa, F., Serrano, B., de Lasa, H., Eight-lamp externally irradiated bench-scale photocatalytic reactor: Scale-up and performance prediction. *Chem. Eng. J.*, 282, 142–151, 2015. https://doi.org/10.1016/j.cej.2015.03.039.
36. Domínguez-Arvizu, J.L., Jiménez-Miramontes, J.A., Salinas-Gutiérrez, J.M., Meléndez-Zaragoza, M.J., López-Ortiz, A., Collins-Martínez, V., Study of NiFe2O4 nanoparticles optical properties by a six-flux radiation model towards the photocatalytic hydrogen production. *Int. J. Hydrogen Energy*, 44, 12455–12462, 2019. https://doi.org/10.1016/j.ijhydene.2018.08.148.
37. Brandi, R.J., Alfano, O.M., Cassano, A.E., Rigorous model and experimental verification of the radiation field in a flat-plate solar collector simulator employed for photocatalytic reactions. *Chem. Eng. Sci.*, 54, 2817–2827, 1999. https://doi.org/10.1016/S0009-2509(98)00355-8.
38. Tolosana-Moranchel, A., Casas, J.A., Carbajo, J., Faraldos, M., Bahamonde, A., Influence of TiO2 optical parameters in a slurry photocatalytic reactor: Kinetic modelling. *Appl. Catal. B Environ.*, 200, 164–173, 2017. https://doi.org/10.1016/j.apcatb.2016.06.063.
39. Hurtado, L., Natividad, R., Torres-García, E., Farias, J., Li Puma, G., Correlating the photocatalytic activity and the optical properties of LiVMoO6 photocatalyst under the UV and the visible region of the solar radiation

spectrum. *Chem. Eng. J.*, 262, 1284–1291, 2015. https://doi.org/10.1016/j.cej.2014.10.052.

40. Manassero, A., Satuf, M.L., Alfano, O.M., Evaluation of UV and visible light activity of TiO2 catalysts for water remediation. *Chem. Eng. J.*, 225, 378–386, 2013. https://doi.org/10.1016/j.cej.2013.03.097.
41. Tolosana-Moranchel, Á., Manassero, A., Satuf, M.L., Alfano, O.M., Casas, J.A., Bahamonde, A., Influence of TIO2-rGO optical properties on the photocatalytic activity and efficiency to photodegrade an emerging pollutant. *Appl. Catal. B Environ.*, 246, 1–11, 2019. https://doi.org/10.1016/j.apcatb.2019.01.054.
42. Grčić, I. and Li Puma, G., Six-flux absorption-scattering models for photocatalysis under wide-spectrum irradiation sources in annular and flat reactors using catalysts with different optical properties. *Appl. Catal. B Environ.*, 211, 222–234, 2017. https://doi.org/10.1016/j.apcatb.2017.04.014.
43. Moreira, J., Serrano, B., Ortiz, A., de Lasa, H., TiO2 absorption and scattering coefficients using Monte Carlo method and macroscopic balances in a photo-CREC unit. *Chem. Eng. Sci.*, 66, 5813–5821, 2011. https://doi.org/10.1016/j.ces.2011.07.040.
44. Trujillo, F.J., Safinski, T., Adesina, A.A., CFD analysis of the radiation distribution in a new immobilized catalyst bubble column externally illuminated photoreactor. *J. Sol. Energy Eng.*, 129, 27–36, 2006.
45. Motegh, M., Ruud van Ommen, J., Appel, P.W., Mudde, R.F., Kreutzer, M.T., Bubbles scatter light, yet that does not hurt the performance of bubbly slurry photocatalytic reactors. *Chem. Eng. Sci.*, 100, 506–514, 2013. https://doi.org/10.1016/j.ces.2013.02.022.
46. Cao, F., Li, H., Chao, H., Zhao, L., Guo, L., Optimization of the concentration field in a suspended photocatalytic reactor. *Energy*, 74, 140–146, 2014.
47. Gaya, U.I. and Abdullah, A.H., Heterogeneous photocatalytic degradation of organic contaminants over titanium dioxide: A review of fundamentals, progress and problems. *J. Photochem. Photobiol. C Photochem. Rev.*, 9, 1–12, 2008.

10

The Development of Self-Powered Nanoelectrocatalytic Reactor for Simultaneous Piezo-Catalytic Degradation of Bacteria and Organic Dyes in Wastewater

Daniel Masekela[1], Nomso C. Hintsho-Mbita[2] and Nonhlangabezo Mabuba[1*]

[1]Department of Chemical Sciences, University of Johannesburg, Doornfontein Campus, Johannesburg, Republic of South Africa
[2]Department of Chemistry, University of Limpopo, Private Bag X, Sovenga, Polokwane, Republic of South Africa

Abstract

Electrochemical advanced oxidation processes (EAOPs) have been extensively used for wastewater treatment. These processes use reactive oxygen species (ROS), such as hydroxyl, superoxide and hydrogen peroxide for the degradation of pollutants (organic, inorganic and pathogenic bacteria). However, in most cases they use external electric power (high energy consumption) derived from the burning of fossil fuels. The electric power derived from fossil fuels has detrimental effects such as emission of greenhouse gases to the atmosphere. Therefore, in order to self-sustain these EAOPs, it is important to power these using renewable sources. In this chapter, we highlight the use of piezoelectric materials to power EAOPs especially for bacterial and organic dye degradation. These piezoelectric materials tend to produce electric energy under the influence of applied pressure; however, they are known to be poor electrocatalysts. Their limitations and possible future perspectives will also be discussed.

***Keywords*:** Advanced oxidation processes, piezoelectric materials, electrocatalyst, bacteria, organic dye

**Corresponding author*: nmabuba@uj.ac.za

Elvis Fosso-Kankeu, Sadanand Pandey, and Suprakas Sinha Ray (eds.) Photoreactors in Advanced Oxidation Processes: The Future of Wastewater Treatment, (305–338) © 2023 Scrivener Publishing LLC

Abbreviations

AOPs	Advanced oxidation processes
BBV	Basic blue violet
BOD	Biological oxygen demand
CFU	Colony forming unit
CVD	Chemical vapor deposition
DBPs	Disinfection by-products
EAOPs	Electrochemical advanced oxidation processes
EO	Electrochemical oxidation
GHGs	Greenhouse gases
Haa	Haloacetic acid
MB	Methylene blue
MDA	Malondialdehyde
MEMs	Microelectromechanical system
MO	Methyl orange
PEHs	Piezoelectric energy harvesters
PENGs	Piezoelectric nanogenerators
PVDF	Polyvinylidene difluoride
PZM	Piezoelectric material
PZT	Lead zirconate
RhB	Rhodamine B
ROS	Reactive oxygen species
SEM	Scanning electron microscope
THMs	Trihalomethanes
TMDs	Transition metal dichalcogenides (TMDs)
WHO	World Health Organization

10.1 Introduction

The consumption of water contaminated with harmful pollutants, such as organic, inorganic, and pathogens, has a negative impact on aquatic living organisms and human health [1]. Some of these organic pollutants include synthetic dyes, such as methylene blue (MB), methyl orange (MO), and basic blue violet (BBV). These dyes are released into rivers and streams from various industries such as textile, chemical, and paper and pulp when used as coloring agents [2]. Furthermore, they are extremely toxic, non-biodegradable and thus they pose a serious threat to the ecosystem [3]. Drinking water containing low levels (<1 ppm) of these toxic dyes can cause serious damage to multiple human organs, such as lungs, liver, and kidney [4]. Other pollutants that have been of great concern in wastewater are bacteria, such as *Escherichia coli* (*E. coli*), *Pseudomonas aeruginosa* (*P. aeruginosa*), *Staphylococcus aureus* (*S. aureus*) *Enterococcus faecalis* (*E. faecalis*), etc. These strains are lethal and can cause waterborne diseases such as diarrhoea, typhoid fever, etc. [5]. It has been reported that approximately more than 3 million people worldwide have died as a result of waterborne diseases [6]. The World Health Organization (WHO) statistics stated that more than 1.1 billion of the world population still do not have access to clean water [7]. Therefore, due to the global over population and an increase in industries, there is a need to develop wastewater treatment technologies that are safe and efficient.

There are several existing physical and chemical wastewater treatment technologies, such as chlorination, ozonation, membrane filtration, reverse osmosis, adsorption, and advanced oxidation processes (AOPs), that have been vastly used [8]. Unfortunately, many of these methods have several setbacks, such as high operation costs, generation of secondary toxic by-products, high energy consumption, and incomplete removal. For instance, chlorination is one of the most commonly applied disinfection processes for bacterial removal; however, it generates toxic disinfection by products (DBPs), such as trihalomethanes (THMs) and haloacetic acid (Haa) [9]. Chemical precipitation requires a large quantity of expensive reagents to treat a small amount of organic and inorganic pollutants. On the other hand, membrane filtration and adsorption suffer from the generation of toxic sludge and regeneration, respectively. Among the wastewater treatment strategies, advanced oxidation processes (AOPs) have been regarded as one of the effective techniques for the degradation of toxic organic dyes and microorganisms present in wastewater.

Over the past years, AOPs, such as Electrochemical Advanced Oxidation processes (EAOPs), have been extensively applied for wastewater treatment. The benefits of using electrochemical EAOPs include the use of a clean reagent (electron), they are easy to operate and environmentally friendly [10]. This process (EAOPs) uses electrogenerated reactive oxygen species (ROS), such as hydroxyl, superoxide, hydrogen peroxide, as strong oxidants for the chemical degradation of waste pollutants to less harmful by-products, such as CO_2 and H_2O. However, there are drawbacks associated with the use of EAOPs for bacterial decomposition, electrocatalytic degradation, and wastewater treatment. One of the main disadvantages is the high energy that is consumed using an external electrical power [11]. In most cases, their electric power is derived from the use of burning of fossils fuels. The electric energy derived from burning of fossil fuels has its own disadvantages, such as emission of greenhouse gases (GHGs) to the atmosphere. These GHGs, such as CO_2, CH_4, and N_2O, have a negative impact on the environment [12]. Therefore, in order to self-sustain any electrochemical technique for various applications, especially for wastewater purification, it is important to power these using renewable energy. The advantage of using renewable energy resources includes production of clean energy, which does not have any negative impact to the atmosphere.

Piezoelectric materials (PZM) have gained momentum as energy-harvesting materials for electrical energy production. These materials generate electric charges under the influence of mechanical vibrations [13]. The electricity produced by these materials is called piezoelectricity. Recently, these materials have been widely used to power electrocatalytic reactors. The most extensively used PZMs are lead based piezoelectric materials, such as lead zirconate (PZT). However, PZT contain about 80% lead (Pb) content, thus limiting their use in various applications. The disposal of toxic lead (Pb) can have a negative impact on our surroundings [14]. Thus, there has been a development of various lead free piezoelectric materials, such as $BaTiO_3$ (BT), $BiNaTiO_3$ (BNT), $KNaNbO_3$ (KNN), etc. [15].

These materials are nontoxic and environmentally friendly compared to lead-based materials. However, many of these lead free piezoelectric materials are considered as poor electrocatalysts, thus have poor electrochemical performance [15]. Several studies have shown that their electrochemical performance can be enhanced by coupling them with excellent electrocatalysts [16]. Thus, in this chapter, we highlight the different types of piezoelectric materials, the synthesis of various PZM materials, their challenges and modifications. Lastly, their use for controlling EAOPs for the degradation of organic dyes and bacteria is discussed.

10.2 Degradation Techniques

Several degradation methods, including physical, chemical, biological, and radiation, have been employed for wastewater treatment. However, a majority of these techniques have several setbacks. For instance, physical strategies suffer from incomplete removal of pollutants and cause secondary pollution [17]. Biological techniques are also not effective in the removal of organic pollutants from wastewater [17]. On the other hand, advanced oxidation process techniques (AOPs), such as electrochemical advanced oxidation process and piezocatalysis, have been reported to effectively decompose organic pollutants into less harmful products (CO_2 and H_2O).

10.2.1 Electrochemical Advanced Oxidation Processes (EAOPs)

Electrochemical advanced oxidation processes, such as electrochemical oxidation (EO), have emerged as promising techniques for the removal of organic pollutants from aqueous solution. The EO technique uses electrogenerated reactive oxygen species (ROS), such as hydroxyl radicals and superoxide radicals at the surface of the anode electrode/material to degrade wastewater pollutants (Figure 10.1) [18]. The electrogenerated ROS are either physically or chemically adsorbed on the surface of the anode. The adsorption of ROS depends on the characteristics of the anode materials.

For wastewater treatment, high stable anodes, which exhibit high anodic potential, are required for water oxidation to produce bulk hydroxyl radicals (•OH) to oxidize organic waste pollutants. Several studies have shown that the nature of the anode tends to have a strong influence on the electrocatalytic oxidation performance [20]. Table 10.1 shows some of the anode materials used for electrochemical processes.

Electrochemical oxidation (EO) as one of the AOPs requires an external electric power source to drive the oxidation-reduction reaction. Approximately 60% of electricity used to power electrocatalytic reactors is produced from burning of fossil fuels, thus resulting in the release of toxic

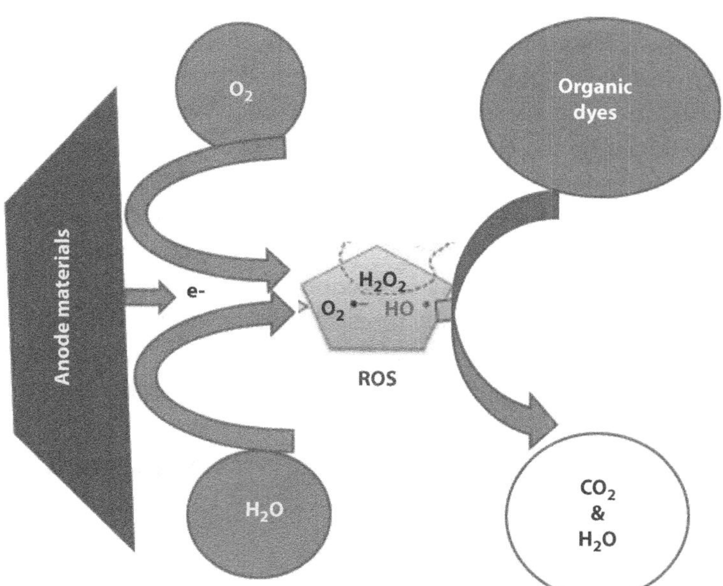

Figure 10.1 EAOPs degradation mechanism for organic dyes [19].

Table 10.1 Types of anode materials with their respective oxidation potential and oxidation power [21].

Anode materials	Oxidation potential	Adsorption of •OH on the surface of electrode	Anodic oxidation power
RuO_2-TiO_2 (DSA-Cl_2)	1.4-1.7	Chemisorption	Low
IrO_2-Ta_2O_5 (DSA-O_2)	1.5-1.8		
Ti/Pt	1.7-1.9		
Ti/PbO_2	1.8-2.0		
Ti/SnO_2-Sb_2O_5	1.9-2.2		
p-Si/BDD	2.2-2.6	Physisorption	High

gases, such as carbon dioxide (CO_2), methane (CH_4), and nitrous oxide (N_2O), in the environment. Due to fossil fuel depletion, there is a need to develop new methods, which are sustainable, produce clean energy, and do not have negative impact on the environment [12]. Alternative ways include the use of renewable resources to generate electricity. One of the major advantages of utilizing renewable resources includes production of clean energy [22]. In this review, we focus on the exploitation of piezoelectric energy harvesting materials as an alternative way to produce electrical energy to drive electrochemical processes, especially for electrocatalytic degradation of organic dyes and bacteria. The overview of the piezoelectric materials for catalytic degradation of wastewater pollutants is discussed thoroughly in section 10.6.

10.3 Characteristics and Properties of Piezoelectric Materials

Piezoelectric materials are known as energy harvesting materials which produce electricity when pressure is applied on them [13]. There are more than 200 piezoelectric materials available for energy harvesting. These piezoelectric materials can be classified into four types as (1) single crystal materials, (2) lead based piezoelectric ceramics, (3) lead free piezoelectric ceramics and (4) piezopolymeric materials [23]. The single crystal piezoelectrical materials are also known as natural existing materials, whereas lead based piezoceramics, lead free piezoceramics and piezopolymeric materials are manmade (synthetic) materials. The most widely used are piezoelectric ceramics and piezopolymer materials. Some of these piezoelectric materials are summarized in Table 10.2.

There are several parameters, which have a great influence on piezoelectric materials. These include piezoelectric deformation constant (d), piezoelectric potential constant (g), acoustic impedance (Z), mechanical quality factor (Q_M), and electrochemical coupling factor (k). The selection choice of piezoelectric materials depends mostly on some of these parameters. Table 10.3 highlights some of these important piezoelectric parameters on selected materials.

Table 10.2 Piezoelectric material properties and characteristics [24].

Piezoelectric materials	Characteristics/properties	Examples
Single crystals	• Mono crystals with weak piezoelectric effect. • Depending on their synthetic routes, nanostructures.	• ZnO, SiO_2, $LiNbO_3$
Lead-based piezoceramics	• Polycrystals with high piezoelectric effect. • Consist of chemical formula ABO3 structure. • Consist of toxic lead (Pb), thus harmful.	• $PbTiO_3$, $PbZrO_3$
Lead free piezoceramics	• Do not contain lead (Pb), thus not harmful. • The most widely used nontoxic-perovskites structure.	• $BaTiO_3$, $BiFeO_3$, $KNbO_3$
Piezopolymeric materials	• Highly flexible electroactive materials. • Nontoxic and biocompatible. • Less expensive production process.	• PVDF, PVDF-TrFE

Table 10.3 Properties of piezopolymeric and ceramic materials [23].

Properties	**Piezoceramic (PZT)**	**Piezopolymeric (PVDF)**
Piezoelectricity	High	Low
Electrochemical factor (k)	$(k_{31})^2 = 30$	$(k_{31})^2 = 12$
Dielectric constant	1180	10-15
Flexibility	Poor	Highly flexible
Acoustic impedance (Z)	Low	High
Curie temperature (°C)	High temp (>300 °C)	Low temp (<100 °C)
Relative permittivity	1200	12

- Piezoelectric deformation constant (d), is mechanical strain (x) generated when electrical field (E) is applied on piezoelectric material. It is defined by equation 10.1.

$$X = dE \tag{10.1}$$

- Piezoelectric potential constant (g), is the electric field (E) produced by piezoelectric material under the influence of applied external stress (X), as indicated in equation 10.2.

$$\mathrm{E} = \mathrm{gX} \tag{10.2}$$

- Electrochemical coupling factor (k), is the ratio of stored mechanical energy to input electrical energy, it is an important parameter, which measures the effectiveness of piezoelectric material to convert mechanical energy into electrical energy or vice versa (equation 10.3).

$$k^2 = \text{Stored mechanical energy/input electrical energy} \tag{10.3}$$

- Mechanical quality factor (Q_M), is one important parameter, which measures mechanical losses of piezoelectric materials. The mechanical losses of piezoelectric materials are a result of internal friction mechanism or cracking of the geometric structure.

- Acoustic impedance (Z) is another important parameter used for assessing the acoustic energy exchange between two materials. It is defined using equation 10.4:

$$Z = \sqrt{\rho c} \tag{10.4}$$

where ρ and c represent the density and elastic stiffness of the material, respectively.

10.3.1 Natural Piezoelectric Materials

In the past few decades, several existing materials have been discovered by Curie brothers, which tend to exhibit piezoelectric effect including Rochelle salt (potassium sodium tartrate), the mineral topaz, quartz, sugar cane, bones, and many more (Figure 10.2) [25].

A majority of these natural piezoelectric materials have several setbacks, which limit their industrial applications. For instance, Rochelle salt, which was the first discovered natural ferroelectric material, tends to lose its ferroelectric properties under high-pressure stress. Ferroelectricity is regarded as one of the significant properties for piezoelectric effect. Therefore, this

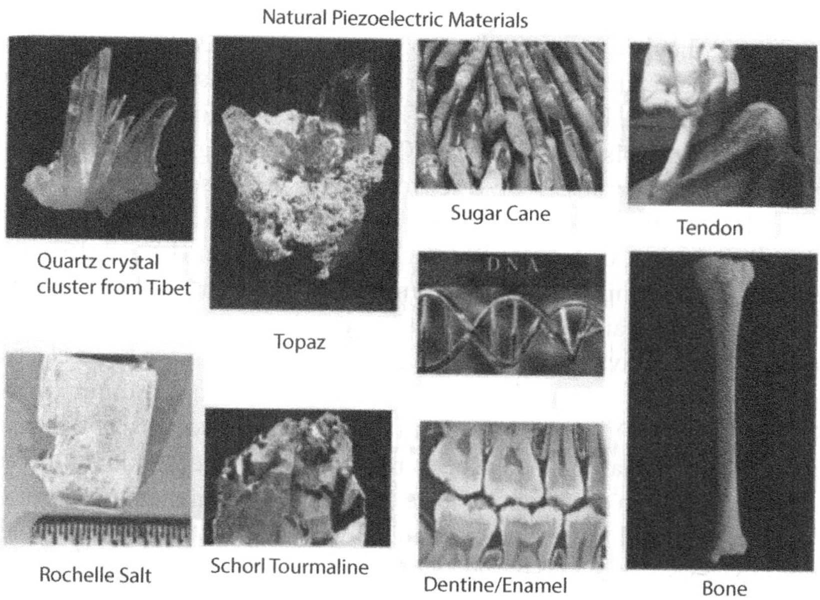

Figure 10.2 Natural existing piezoelectric materials [25].

has led to the development of synthetic piezoelectric materials, which are more effective as compared with natural existing piezoelectric materials.

10.3.2 Synthetic Piezoelectric Materials

Synthetic piezoelectric materials are fabricated in laboratories using several synthetic methods, such as hydrothermal (e.g., barium titanate), sol gel (e.g., lead zirconate titanate), microemulsion (e.g., Bismuth ferrite), and solid state reaction (e.g. lead zirconate titanate, bismuth ferrite) [26]. Lead-based piezoelectric materials, such as lead zirconate titanate ($PbZrO_3$), lead titanate ($PbTiO_3$), lead metaniobate ($Pb_2Nb_2O_6$), are the mostly widely used since they exhibit exceptional ferroelectric, optical and piezoelectric properties [27]. However, these materials consist of a large content of the toxic element lead (Pb). About 80% of lead content is used when manufacturing these synthetic materials. The disposal of these materials after their application can have a negative impact on the environment. Long term exposure to lead via inhalation or ingestion can cause serious multiple organ damage, including kidney, liver and lung damage [28]. Despite their toxicity, the lead containing piezoelectric materials are more effective compared to others. Recently, several researchers have deviated from the use of lead containing piezoelectric materials to lead free ones [27]. The main advantage of using lead free piezoelectric materials is that they are environmentally friendly. While not as effective as the lead-based ones, they are now popular replacements due their biocompatibility (nontoxicity). Lead free PZMs materials include potassium sodium niobate (KNN) based, barium titanate (BT) based, sodium bismuth titanate (BNT) based and bismuth ferrite (BFO) based ones (Figure 10.3) [29]. The first lead free materials to be discovered were BT based ceramics; these materials are known as perovskites with the general chemical formula ABO_3 [in which A denotes either barium (Ba), lithium (Li), or strontium (Sr), and B is titanium (Ti), niobium (Nb), or tantalum (Ta)] [30]. The BT-based piezoceramics are the most frequently used lead-free PZMs due to the following advantages: they are electrically stable, have good electrochemical coupling, and low dielectric loss [31]. Potassium sodium niobate [KNN, $(K,Na)NbO_3$] was reported in 1960s after the discovery of BT-based ceramics. These materials have been reported to have exceptional electrical properties, high curie temperature (420°C), high remanent polarization (33 $\mu C\ cm^{-2}$), and large electrochemical coupling factor (approximately 0.454) [32]. Usually KNN-based ceramics, including $KNbO_3$, are prepared using the solid-state reaction method due to its simplicity and high yield. However, this method produces KNN based ceramics with high level of

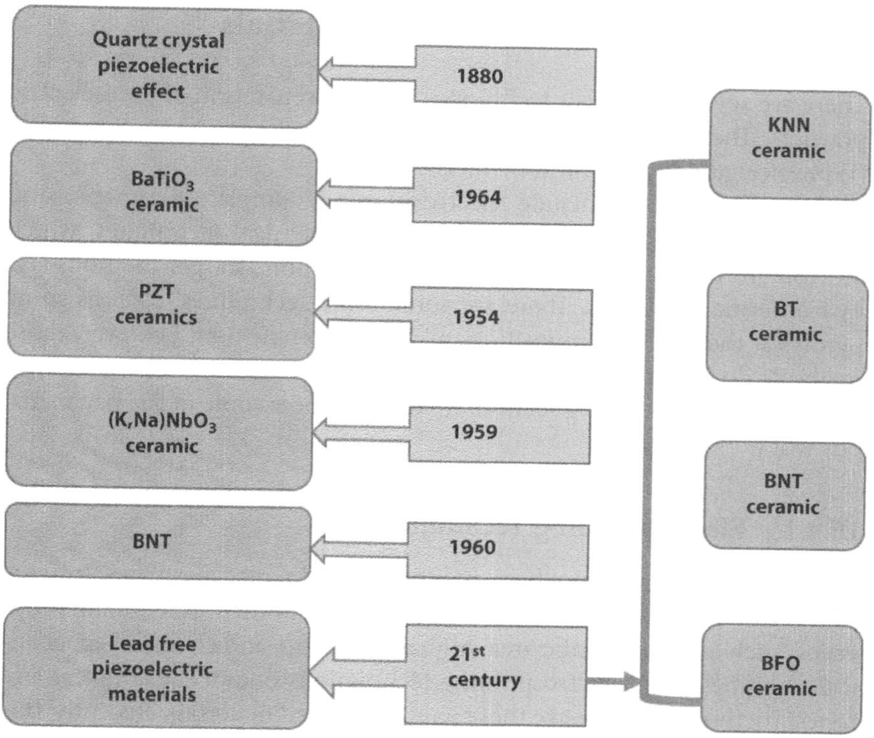

Figure 10.3 Historic development of lead free piezoelectric materials [29].

impurities and also uses high temperatures of above 1000°C. In order to reduce formation of impurities, several other methods, including normal sintering and wet chemistry methods, have been implemented [31]. For example, Zhang *et al.* [33] synthesized highly pure lead-free piezoelectrics $(_{0.96}(K_{0.5}Na_{0.5})_{0.95}Li_{0.05}Nb_{1-x}Sb_xO_{3\text{-}0.04}BaZrO_3)$ thorough solid state reaction method followed by normal sintering; the purity level of the product was greater than 99.5%. Sodium bismuth titanates (BNT) were also discovered in 1960s, as lead-free based piezoelectric materials, and they are known for their large remanent polarization and high Curie temperature [34]. Due to these properties, BNT piezoelectric materials, such as $Bi_{0.5}Na_{0.5}TiO_3$-$BaTiO_3$ (BNT-BT), $Bi_{0.5}Na_{0.5}TiO_3$-$Bi_{0.5}K_{0.5}TiO_3$ (BNT-BKT), and $(1\text{-}x)(Na_{0.5}Bi_{0.5})TiO_3{-}xBaTiO_3$ (NBBT), have attracted a lot of attention in various applications, including sensors and actuators [35]. However, the modified-NBBT has limited application due to their low depolarization temperature, serious dielectric loss, and large leakage current.

10.4 Synthesis of Piezoelectric Materials

There are several synthesis techniques for the preparation of piezoelectric powders. These synthetic techniques are divided into two main categories: (i) physical and (ii) chemical methods.

Physical methods include electrospinning, template synthesis, and mixed oxide synthesis. On the other hand, chemical techniques, which are usually used, include chemical-coprecipitation, sol-gel method, and hydrothermal synthesis. There are some other techniques, such as spray pyrolysis, thermal decomposition, and auto combustion; however, these methods do not have a significant role in the formation of piezoelectric powders. In the following section, we describe how some of the key methods work.

10.4.1 Electrospinning Technique

Over past few years, nanofibers, nanowires, and nanotubes have attracted a lot of attention as nanostructure materials due to their exceptional properties, such as large specific area, high aspect ratio, and chemical/mechanical stabilities [36]. Electrospinning technique is one of the simplest and fastest method to synthesize these nanostructured materials. Recently, this physical approach techniques have been used to prepare 1D perovskites nanofibers and/nanofibers [37]. In this approach, the perovskites are synthesized in the presence of homogenous polymer solution. The polymer

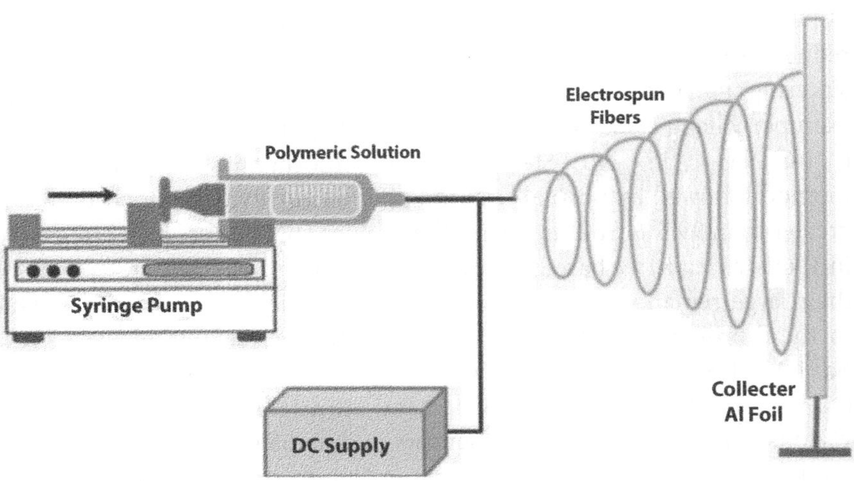

Figure 10.4 Schematic representation of electrospinning process [39].

is dissolved in an organic solvent, such as ethanol to form a homogenous solution. Thereafter, the sol-gel precursor is then mixed with the polymer solution. The mixture is then transferred into a syringe and pumped at the electrode (or tip of the needle). When high voltage is applied in the system, an electric field is created between the syringe and a collector plate thus electrifying the pumped solution. As a result, the surface tension of the liquid droplet tends to be distorted and forms long and thin threads (Figure 10.4) [38].

10.4.2 Template Synthesis

The template method uses nanoporous membrane materials, such as silicone, aluminium, polycarbonate, etc., as templates to grow new nanostructured materials of nanofibers [40]. Fabrication of ferroelectric nanofibers via templating has the following benefits: (i) high yield of nanofibers or nanotubes, (ii) their dimension can be controlled on the template, (iii) aggregation of nanofibers can be prevented by pore walls, and (iv) nanoarray structures are subjected to template structures [37]. The templating method combined with the sol-gel method has been applied to prepare the following piezoelectric materials, $BaTiO_3$ [41], $PbTiO_3$ [42], $PbZrO_3$ [43], and $BiFeO_3$ [44]. In another study, $BiFeO_3$ nanotubes were fabricated using the combination of sol-gel and template method [44]. The templates used had a significant role on the size, shape, and structure of the resultant nanotubes. Other parameters that affected geometrical properties of fabricated nanotubes were capillary force and sol-gel concentration [45].

10.4.3 Mixed Metal Oxide (MMO)/Solid State Synthesis

The solid state method involves the chemical decomposition reaction between two solid materials at high temperatures to form a new material. Several factors, including surface area, chemical structure and composition, stoichiometry ratio of the starting solid materials, and reaction conditions, such as temperature and pressure, have an impact on the formation of new ferroelectric materials. The benefits of using this technique include its simplicity and high yield; however, the requisite high temperatures limit its synthetic application. Simon-Seveyra *et al.* [46] used solid state reaction to fabricate lead-free piezoelectric material ($BaTiO_3$) using $BaCO_3$ and TiO_2 as staring materials. The reaction steps involved were ball milling and calcining. The same method was used to fabricate piezoelectric ceramics, such as $KNbO_3$ [47], $(Na0.5Bi0.5)_1-_xBa_xTiO_3$ [48], $(Bi_{0.5}Na_{0.5})TiO_3$ [49], and $(Na_{0.5}K_{0.5})NbO_3$ [50] for different applications.

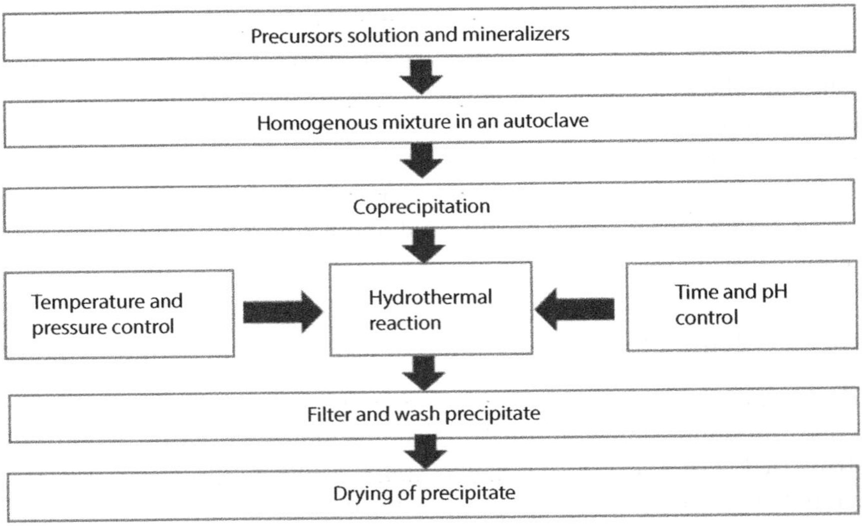

Figure 10.5 Hydrothermal synthetic flowchart [26].

10.4.4 Hydrothermal/Solvothermal Method

Hydrothermal and solvothermal synthesis are similar, the only difference is based on the solvent used during chemical reaction. The solvent used in hydrothermal method is an aqueous solution whereas for solvothermal it is a nonaqueous solution. Both chemical reactions take place in an autoclave at temperatures above 100°C [51]. These techniques are mostly used to synthesize piezoelectric or perovskite nanomaterials due to the following advantages: they are less expensive and form pure and stable materials. Hydrothermal reaction can be influenced by various reaction parameters, such as temperature, pressure, time, and pH of the solution. Figure 10.5 shows synthetic flow charts by hydrothermal routes.

10.4.5 Sol-Gel Method

The sol-gel method is a fabrication technique known to prepare gels, which can be easily broken into fine powders during grinding or the drying process. This technique involves metal alkoxides as starting precursors being dissolved in appropriate organic solvents. Thereafter, hydrolysis takes place when a small amount of water is added to the solution. The resultant gel from a hydrolysis reaction undergoes thermal treatment at high temperature between 600°C and 1000°C for liquid removal and to prevent formed product/material from densification

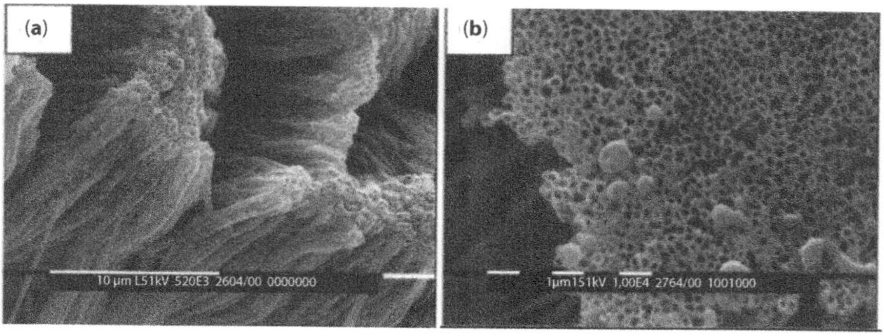

Figure 10.6 SEM images for side view of (a) $BaTiO_3$ and top view of (b) $PbTiO_3$ formed by sol-gel template synthesis [56].

process [52]. This method has been employed in the synthesis of various piezoelectric materials including as $BaTiO_3$ [53], $PbTiO_3$ [54] and $BiFeO_3$ [55]. Hernandez *et al.* [56] fabricated nanotubes of $PbTiO_3$ and $BaTiO_3$ using synergic synthesis of sol-gel and template method. Firstly, the gels were prepared from metal acetates ($Ba(OAc)_2$ and $Pb(OAc)_2$) separately with acetic acid, titanium isopropoxide and ethanol. Thereafter, a template method was used to prepare piezoelectric nanotubes. The resultant gels were dipped into alumina membrane template with 200 nm pore size. The sol-gel templates were calcined at different temperatures, for barium (Ba) and lead (Pb) 700 °C and 650 °C, respectively. Figure 10.6 shows top and side view SEM images for $PbTiO_3$ and $BaTiO_3$, respectively. The resultant perovskite nanotubes were 50-μm-long with diameter of 200-nm.

Though there are several methods used to synthesize these materials, there are several limitations with the formed materials, which are discussed in the next section.

10.5 Challenges of Piezoelectric Nanomaterials/ Nanogenerators

The research and development on piezoelectric energy harvesters (PEHs) is currently growing. Figure 10.7 shows a growing trend on PEHs and some of the vibrational energy harvesters in the past few years (2003–2013). The most widely used piezoelectric materials are PZT and have been applied to power small electronics including biomedical implant devices, wireless sensors and cell phone batteries. The electric energy produced by these

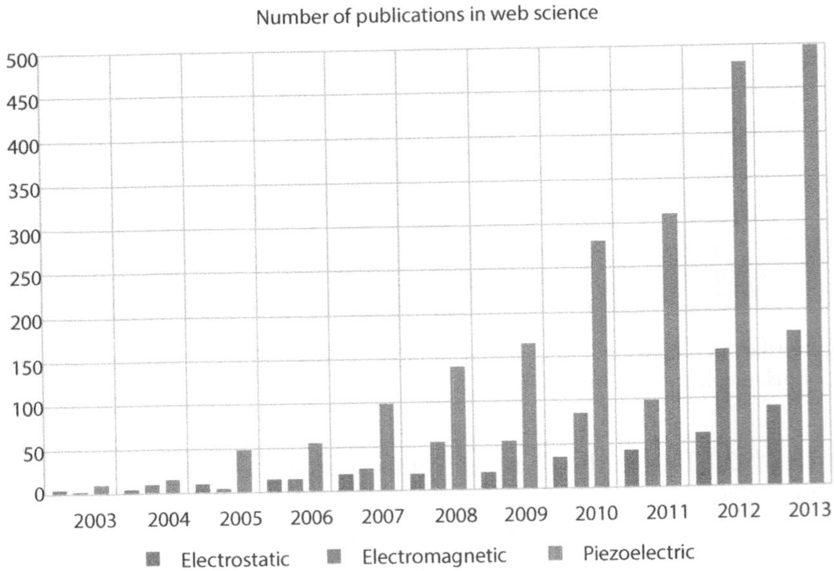

Figure 10.7 Number of publications on piezoelectric, electrostatic and electromagnetic energy harvesters in web of science from 2003 to 2013 [58].

materials ranges between mW and μW. Unlike other energy harvesters, such as photovoltaics, which produce large quantities of electric energy up to thousands of watts, piezoelectric effect harvests a small quantity of electric energy [57].

It is very important to note there is no straightforward/fair comparison of energy harvesters with different architecture, size, material and operational mode. The problem associated with unfair comparison is based on lack of universal merit of figures and on the other hand, there is a lot of data in the literature which is not consistent. The performance of energy harvesters greatly depends on several factors, such as materials, dimensions, operational conditions, and circuits, employed. Several energy harvesters operate under different vibrational conditions, therefore it would be unfair to compare various designs based on limited factors or one standard. Hence, it is very important for researchers to clearly demonstrate parameters, such as power output, volume, and input excitation (acceleration) for various designs. Table 10.4 [59–62] illustrates output performance of piezoelectric materials (PZT) under different conditions in the last few years.

The benefits of PEHs when compared to other vibration based energy harvesting materials includes higher energy density and the ability to operate at higher vibrational frequencies [63, 64]. PZTs contain a high content

Table 10.4 Performance comparison of PZT energy harvesters under different conditions.

Materials	Input excitation	Volume (mm^3)	Power (μW)	Frequency (Hz)	Normalized power density (μW/ mm^3/g^2)	Ref
Bulk PZT	0.255 g	535	375	120	10.80	[59]
Bulk PZT	0.100 g	19.7	2.74	167	13.91	[60]
Bulk PZT	0.250 g	613	4000	100	104.7	[61]
Sol-gel PZT	1.000 g	613	461	10.80		[62]

of lead (Pb), thus these materials are not environmentally friendly and are also not biocompatible. The most utilized piezomaterials in microelectromechanical system (MEMS) are PZT thin films. There are several techniques used to fabricate PZT thin films, these include sol-gel, chemical vapor deposition (CVD), sputtering, aerosol and epitaxial method [65]. The sol-gel method is commonly used to deposit sol-gel on the thin film substrate. The coated PZT thin film is then annealed at higher temperatures between 450°C and 700°C for crystallization growth. The PZT thin films fabricated using the sol-gel technique tends to have less impurities and thicker films above 1 μm. However, this process suffers from producing thin films with some irregular surface morphology with cracks and holes [66, 67]. These defects can lower the piezoelectric effect of the PZT thin films. To overcome the challenges on defects, the sputtering technique has been used to form uniform thin films of PZT. As opposed to other techniques, such as aerosol and epitaxial method, which requires high temperatures for excellent crystallization growth, the sputtering process operates at low temperature (<150°C). The main challenge associated with the sputtering process is to produce thick layers of PZT above 1 μm [65]. Another promising deposition technique, which tends to fabricate PZT thin films with excellent surface uniformity, is chemical vapor deposition (CVD). This process uses moderate temperature (approximately 225°C) to deposit evaporated PZT on the substrate [68]. The thickness and shape of the PZT film are more controllable using this process. The deposited PZT layers on the substrate can be positioned to be desirable by either lift off, wet or dry etching. The most favorable method is dry etching to remove

PZT thin films from the substrate. However, during the etching process, there is a high risk of cross-contamination from the surrounding environment. Other ways of avoiding contamination is to apply the wet etching process. This process uses liquid chemicals (HCl and BHF) to remove PZT layers from the substrate/wafer [69]. The main limitations include cutting of the PZT layers during the etching process and types of etchants used. Some etchants are compatible with certain wafers, for instance high concentration of HCl would selectively etch silicon dioxide and titanium layers. Table 10.5 illustrates the different deposition techniques with their strengths and limitations/challenges.

Some of the other methods for the preparation of PZT thin films include aerosol deposition, sputtering and epitaxial method. However, aerosol and epitaxial techniques require higher temperatures for excellent crystallization growth. Though there have been several challenges in the formation of these materials, they have been used in several applications, such as the electrodegradation of dyes and bacteria.

Table 10.5 Deposition techniques used to prepare piezoelectric thin films and their challenges [65].

Deposition techniques	Specifications (Advantages and disadvantages)
Sputtering method	• Produces very good thin layers with a thickness of 1 μm. • Smooth and uniform films are producible. • Process occurs at low temperature, below 150 °C. • High risk of cross contamination.
Sol-gel technique	• Thicker films above 1 μm are producible • Process occurs at high temperature, above 450 °C. • Deposition can produce uneven PZT surface/ pin-holes. • Low risk of cross contamination.
Chemical vapor deposition	• Process operate at moderate temperature. • Suitable step coverage for 3D applications. • Good composition control. • Require specialized equipment and materials.

10.6 Application of Piezoelectric Materials for Piezo-Electrocatalytic Degradation of Dyes and Bacteria in Wastewater

Dyes, such as methylene blue (MB), methyl orange (MO), rhodamine B (RhB), etc., have been used extensively as coloring agents in particular in the textile, dying and pulp industries [2]. These industries, contribute to huge amounts of toxic dyes in the environment, as shown in Figure 10.8, the textile industry is the major contributor of dye effluents.

It has been estimated that approximately more than ten thousands (10, 0000) tons of dyes per year are produced from the textile industries, and about one hundred (100) tons per year are released into industrial wastewater streams [71]. The presence of these dyes even at lower level concentration (1ppm) tends to influence water quality negatively [72]. Industrial wastewater containing these dyestuffs have a serious health problem to human health, aquatics and plants due to their toxicity [2]. According to international standards, there's a certain allowed permissible limit levels of dye effluents in the environment. Table 10.6 shows the allowed level limit of chemical oxygen demand (COD), biological oxygen demand (BOD), temperature, solid particles, and concentration color.

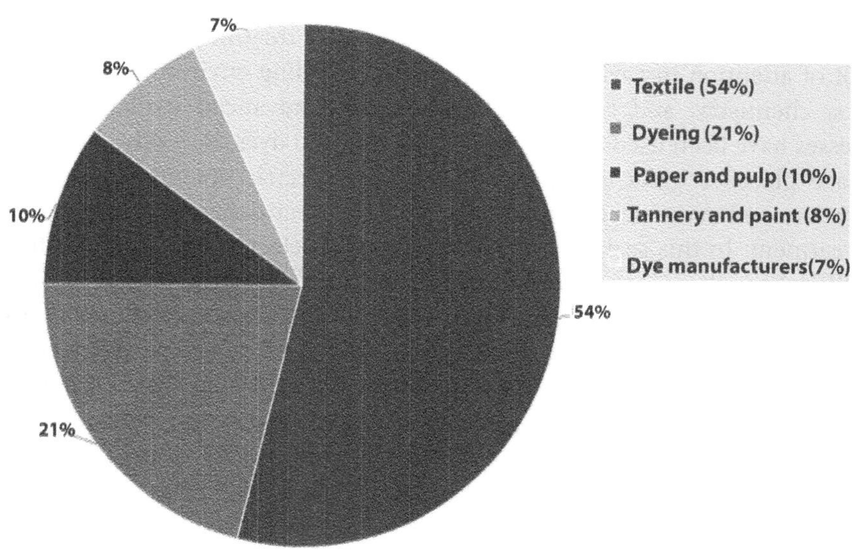

Figure 10.8 Percentage contribution of organic dyes by various industries [70].

Table 10.6 Allowed limits of dye effluents in the environment according to international standards [4].

Parameters	Allowed permissible limit
Chemical Oxygen Demand	< 50.0 mg/L
Biological Oxygen Demand	< 30.0 mg/L
Concentration of color	< 1.00 mg/L
pH solution	Neutral (6-7)
Temperature	< 42 °C

Wastewater from industries consist of pathogenic bacteria, such as *E. coli, P. aeruginosa, E. faecalis, S. aurea,* etc., which can cause several water-borne diseases, such as diarrhea, cholera, malaria, typhoid, and many more [73]. It has been reported that about 30,000 people die daily from these diseases. These bacteria have shown to be resistant to several antibiotics including ciproflaxin, quinolone, evofloxacin, *β*-lactams, and piperacillin, amoxicillin, and many more [74, 75]. According to the World Health Organization (WHO), the colony forming units for these bacteria (*E. coli, P. aeruginosa, E. faecalis*) in drinking water should be below 0 CFU/mL [76]. This means we should not detect any of these pathogenic bacteria in drinking water.

Bacterial and dye degradation for wastewater treatment has attracted a lot of attention in various research fields, including environmental physics, chemistry, and biology. Electrochemical advanced oxidation processes have been used to decompose bacteria and dyes. Recently, scientists have placed a great effort in utilizing piezoelectric-potential generated by strained piezoelectric materials to drive electrochemical wastewater treatment. In this section, we will explore piezoelectric energy harvesters (PEHs) as self-powered electrocatalytic reactor for degradation of organic pollutants and bacteria. It has been reported that piezoelectric materials (PZMs) under mechanical stress behave in a similar manner as electrocatalytic reactors. When mechanical stress is applied on PZMs, internal electric charges (positive or negative charges) move on opposite edges of the material. These free electric charges are responsible to induce redox reaction thus resulting in ROS, such as O_2^- and •OH. The free electric charges (electrons and holes) at opposite sides of these materials tend to act as anodes and cathodes. The reaction equations (10.5) and (10.7) show the formation of reactive oxygen species from free electric charges on the opposite sites of piezoelectric materials. As shown in equation (10.6) and (10.7), the free positive charges react with water to form hydroxyl radicals (•OH), whereas negative charges react with free oxygen molecules to form superoxide radicals ($^{\bullet}O_2^-$). These reactive oxygen species ($^{\bullet}O_2^-$ in equation 8 and •OH in equation 10.7) are regarded as strong oxidants and are responsible for the degradation of organic dyes and bacterial disinfection.

$$\text{BTO (Piezoelectric material)} + \text{mechanical vibration} \rightarrow \text{BTO } (e^- + h^+) \quad (10.5)$$

Negatively charged surface of piezoelectric material

$$h^+ + H_2O \rightarrow \bullet OH + H^\bullet \quad (10.6)$$

$$h^+ + OH^- \rightarrow \bullet OH \quad (10.7)$$

Positively charged surface of piezoelectric material

$$e^- + O_2 \rightarrow O_2^{\bullet -} \quad (10.8)$$

Piezocatalytic degradation of organic pollutants

$$\text{ROS } \bullet OH \text{ and } O_2^{\bullet -} + \text{organic pollutants} \rightarrow \text{Organic degradation} \rightarrow CO_2 + H_2O$$

10.6.1 Piezo-Electrocatalytic Degradation of Organic Dyes and Bacteria in Wastewater

There is limited research on the utilization of piezoelectric nanogenerators (PENGs) for self-powered wastewater treatment, especially for the degradation of organic dyes from aqueous solution. Hong *et al.* [77] investigated piezo-electrocatalytic activity of $BaTiO_3$ dendrites for the removal of Azo dyes from wastewater, especially acid orange 7 (AO7) dye. The ultrasonication was applied as a source of mechanical vibration to trigger piezoelectric effect of $BaTiO_3$. The strained $BaTiO_3$ was able to degrade about 80% of the AO7 after 90 minutes. Figure 10.9 illustrates the piezocatalytic mechanism of AO7 by induced free electric charges. As seen in the figure, the negatively charged side of the strained $BaTiO_3$ acts as an anode, at this side (anode) oxidation of water occurs which generates hydroxyl radicals

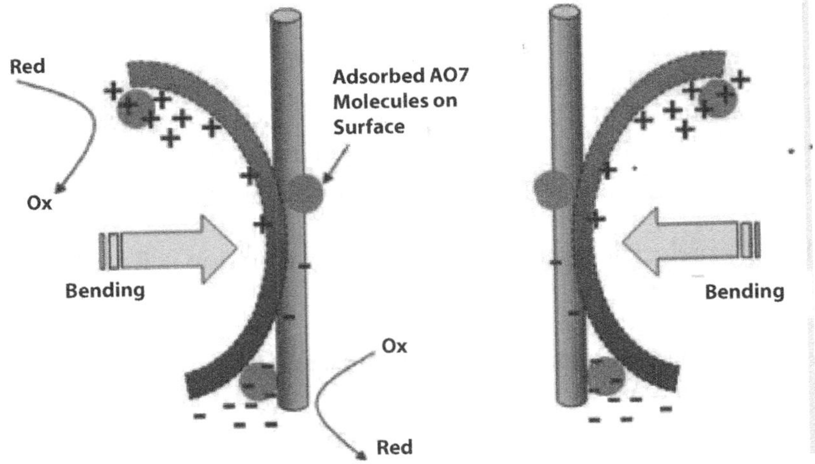

Figure 10.9 Pieozocatalytic mechanism of AO7 using strained $BaTiO_3$ [77].

(•OH). On the other side, (positively charged side), reduction reaction induces other hydroxyl radicals. The generated hydroxyl radicals (•OH) were responsible for the decomposition of AO7.

The piezo-photocatalytic degradation of Rhodamine B (RhB) was achieved using bismuth oxychloride (BiOCl) [78]. The hydrothermally synthesized BiOCl behaved as piezocatalyst and photocatalyst, BiClO as a piezocatalyst degraded 26% of RhB, whereas when BiClO was applied as a photocatalyst, 72% of RhB was decomposed from the aqueous solution. The synergic effect of photocatalysis and piezocatalysis (Piezo-photocatalysis) showed significant improvement of RhB degradation. When BiClO was employed as a piezo-photocatalyst, about 99% of RhB was degraded. This showed that coupling vibration and light source increased the piezoelectrocataytic performance of BiClO. A similar study was conducted by Xue *et al.* [79], the fabricated ZnO nanowires were investigated as piezocatalyst and photocatalyst for the degradation of methylene blue (MB). The piezoelectric effect enhanced the photocatalytic performance of ZnO nanowires. ZnO nanowires under light source only (without any mechanical vibration) degraded 64% of MB, the percentage removal increased to 96% when both UV irradiation and mechanical vibration was applied. Li *et al.* [80] conducted piezoelectrocatalytic degradation of RhB and tetracycline using MoS_2, WS_2, and WSe_2. Under ultrasonication vibration, the RhB degradation percentages were 43.5%, 65.7%, and 96% within 60 min of vibration for WSe_2, WS_2, and MoS_2, respectively (Figure 10.10(a)). In terms of tetracycline, the degradation were 62% (WSe_2), 72% (WS_2), and 93%

(MoS_2) within 120 minutes. MoS_2 showed a better piezocatalytic activity as compared to other TMDs samples, the rate of piezocatalytic degradation for RhB and tetracycline was faster when MoS_2 was employed as a piezocatalyst (Figure 10.10(b)).

Feng *et al.* [81] reported on the deactivation of *Escherichia coli (E. coli)* from synthetic water using hydrothermally synthesized tetragonal-barium titanate (*t*-$BaTiO_3$). Thiobarbituric acid method was used to measure malondialdehyde (MDA) which indicates oxidation process. Usually, inactivation of microbes starts by a physical or chemical damage of cell membranes either by chemical oxidation. The MDA concentration increased under the exposure of *t*-$BaTiO_3$. Physical damages of *E. coli* were observed through SEM (Figures 10.11(a)-(f)). The cell structure of *E. coli* changed from a rod shape to a rough surface with several potholes after 60 minutes

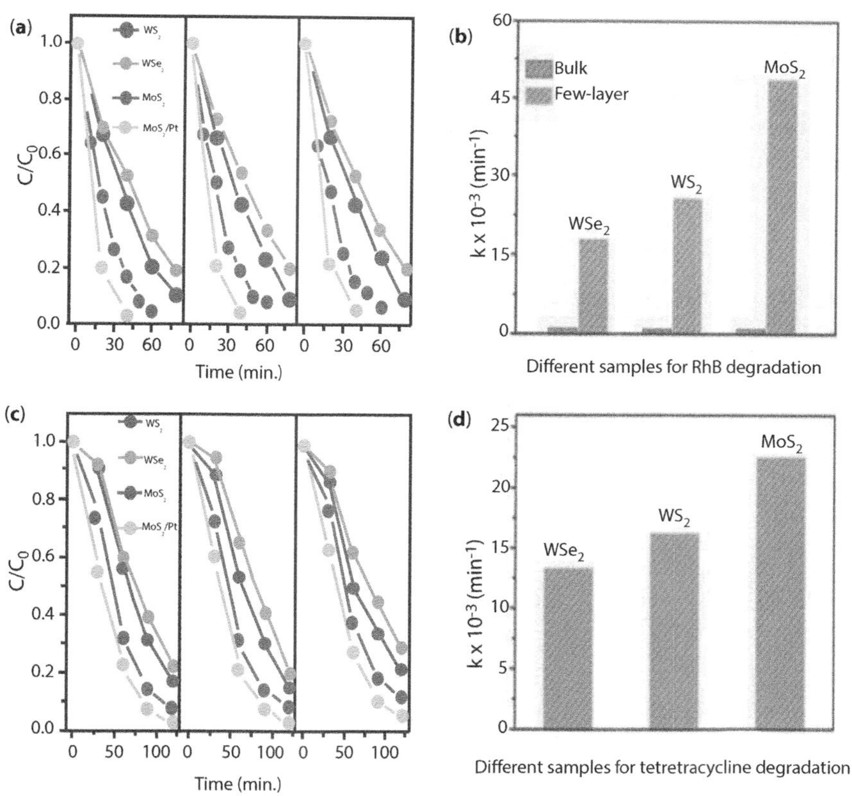

Figure 10.10 Three recycling tests of piezocatalytic activity for (a) RhB and (c) tetretracycline. Piezocatalytic degradation kinetics for (b) RhB and (d) tetretracycline using different few layered TMDs samples [80].

of treatment. The structural change to rough surface indicated that the cell membrane was tempered with. After 120 minutes of treatment, only debris were observed, which indicated its decomposition (Figure 10.11(d)). The strained *t*-$BaTiO_3$ achieved 95% inactivation efficiency for *E. coli* under 30 minutes of sono-pieozocatalysis.

Kumar *et al.* [82], investigated the synergistic effect of piezocatalysis and photocatalysis for bacterial degradation by unpoled and poled $BaTiO_3$ ceramic. A low frequency of 8 Hz was used for deformation of $BaTiO_3$ (BTO) to induce piezo-potential responsible for the degradation of bacteria. Under ultrasonic vibration, unpoled $BaTiO_3$ ceramic destroyed 56% of bacteria within 30 minutes (Figure 10.12). It was evident that the UV light exposure improved the degradation rate of bacteria, the bacterial efficiency was increased to 70% within the same time (30 minutes). The poled $BaTiO_3$ ceramic had a better piezocatalytic degradation activity as compared to unpoled $BaTiO_3$. Under vibrational exposure only, poled $BaTiO_3$ achieved piezocatalytic bacterial degradation efficiency of 97%. When light and vibration excitation was applied to BTO, piezocatalytic degradation efficiency reached 99.99% within 20 minutes. The results showed that the combination of a piezocatalysis and photocatalysis enhanced the catalytic degradation activity.

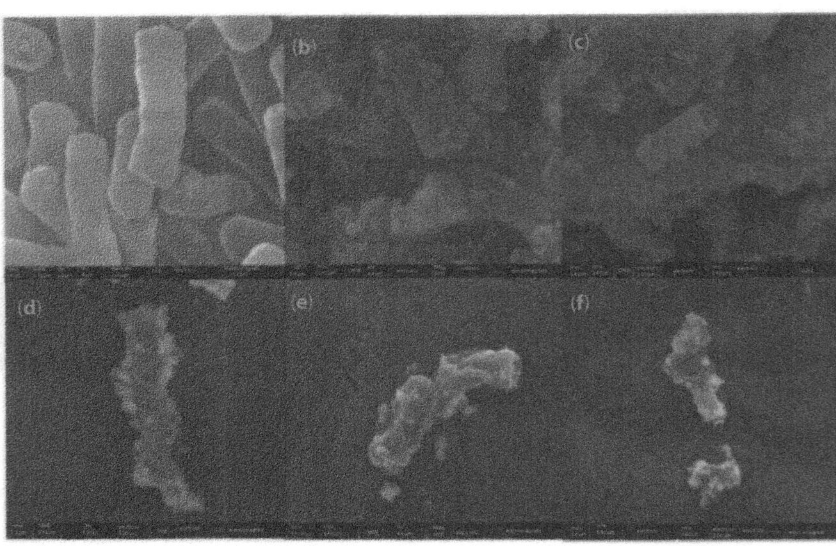

Figure 10.11 SEM images of *E. coli* after piezocatalytic treatment for (a) 0 minute, (b) 30 minutes, (c) 60 minutes, (d) 120 minutes, (e) 150 minutes, and (f) 180 minutes [81].

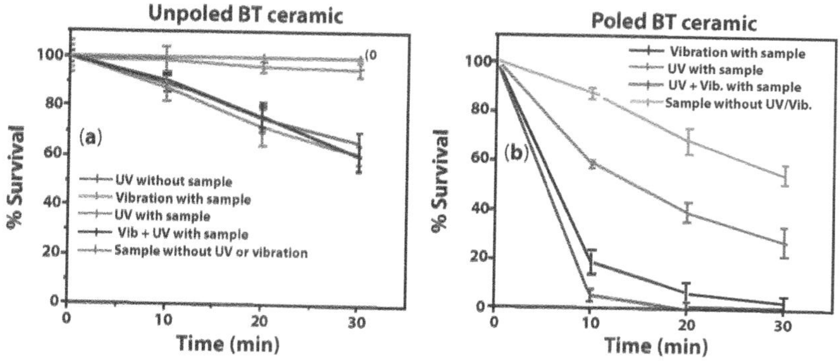

Figure 10.12 Survival percentages of bacteria during photocatalytic and piezo-photocatalytic treatment [82].

Sharma *et al.* [83] investigated piezocatalytic, pyrocatalytic, and photocatalytic activity of $Ba_{0.85}Ca_{0.15}Ti_{0.9}Zr_{0.1}O_3$ (BCZTO) toward catalytic degradation of pharmaceuticals, dyes, and bacteria. In terms of bacterial degradation, the rate of piezocatalytic degradation was faster when poled BCZTO was employed as compared to unpoled BCZTO. Under ultrasonic vibrations, the poled BCZTO as piezocatalyst was able to degrade almost 100% of the bacteria (Figures 10.13(a)-(b)). The bacterial disinfection mechanism was based on the ROS species generated via a redox reaction when BCZTO underwent mechanical vibrations. These ROS, such as hydroxyl radicals (•OH), are known to be strong oxidants and can react with amino acids and DNA molecules thus damaging the bacterial cells [84]. The same authors [85], used different piezocatalyst ($BaZr_{0.02}Ti_{0.98}O_3$ (BZT)) for bacterial disinfection of *E. coli* from stimulated water. In that study, photocatalysis was combined with the piezoelectric effect to enhance the catalytic activity of BZT. A similar behavior was observed when unpoled and poled BZT were applied to degrade bacteria. The poled BZT under both UV irradiation and vibrations was more effective as compared to unpoled BZT. The maximum degradation of *E. coli* achieved by poled BZT under simultaneous exposure of light and vibration was also almost 100%.

Chou *et al.* [86] fabricated new piezo-nanocatalyst to inactive *E. coli* from water by using single-layered and multilayered molybdenum disulfide (MoS_2) nanosheets (NNs). The MoS_2 as a disinfectant destroyed 99.99% of the *E. coli* after 60 min of simultaneous exposure to light source and vibration. Figure 10.14 shows a schematic illustration of bacterial mechanism when (MoS_2) nanosheets were exposed to UV light source and mechanical vibration. The deformed MoS_2 NNs created free electric charges to induce

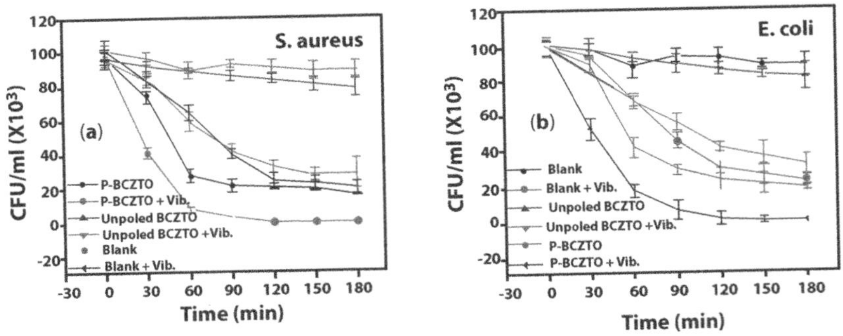

Figure 10.13 Number of colony forming unit (CFU)/mL of (a) *S. aureus* and (b) *E. coli* before and after piezo-photocatalytic treatment with BCZTO [83].

redox-reaction, thus creating reactive oxygen species, such as hydroxyl radicals (•OH) and superoxide ($O_2^{-•}$), through water splitting. These ROS interacted with bacteria by damaging their cell membranes. Masimukku *et al.* [87] used single layer tungsten nanoflowers composite (PDMS/WS_2 NFs) to degrade 99.99% of *E. coli* after 60 minutes of ultrasonic vibration.

Tan *et al.* [88] developed lead free piezoceramics of potassium sodium niobate (KNN), as a self-powered electrocatalytic reactor to degrade

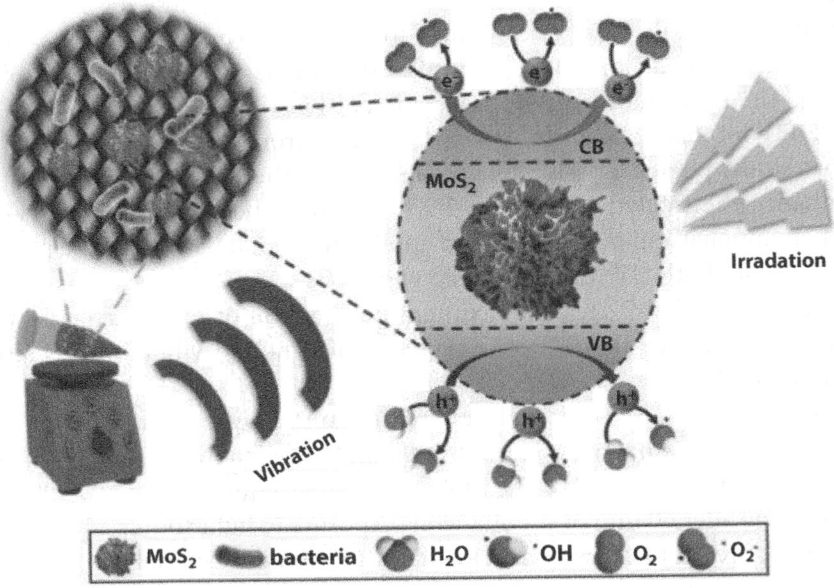

Figure 10.14 Piezo-photocatalytic degradation mechanism of *E. coli* by MoS_2 [86].

bacteria. The polarized KNN under ultrasonic vibration positive and negative charges on opposite surfaces of the KNN material. The results obtained from DCFH fluorescence showed that the highest amount level of ROS were produced on the positively charged surface (KNN+70) (Figure 10.15(a)). These ROS generated by KNN+70 played a major role in bacterial disinfection as compared to KNN-70. Figure 10.15(b) shows that KNN+70 was more effective in destroying *S. aureus* since the number of bacterial strains were reduced; however, nonpolarized KNN and polarized KNN-70 did not show any antimicrobial properties against *S. aureus*. The antimicrobial results (Figure 10.15(c)) correlated well with ROS generation (Figure 10.15(a)). The higher the number of ROS, the more effective they are against bacteria. It is very important that piezoceramics should destroy bacteria without tampering with normal mammalian cells. CCK-8 investigations were conducted to identify proliferation of mouse mesenchymal stem cells (MSCs) on individual KNN surfaces (KNN, KNN+70, KNN-70, and HA). The results showed no significant difference in proliferating cells

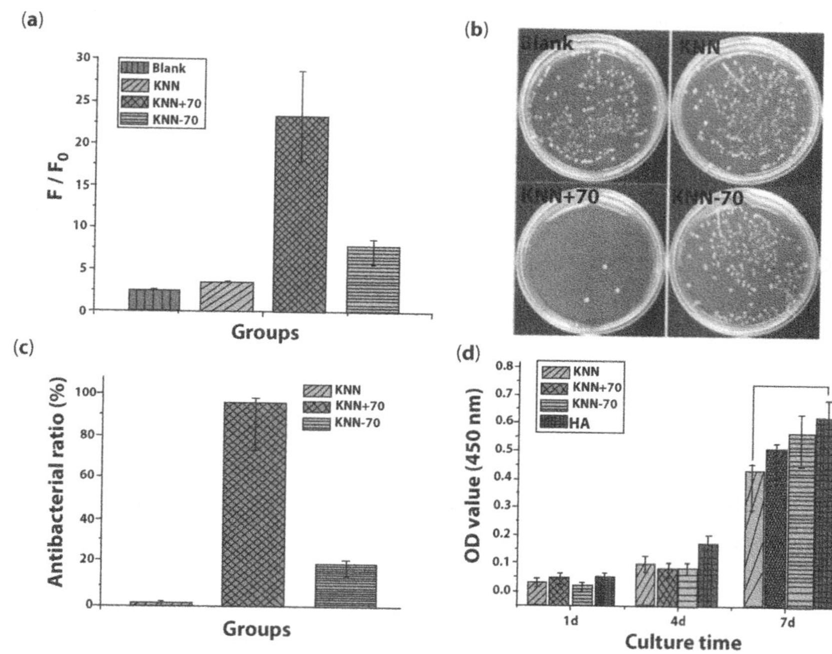

Figure 10.15 (a) Quantification of ROS levels (intensity of F/F_0 represents ROS). (b) Number of colony forming unit (CFU)/mL on the plates after piezocatalytic treatment. (c) Pieozocatalytic treatment ratio of bacteria and (d) cell proliferation (CCK-8 assay) after several days of incubation [88].

for strained KNN (Figure 10.15(d)), indicating that the polarized material had excellent biocompatibility. The polarized KNN can safely kill bacterial cell without tempering with mammalian cells.

10.7 Conclusion and Future Perspectives

Synthetic procedures and challenges faced by piezocatalyst as self-powered electrocatalytic reactors were described in this chapter. Lead-based piezoelectric materials are the most widely used since they have a high piezoelectric effect; however, they are toxic to the environment due to their high lead (Pb) content. These materials have also been shown to be of importance in different applications, including bacterial disinfection, hydrogen production, sensors, and wastewater treatment. However, such materials have been considered as poor electrocatalysts to drive electrochemical reaction processes. To enhance their electrochemical performance, these materials have to be coupled with electroactive materials or excellent electrocatalysts. Their application in dye degradation and bacterial disinfection was also discussed in details. In wastewater treatment, the synergic effect of piezocatalysis and photocatalysis was shown to enhance the catalytic degradation of organic dyes and bacteria. In conclusion, piezoelectric materials can be used in the future for wastewater treatment provided they are coupled with other catalysts.

Acknowledgments

This work was supported by Global Excellence Stature (GES), University of Johannesburg; FRC/URC Faculty of Science, University of Johannesburg, the Centre for Nanomaterials Science Research (University of Johannesburg); National Research Foundation of South Africa (Thuthuka Grant No. TTK 107066, Thuthuka grant no. TTK-117999) and University of Johannesburg (UJ), the Faculty of Science, South Africa and University of Limpopo for financial support.

References

1. Owa, F.D., Water pollution: Sources, effects, control and management. *Mediterr. J. Soc. Sci.*, 4, 8, 65, 2013.

2. Silveira, E., Marques, P.P., Silva, S.S., Lima-Filho, J.L., Porto, A.L.F., Tambourgi, E.B., Selection of Pseudomonas for industrial textile dyes decolourization. *Int. Biodeterior. Biodegradation*, 63, 2, 230–235, 2009.
3. Fernández, C., Larrechi, M.S., Callao, M.P., An analytical overview of processes for removing organic dyes from wastewater effluents. *Trends Analyt. Chem.*, 29, 10, 1202–1211, 2010.
4. Katheresan, V., Kansedo, J., Lau, S.Y., Efficiency of various recent wastewater dye removal methods: A review. *J. Environ. Chem. Eng.*, 6, 4, 4676–4697, 2018.
5. Malato, S., Fernández-Ibáñez, P., Maldonado, M.I., Blanco, J., Gernjak, W., Decontamination and disinfection of water by solar photocatalysis: Recent overview and trends. *Catal. Today*, 147, 1, 1–59, 2009.
6. Cruvinel, V.R.N., Zolnikov, T.R., Bashash, M., Marques, C.P., Scott, J.A., Waterborne diseases in waste pickers of estrutural, Brazil, the second largest open-air dumpsite in world. *Waste Manage.*, 99, 71–78, 2019.
7. Alrumman, S.A., El-kott, A.F., Keshk, S.M.A.S., Water pollution: Source and treatment. *Am. J. Environ. Eng.*, 6, 3, 88–98, 2016.
8. Sirés, I., Brillas, E., Oturan, M.A., Rodrigo, M.A., Panizza, M., Electrochemical advanced oxidation processes: Today and tomorrow. A review. *Environ. Sci. Pollut. Res.*, 21, 14, 8336–8367, 2014.
9. Xiang, H., Shao, Y., Gao, N., Lu, X., An, N., Tan, C., Zheng, Z., Degradation of diuron by chlorination and UV/chlorine process: Degradation kinetics and the formation of disinfection by-products. *Sep. Purif. Technol.*, 202, 365–372, 2018.
10. Fornazari, A.L.T., Malpass, G.R., Miwa, D.W., Motheo, A.J., Application of electrochemical degradation of wastewater composed of mixtures of phenol–formaldehyde. *Water, Air, Soil Pollut.*, 223, 8, 4895–4904, 2012.
11. Martínez-Huitle, C.A., Rodrigo, M.A., Sires, I., Scialdone, O., Single and coupled electrochemical processes and reactors for the abatement of organic water pollutants: A critical review. *Chem. Rev.*, 115, 24, 13362–13407, 2015.
12. Goldemberg, J., Ethanol for a sustainable energy future. *Science*, 315, 5813, 808–810, 2007.
13. Safaei, M., Sodano, H.A., Anton, S.R., A review of energy harvesting using piezoelectric materials: State-of-the-art a decade later (2008–2018). *Smart Mater. Struct.*, 28, 11, 113001, 2019.
14. Panda, P.K. and Sahoo, B., PZT to lead free piezo ceramics: A review. *Ferroelectrics*, 474, 1, 128–143, 2015.
15. Aksel, E. and Jones, J.L., Advances in lead-free piezoelectric materials for sensors and actuators. *Sensors*, 10, 3, 1935–1954, 2010.
16. Rao, A.N.S. and Venkatarangaiah, V.T., Metal oxide-coated anodes in wastewater treatment. *Environ. Sci. Pollut. Res.*, 21, 5, 3197–3217, 2014.
17. Saien, J. and Nejati, H., Enhanced photocatalytic degradation of pollutants in petroleum refinery wastewater under mild conditions. *J. Hazard. Mater.*, 148, 1-2, 491–495, 2007.

18. Fajardo, A.S., Seca, H.F., Martins, R.C., Corceiro, V.N., Freitas, I.F., Quinta-Ferreira, M.E., Quinta-Ferreira, R.M., Electrochemical oxidation of phenolic wastewaters using a batch-stirred reactor with NaCl electrolyte and Ti/RuO2 anodes. *J. Electroanal. Chem.*, 785, 180–189, 2017.
19. Hossain, M.S., Mollah, M.Y.A., Susan, M.A.B.H., Islam, M.M., Role of in situ electrogenerated reactive oxygen species towards degradation of organic dye in aqueous solution. *Electrochim. Acta*, 344, 136146, 2020.
20. Patel, P.S., Bandre, N., Saraf, A., Ruparelia, J.P., Electro-catalytic materials (electrode materials) in electrochemical wastewater treatment. *Proc. Eng.*, 51, 430–435, 2013.
21. Martínez-Huitle, C.A. and Andrade, L.S., Electrocatalysis in wastewater treatment: Recent mechanism advances. *Quim. Nova*, 34, 5, 850–858, 2011.
22. Jacobson, M.Z., Delucchi, M.A., Bazouin, G., Bauer, Z.A., Heavey, C.C., Fisher, E., Morris, S.B., Piekutowski, D.J., Vencill, T.A., Yeskoo, T.W., 100% clean and renewable wind, water, and sunlight (WWS) all-sector energy roadmaps for the 50 United States. *Energy Environ. Sci.*, 8, 7, 2093–2117, 2015.
23. Mishra, S., Unnikrishnan, L., Nayak, S.K., Mohanty, S., Advances in piezoelectric polymer composites for energy harvesting applications: A systematic review. *Macromol. Mater. Eng.*, 304, 1, 1800463, 2019.
24. Covaci, C. and Gontean, A., Piezoelectric energy harvesting solutions: A review. *Sensors*, 20, 12, 3512, 2020.
25. Pramanik, S., Pingguan-Murphy, B., Osman, N.A., Developments of immobilized surface modified piezoelectric crystal biosensors for advanced applications. *Int. J. Electrochem. Sci.*, 8, 8863–8892, 2013.
26. Miclea, C., Preparation of piezoelectric nanoparticles, in: *Piezoelectric Nanomaterials for Biomedical Applications*, pp. 29–61, Springer, Berlin, Heidelberg, 2012.
27. Rödel, J., Jo, W., Seifert, K.T., Anton, E.M., Granzow, T., Damjanovic, D., Perspective on the development of lead-free piezoceramics. *J. Am. Ceram. Soc.*, 92, 6, 1153–1177, 2009.
28. Ara, A. and Usmani, J.A., Lead toxicity: A review. *Interdiscip. Toxicol.*, 8, 2, 55–64, 2015.
29. Wu, J., Perovskite lead-free piezoelectric ceramics. *J. Appl. Phys.*, 127, 19, 190901, 2020.
30. Yasseri, M., Abdizadeh, H., Shakeri, A., Golobostanfard, M.R., Low temperature synthesis of barium titanate powder by a modified sol-gel method, in: *Advanced Materials Research*, vol. 829, pp. 727–731, Trans Tech Publications Ltd, 2014.
31. Wei, H., Wang, H., Xia, Y., Cui, D., Shi, Y., Dong, M., Liu, C., Ding, T., Zhang, J., Ma, Y., Wang, N., An overview of lead-free piezoelectric materials and devices. *J. Mater. Chem. C*, 6, 46, 12446–12467, 2018.

32. Wang, X., Wu, J., Xiao, D., Zhu, J., Cheng, X., Zheng, T., Zhang, B., Lou, X., Wang, X., Giant piezoelectricity in potassium–sodium niobate lead-free ceramics. *J. Am. Chem. Soc.*, 136, 7, 2905–2910, 2014.
33. Zhang, B., Wu, J., Cheng, X., Wang, X., Xiao, D., Zhu, J., Wang, X., Lou, X., Lead-free piezoelectrics based on potassium–sodium niobate with giant d 33. *ACS Appl. Mater. Interfaces*, 5, 16, 7718–7725, 2013.
34. Zheng, T., Wu, J., Xiao, D., Zhu, J., Recent development in lead-free perovskite piezoelectric bulk materials. *Prog. Mater. Sci.*, 98, 552–624, 2018.
35. Jiang, C., Zhou, X., Zhou, K., Wu, M., Zhang, D., Synthesis of Na0. 5Bi0. 5TiO3 whiskers and their nanoscale piezoelectricity. *Ceram. Int.*, 43, 14, 11274–11280, 2017.
36. Dzenis, Y., Spinning continuous fibers for nanotechnology. *Science*, 304, 5679, 1917–1919, 2004.
37. Nuraje, N. and Su, K., Perovskite ferroelectric nanomaterials. *Nanoscale*, 5, 19, 8752–8780, 2013.
38. Li, D. and Xia, Y., Electrospinning of nanofibers: Reinventing the wheel? *Adv. Mater.*, 16, 14, 1151–1170, 2004.
39. Alharbi, A.R., Alarifi, I.M., Khan, W.S., Asmatulu, R., Highly hydrophilic electrospun polyacrylonitrile/polyvinypyrrolidone nanofibers incorporated with gentamicin as filter medium for dam water and wastewater treatment. *J. Memb. Separ. Tech.*, 5, 2, 38–56, 2016.
40. Ugo, P. and Moretto, L.M., Template deposition of metals, in: *Handbook of Electrochemistry*, pp. 678–709, Elsevier, 2007.
41. Luo, Y., Szafraniak, I., Zakharov, N.D., Nagarajan, V., Steinhart, M., Wehrspohn, R.B., Wendorff, J.H., Ramesh, R., Alexe, M., Nanoshell tubes of ferroelectric lead zirconate titanate and barium titanate. *Appl. Phys. Lett.*, 83, 3, 440–442, 2003.
42. Hsu, M.C., Leu, I.C., Sun, Y.M., Hon, M.H., Template synthesis and characterization of PbTiO3 nanowire arrays from aqueous solution. *J. Solid State Chem.*, 179, 5, 1421–1425, 2006.
43. Singh, S. and Krupanidhi, S.B., Synthesis and structural characterization of the antiferroelectric lead zirconate nanotubes by pulsed laser deposition. *Appl. Phys. A*, 87, 1, 27–30, 2007.
44. Park, T.J., Mao, Y., Wong, S.S., Synthesis and characterization of multiferroic BiFeO3 nanotubes. *Chem. Commun.*, 23), 2708–2709, 2004.
45. Mao, Y., Park, T.J., Wong, S.S., Synthesis of classes of ternary metal oxide nanostructures. *Chem. Commun.*, 46), 5721–5735, 2005.
46. Simon-Seveyrat, L., Hajjaji, A., Emziane, Y., Guiffard, B., Guyomar, D., Re-investigation of synthesis of BaTiO3 by conventional solid-state reaction and oxalate coprecipitation route for piezoelectric applications. *Ceram. Int.*, 33, 1, 35–40, 2007.
47. Wada, T., Suzuki, A., Saito, T., Fabrication of lead-free piezoelectric KNbO3 ceramics by modified solid state reaction method. *Jpn. J. Appl. Phys. 1*, 45, 9S, 7431, 2006.

48. Xu, Q., Chen, S., Chen, W., Wu, S., Zhou, J., Sun, H., Li, Y., Synthesis and piezoelectric and ferroelectric properties of (Na0.5Bi0.5) 1–xBaxTiO3 ceramics. *Mater. Chem. Phys.*, 90, 1, 111–115, 2005.
49. Pourianejad, S. and Movahedi, B., Solid-state synthesis of modified (Bi0.5Na0.5) TiO3 piezoelectric nanocrystalline ceramics and evaluating their properties. *Mater. Sci. Semicond. Process.*, 29, 337–344, 2015.
50. Hreščak, J., Malič, B., Cilenšek, J., Benčan, A., Solid-state synthesis of undoped and Sr-doped K0.5Na0.5NbO3. *J. Therm. Anal. Calorim.*, 127, 1, 129–136, 2017.
51. Zhong, H., Mirkovic, T., Scholes, G.D., Nanocrystal synthesis, in: *Comprehensive Nanoscience and Technology*, pp. 153–201, Elsevier Inc, 2011.
52. Neacşu, I.A., Nicoară, A.I., Vasile, O.R., Vasile, B.Ş., Inorganic micro-and nanostructured implants for tissue engineering, in: *Nanobiomaterials in Hard Tissue Engineering*, pp. 271–295, William Andrew Publishing, 2016.
53. Harizanov, O., Harizanova, A., Ivanova, T., Formation and characterization of sol–gel barium titanate. *Mater. Sci. Eng. B*, 106, 2, 191–195, 2004.
54. Bao, D., Yao, X., Wakiya, N., Shinozaki, K., Mizutani, N., Structural, dielectric, and ferroelectric properties of PbTiO3 thin films by a simple sol–gel technique. *Mater. Sci. Eng. B*, 94, 2-3, 269–274, 2002.
55. Xu, J.H., Ke, H., Jia, D.C., Wang, W., Zhou, Y., Low-temperature synthesis of BiFeO3 nanopowders via a sol–gel method. *J. Alloys Compd.*, 472, 1-2, 473–477, 2009.
56. Hernandez, B.A., Chang, K.S., Fisher, E.R., Dorhout, P.K., Sol–gel template synthesis and characterization of BaTiO3 and PbTiO3 nanotubes. *Chem. Mater.*, 14, 2, 480–482, 2002.
57. Glaser, P.E., Solar energy—An alternative source for power generation. *Futures*, 1, 4, 304–313, 1969.
58. Toprak, A. and Tigli, O., Piezoelectric energy harvesting: State-of-the-art and challenges. *Appl. Phys. Rev.*, 1, 3, 031104, 2014.
59. Roundy, S. and Wright, P.K., A piezoelectric vibration based generator for wireless electronics. *Smart Mater. Struct.*, 13, 5, 1131, 2004.
60. Aktakka, E.E., Peterson, R.L., Najafi, K., Thinned-PZT on SOI process and design optimization for piezoelectric inertial energy harvesting, in: *2011 16th International Solid-State Sensors, Actuators and Microsystems Conference*, IEEE, pp. 1649–1652, June 2011.
61. Reilly, E.K., Burghardt, F., Fain, R., Wright, P., Powering a wireless sensor node with a vibration-driven piezoelectric energy harvester. *Smart Mater. Struct.*, 20, 12, 125006, 2011.
62. Fang, H.B., Liu, J.Q., Xu, Z.Y., Dong, L., Wang, L., Chen, D., Cai, B.C., Liu, Y., Fabrication and performance of MEMS-based piezoelectric power generator for vibration energy harvesting. *Microelectron. J.*, 37, 11, 1280–1284, 2006.
63. Invernizzi, F., Dulio, S., Patrini, M., Guizzetti, G., Mustarelli, P., Energy harvesting from human motion: Materials and techniques. *Chem. Soc. Rev.*, 45, 20, 5455–5473, 2016.

64. Zhang, K., Wang, X., Yang, Y., Wang, Z.L., Hybridized electromagnetic–triboelectric nanogenerator for scavenging biomechanical energy for sustainably powering wearable electronics. *ACS Nano*, 9, 4, 3521–3529, 2015.
65. Rezaei, M., Lueke, J., Raboud, D., Moussa, W., Challenges in fabrication and testing of piezoelectric MEMS with a particular focus on energy harvesters. *Microsyst. Technol.*, 19, 8, 1195–1219, 2013.
66. Fu, R.X., Mamazza, R., Zheleva, T.S., Kirchner, K.W., Piekarski, B.H., Correlation of materials and ferroelectric properties of sputtered and sol-gel PZT films, in: *2009 18th IEEE International Symposium on the Applications of Ferroelectrics*, IEEE, pp. 1–4, August 2009.
67. Bathurst, S.P. and Kim, S.G., Designing direct printing process for improved piezoelectric micro-devices. *CIRP Ann.*, 58, 1, 193–196, 2009.
68. Ashrit, P., Introduction to transition metal oxides and thin films, in: *Transition Met Oxide Thin Film Based Chromogenics Devices*, vol. 73, p. 151, 2017.
69. Yeh, T.H., Suresh, M.B., Shen, J.N., Yu, J.C., Chou, C.C., Chemical reactions during wet-etching process of LSMO/PZT/LSMO-structured device fabrication. *Ferroelectrics*, 380, 1, 97–101, 2009.
70. Rauf, M.A. and Ashraf, S.S., Survey of recent trends in biochemically assisted degradation of dyes. *Chem. Eng. J.*, 209, 520–530, 2012.
71. Gupta, V.K., Kumar, R., Nayak, A., Saleh, T.A., Barakat, M.A., Adsorptive removal of dyes from aqueous solution onto carbon nanotubes: A review. *Adv. Colloid Interface Sci.*, 193, 24–34, 2013.
72. Chiu, Y.H., Chang, T.F.M., Chen, C.Y., Sone, M., Hsu, Y.J., Mechanistic insights into photodegradation of organic dyes using heterostructure photocatalysts. *Catalysts*, 9, 5, 430, 2019.
73. Hofkes, E.H. (Ed.), *Small Community Water Supplies*, International Reference Centre for Community Water Supply and Sanitation and Wiley, 1981.
74. Masekela, D., Hintsho-Mbita, N.C., Mabuba, N., Diethylamine functionalised Moringa oleifera leaves for the removal of chromium (VI) and bacteria from wastewater. *Int. J. Environ. Anal. Chem.*, 1–21, 2020.
75. Mahlaule-Glory, L.M., Mbita, Z., Mathipa, M.M., Tetana, Z.N., Hintsho-Mbita, N.C., Biological therapeutics of AgO nanoparticles against pathogenic bacteria and A549 lung cancer cells. *Mater. Res. Express*, 6, 10, 105402, 2019.
76. Edition, F., Guidelines for drinking-water quality. *WHO Chron.*, 38, 4, 104–8, 2011.
77. Hong, K.S., Xu, H., Konishi, H., Li, X., Piezoelectrochemical effect: A new mechanism for azo dye decolorization in aqueous solution through vibrating piezoelectric microfibers. *J. Phys. Chem. C*, 116, 24, 13045–13051, 2012.
78. Ismail, M., Wu, Z., Zhang, L., Ma, J., Jia, Y., Hu, Y., Wang, Y., High-efficient synergy of piezocatalysis and photocatalysis in bismuth oxychloride nanomaterial for dye decomposition. *Chemosphere*, 228, 212–218, 2019.

79. Xue, X., Zang, W., Deng, P., Wang, Q., Xing, L., Zhang, Y., Wang, Z.L., Piezo-potential enhanced photocatalytic degradation of organic dye using ZnO nanowires. *Nano Energy*, 13, 414–422, 2015.
80. Li, S., Zhao, Z., Yu, D., Zhao, J.Z., Su, Y., Liu, Y., Lin, Y., Liu, W., Xu, H., Zhang, Z., Few-layer transition metal dichalcogenides (MoS2, WS2, and WSe2) for water splitting and degradation of organic pollutants: Understanding the piezocatalytic effect. *Nano Energy*, 66, 104083, 2019.
81. Feng, J., Fu, Y., Liu, X., Tian, S., Lan, S., Xiong, Y., Significant improvement and mechanism of ultrasonic inactivation to escherichia coli with piezoelectric effect of hydrothermally synthesized t-BaTiO3. *ACS Sustain. Chem. Eng.*, 6, 5, 6032–6041, 2018.
82. Kumar, S., Sharma, M., Kumar, A., Powar, S., Vaish, R., Rapid bacterial disinfection using low frequency piezocatalysis effect. *J. Ind. Eng. Chem.*, 77, 355–364, 2019.
83. Sharma, M., Singh, V.P., Kumar, S., Vaish, R., Multicatalytic behavior of Ba0.85Ca0.15Ti0.9Zr0.1O3 ceramics for pharmaceutical/dye/bacterial treatments. *J. Appl. Phys.*, 127, 13, 135103, 2020.
84. Simic, M.G., Bergtold, D.S., Karam, L.R., Generation of oxy radicals in biosystems. *Mutat. Res. Fundam. Mol. Mech. Mutagen.*, 214, 1, 3–12, 1989.
85. Sharma, M., Singh, G., Vaish, R., Dye degradation and bacterial disinfection using multicatalytic BaZr0.02Ti0.98O3 ceramics. *J. Am. Ceram. Soc.*, 103, 4774–4784, 2020.
86. Chou, T.M., Chan, S.W., Lin, Y.J., Yang, P.K., Liu, C.C., Lin, Y.J., Wu, J.M., Lee, J.T., Lin, Z.H., A highly efficient Au-MoS2 nanocatalyst for tunable piezocatalytic and photocatalytic water disinfection. *Nano Energy*, 57, 14–21, 2019.
87. Masimukku, S., Hu, Y.C., Lin, Z.H., Chan, S.W., Chou, T.M., Wu, J.M., High efficient degradation of dye molecules by PDMS embedded abundant single-layer tungsten disulfide and their antibacterial performance. *Nano Energy*, 46, 338–346, 2018.
88. Tan, G., Wang, S., Zhu, Y., Zhou, L., Yu, P., Wang, X., He, T., Chen, J., Mao, C., Ning, C., Surface-selective preferential production of reactive oxygen species on piezoelectric ceramics for bacterial killing. *ACS Appl. Mater. Interfaces*, 8, 37, 24306–24309, 2016.

Index

Also of Interest

Check out these other books by Elvis Fosso-Kankeu published by Scrivener Publishing

Photoreactors in Advanced Oxidation Processes
The Future of Wastewater Treatment
Edited by Elvis Fosso-Kankeu, Sadanand Pandey and Suprakas Sinha Ray
Published 2023. ISBN 978-1-394-16629-9

Application of Nanotechnology in Mining Processes
Beneficiation and Sustainability
Edited by Elvis Fosso-Kankeu, Martin Mkandawire and Bhekie Mamba
Published 2022. ISBN 978-1-119-86499-8

Photocatalysts in Advanced Oxidation Processes for Wastewater Treatment
Elvis Fosso-Kankeu, Sadanand Pandey and Suprakas Sinha Ray
Published 2020. ISBN 978-1-119-63139-2

Recovery of Byproducts from Acid Mine Drainage Treatment
Edited by Elvis Fosso-Kankeu, Christian Wolkersdorfer and Jo Burgess
Published 2020. ISBN 978-1-119-62007-5

Nano and Bio-Based Technologies for Wastewater Treatment
Prediction and Control Tools for the Dispersion of Pollutants in the Environment
Edited by Elvis Fosso-Kankeu
Published 2019. ISBN 978-1-119-57709-6

www.Scrivenerpublishing.com